Ruthenium Chemistry

Ruthenium Chemistry

edited by

Ajay Kumar Mishra
Lallan Mishra

Pan Stanford Publishing

Published by

Pan Stanford Publishing Pte. Ltd.
Penthouse Level, Suntec Tower 3
8 Temasek Boulevard
Singapore 038988

Email: editorial@panstanford.com
Web: www.panstanford.com

British Library Cataloguing-in-Publication Data
A catalogue record for this book is available from the British Library.

Ruthenium Chemistry

ISBN 978-981-4774-39-0 (Hardcover)
ISBN 978-1-315-11058-5 (eBook)

Contents

Preface

Ruthenium is a Group 8 transition metal and is inert to most chemicals. The chemistry of ruthenium is currently receiving a lot of attention in the area of inorganic electrochemistry and photochemistry primarily because its complexes have good electron transfer and energy transfer properties. The synthetic versatility of ruthenium complexes has led to the synthesis of a large number of structurally diverse complexes. Some of these compounds inhibit the proliferation of cancer cells and tumor growth via interaction with a variety of intracellular and extracellular targets. Due to less toxicity and effective bio-distribution, ruthenium complexes are of much current interest. Additionally, luminescent ruthenium complexes can be synthesized at the nanoscale and can be directly traced at the tissue level.

Nanotechnology has become more evident nowadays owing to the constant broadening of the borders and complexities. Nanomaterials can be generated from any material, such as metals, semiconductors, and polymers, with specific sizes, shapes, surface properties, defects, and self-assembly properties by utilizing different methodologies from synthetic and materials chemistry.

One of the targets for cancer chemotherapy is DNA—the classic target—which is believed to be the dominant mechanism for cytotoxicity with certain metal-based drugs, while for others such as topo-isomerase, non-classical targets are thought to be more important. Due to effective bio-distribution and multimodal cellular actions, recently ruthenium complexes have drawn much attention as next-generation anticancer agents. In this regard, DNA, considered their main target, is now evident to be highly unselective, and therefore, there is an evolving concept to evaluate whether metal complexes could be able to attenuate certain tumor growth–associated biochemical events at the cellular level. Thus, in ruthenium complexes, the metal center of ruthenium has been shown to interact with a variety of ligands and thereby enabling these complexes to affect various cellular activities.

This book describes ruthenium complexes as chemotherapeutic agents for targeting tumor sites, although it has been the most challenging task in the area of cancer therapy. Nanoparticles are now emerging as the most effective alternative to the traditional chemotherapeutic approach. However, in view of organ system complicacies, instead of using nanoparticles as a delivery tool, it will be more appropriate to synthesize a nanoparticle-size drug that can use the blood transport mechanism to reach the tumor site and result in cancer regression.

The book describes the synthesis, characterization, and applications of various ruthenium complexes as chemotherapeutic agents and also covers the introduction to chemotherapy, classification of ruthenium complexes with respect to their oxidation states and geometry, nano-size ruthenium complexes and their shape and binding selectivity, binding of ruthenium complexes with DNA, DNA cleavage studies, and cytotoxicity. It will be useful for researchers, scientists, biomedical professionals, and research laboratories and academia interested in the area of coordination chemistry, especially ruthenium compounds and their applications. The book will empower the younger generation to create a new world of ruthenium chemistry in materials science as well as medicine.

Ajay Kumar Mishra
Lallan Mishra
Editors

Chapter 1

An Introduction to Ruthenium Chemistry

Lallan Mishra[a] and Ajay Kumar Mishra[b]

[a]*Department of Chemistry, Institute of Science, Banaras Hindu University, Varanasi, India*
[b]*Nanotechnology and Water Sustainability Research Unit, College of Science, Engineering and Technology, University of South Africa, Florida Campus, Johannesburg, South Africa*

lmishrabhu@yahoo.co.in

Ruthenium occupies a middle position of the d-block elements in the periodic table with atomic number 44 and atomic weight 101.07. It has extensive variation of isotopes, but key isotopes have mass 101, 102, and 104. Two of these isotopes, ^{99}Ru and ^{101}Ru, possess a nuclear spin of 5/2. The element ruthenium was first discovered by a Russian Chemist Karl Karlovich Klaus in 1844 while analyzing a residue of platinum ore from Ural Mountains. Its name was derived from Latin name of Russia as Ruthenia. It was partly in tribute to Klaus's homeland and partly in recognition of the earlier efforts of Gottfried Osann, who had suggested the existence of new elements in these residues, one of which he had named ruthenium. It is a hard metallic element of grayish-white color. Its occurrence in the earth crust is rare (it comprises roughly 10^{-7}% of the Earth's crust). It is usually found together with its five fellow "platinum group" metals (Os, Rh,

Ruthenium Chemistry
Edited by Ajay Kumar Mishra and Lallan Mishra

ISBN 978-981-4774-39-0 (Hardcover), 978-1-315-11058-5 (eBook)
www.panstanford.com

Ir, Pd, and Pt). It has a wider span of its oxidation states ranging from −2 to +8. It is hardened with platinum and palladium and the alloy thus obtained is used in the electronics industry as a water-resistant material. Moreover, titanium's resistance to corrosion can be greatly increased by alloying it in small amounts. Ruthenium shows many of the properties of transition metals such as forming complexes in a wide range of formal oxidation states.

The book *Chemistry of Ruthenium*, by Seddon and Seddon (1978), offered complete coverage of ruthenium, which exhibits the highest known formal oxidation state (RuO_4) together with osmium and xenon, in the periodic table. The compound usually used as a starting material in ruthenium chemistry is "hydrated ruthenium trichloride," an almost black, reflective solid that is obtained industrially by dissolving RuO_4 in aqueous HCl and evaporating to dryness. It is soluble in a wider range of solvents and is comparatively reactive. A landmark discovery in coordination chemistry made in 1965 by Allen and Senoff was the first synthetic complex of N_2 as $[Ru(NH_3)_5(N_2)]X_2$ (X = anion). This complex was originally made by the treatment of hydrated ruthenium trichloride with hydrazine and showed a weak band in their infrared spectra at 2,100 cm^{-1}, owing to the N≡N stretching vibration. At present, ruthenium is at the forefront of various important areas of science, including the development of air and moisture-tolerant homogeneous ruthenium catalysts for alkene metathesis. It had a major impact on total synthesis and on materials chemistry, which resulted in the form of Nobel Prize in Chemistry for Grubbs in 2005. Ruthenium complexes are also extensively used in enantioselective hydrogenation reactions in organic synthesis (exemplified by the work of Ryoji Noyori, who was one of the winners of the 2001 Nobel Prize in Chemistry) and are now being investigated in chemotherapy.

Most probably, the most-studied ruthenium complex of the recent time has been $[Ru(2,2'\text{-bipyridine})_3]^{2+}$, sometimes called rubipy, and its derivatives. On irradiation with visible light, rubipy produces a long-lived photoexcited triplet state, essentially consisting of $[Ru(III)(2,2'\text{-bipyridine})_2(2,2'\text{-bipyridine}^{\bullet -})]^{2+}$. The Ru(III) is a strong oxidant, whereas the 2,2′-bipyridine radical anion is a powerful reductant. It was pointed out in the mid-1970s that in principle this photoexcited state is capable of both oxidizing and reducing water to O_2 and H_2, respectively, giving rise to a wider scope of research. Over 150 years of ruthenium's discovery, its chemistry continues to provide fascinating new findings and potential technologies. No doubt its lure will continue to attract others to its study.

Over the past couple of years, the field of coordination chemistry and organometallic chemistry of ruthenium has grown and evolved at unprecedented rates. Recent publications largely highlight key advances in ruthenium complexes and in their chemically addressable applications in challenging areas such as biology, medicine, catalysis, nanoscience, redox, and photoactive materials, etc. (Storr et al., 2014). This explosive expansion is mainly due to the unique ability of the ruthenium core to permit multiple oxidation states, hence versatile electron transfer pathways. Both Ru(II) and Ru(III) oxidation states accommodate six-coordinated octahedral configurations. The axial ligands provide an opportunity to tune the electronic properties of the complexes. Proper variations of ancillary ligands, allowing the modulation of redox properties and ligand interactions, have resulted in a large platform of ruthenium complexes endowed with achiral or chiral configurations (Gunanathan et al., 2014).

The chemistry of ruthenium complexes, with special attention to their electron transfer properties, has received constant attention in the recent decades. Ruthenium offers a wide range of oxidation states, which are accessible chemically and electrochemically (from oxidation state −2 in $[Ru(CO)_4]^{2-}$ to +8 in RuO_4). Therefore, the complexes of ruthenium are redox-active and their application as redox reagents in different chemical reactions is of much current interest. The kinetic stability of ruthenium in several different oxidation states, the often reversible nature of its redox couples, and the relative ease with which mixed-ligand complexes can be prepared by controllable stepwise methods make ruthenium complexes particularly attractive targets of study. Ruthenium complexes exhibit a great deal of applications in many fields of chemical science. Clear correlations can be observed between their properties and the nature of the ligands bound to the central ion. Thus, ruthenium sulfoxide complexes have been extensively investigated in the past two decades because of their properties and usefulness, particularly in catalysis and chemotherapy. Ruthenium complexes with polypyridyl ligands have received much attention owing to their interesting spectroscopic, photophysical, photochemical, and electrochemical properties. These properties lead to their potential use in diverse areas such as photosensitizers for photochemical conversion of solar energy and molecular electronic devices and as photoactive DNA cleavage agents for therapeutic purposes.

Ruthenium complexes are also known to perform a variety of inorganic and organic transformations. Their synthetic versatility, high catalytic performance under relatively mild reaction conditions, and high selectivity make them particularly well suited for this purpose. A wide range of ruthenium-based agents have been synthesized and tested for their antitumor properties in the past 30 years. Most of these agents, independent of the ligands attached to the ruthenium ion, have shown relatively low cytotoxicity and are less toxic than *cis*platin, correspondingly requiring a higher therapeutic dose. Extensive binding to cellular and extracellular components may explain this observation. Despite their low cytotoxic potential, many ruthenium complexes increase the lifetime expectancy in tumor-bearing hosts. Ru(III) complexes remain most likely in their (relatively inactive and unreactive) Ru(III) oxidation state until they reach the tumor site. In this environment, with its lower oxygen content and pH than the healthy tissue, reduction to the more reactive Ru(II) oxidation state may take place. In this manner (termed "activation by reduction"), Ru(III) compounds may provide selective toxicity. However, to be active in vivo, the complexes must have a biologically accessible reduction potential, which can vary considerably with the type of ligands present in their coordination sphere. A second mechanism that could explain the observed antitumor activity could be the high affinity of Ru(III) for the transferrin iron-binding sites. This binding capacity provides a possibility to target Ru(III) complexes to tumors with high transferring receptor densities. A large number of ruthenium complexes (in both the II and III oxidation states) have been synthesized and showed antitumor activity in vitro. Further applications of ruthenium complexes are widely displayed in photochemistry and photophysics, pharmaceuticals, and agrochemicals.

Although it started over 50 years ago, the development of metal-based anticancer drugs has traditionally focused on the cytotoxicity of platinum compounds, several of which have reached clinical application (Wilson et al., 2014). The deeper understanding of cancer biology triggered the introduction of targeted chemotherapies, using other metal-based drugs, able to address specific cancer physiologies or disease states. As most of the ruthenium complexes are found to be less toxic than their platinum counterparts, this offers considerable added value for their medical implementation. The remarkable success in clinical trials of NAMI-A, KP1019, and KP1339, combined with abundant reports on the enhanced in vitro and in vivo activity of other types of ruthenium complexes, prompted ruthenium chemotherapeutics to rapidly become a major area in anticancer drug advancement (Bergamo et al., 2015). In the search for superior metallo-drugs, ruthenium chemistry is gaining momentum through the introduction of ruthenium containing macromolecular complexes (dendrimers, dendronized polymers, protein conjugates, intelligent nanoparticles for advanced drug delivery, and polymeric ruthenium complex conjugates). These complexes may differentiate between tumor cells and healthy cells (Valente et al., 2014).

Currently, among the most promising alternatives to platinum drugs, ruthenium complexes also operate via different mechanisms and are active by interaction with DNA (Nazarov et al., 2015). Both Ru(II) and Ru(III) oxidation states are stable in physiological solutions, with the latter considered to be less reactive; therefore, ruthenium complexes are generally reported to act as redox-activatable prodrugs. Recent research largely illustrates that in vitro and in vivo properties of ruthenium compounds can be finely tuned by ligand variation (Huang et al., 2015). Some arene-functionalized di-nuclear organometallic Ru(II) complexes are capable of cross-linking model peptide and oligonucleotide sequences while cytotoxicities are linked to their more rigid or more flexible conformations (Murray et al., 2014). The mechanism of action of ruthenium-based antitumor drugs was studied, especially with regard to the capacity of ruthenium to mimic iron in its binding to biological molecules. Incorporation of biologically derived ligands that primarily aim at minimizing toxicity toward normal cells is a viable avenue for engineering ruthenium complexes since they provide ways to improve anti-proliferative activity of metal-based drugs. Such ligands display different coordination modes, facilitate compatibility of the complex with the biological environment, and promote a higher cellular uptake.

Photoactivation of ruthenium complexes for triggering and/or modulating their antitumor activity (Saraf et al., 2014) is of great interest. The process enables the transformation of unreactive Ru(II) complexes into light-driven cytotoxic species (Dickerson et al., 2014) that can subsequently interact with proteins and DNA at the cellular level. Several ruthenium(II)polypyridyl complexes have been elaborated to act as two-photon absorption (TPA) agents in mitochondria-targeted photodynamic therapy, a promising new technique for resolving tumors selectively and subduing resistance to alternative anticancer therapies (Liu et al., 2015). Cellular toxicity induced by photo release of certain bioactive molecules has been proved to be a valid alternative for ruthenium photodynamic therapeutic agents (PDT) with dual-action (Albani et al., 2014). Ruthenium complexes additionally find important medicinal applications as NO carriers and donors for the induction of vascular relaxation (some Ru(III) complexes can act as nitric oxide scavengers thus improving graft survival).

Environment-friendly homogeneous and immobilized ruthenium catalysts enabled indispensable synthetic methods inaccessible with other metal-based catalytic systems. With the aim of upgrading the versatility of ruthenium-based catalysis, new mononuclear, binuclear, and polynuclear ruthenium complexes with arene, phosphine, phosphite, alkylidene, vinylidene, allenylidene, indenylidene, N-heterocyclic carbene (NHC), cyclic alkyl amino carbene (CAAC), porphyrin, pincer, Schiff-base, polydentate, and polycyclic ligands have been continuously prepared, minutely characterized and explored in chemo-selective catalytic processes (Baraut et al., 2015; Okamura et al., 2015; Di Giovanni et al., 2013; Laine et al., 2015; Malineni et al., 2015; Marx et al., 2015; Dragutan et al., 2009; Troian-Gautier et al., 2014). A simple method for hydrogen storage has been disclosed. The process, based on a $CO_2(HCO_3^-)/H_2$ and formate equilibrium within an amine-free reversible system, is catalyzed by commercially available ruthenium pincer complexes and produces a mixture of H_2(CO free) and CO_2 (Kothandaraman et al., 2015). A subject of renewed interest is presently the production of oxygen through water splitting with the help of ruthenium catalysts as showcased in excellent reviews (Duan et al., 2015). One recent refined approach applies to $Ru(bpy)_3Cl_2$ as photoredox catalyst and p-toluidine as redox mediator in the visible light photocatalytic thiolene "click" reaction for fast polymer post-functionalization and step-growth polymerization, under aerobic conditions (Xu et al., 2015).

Schiff base (e.g., Salen) ruthenium complexes enable versatile asymmetric catalysis for a range of chemical transformations, especially for carbene, nitrene, and oxene transfer reactions (Dragutan et al., 2014; Ding et al., 2014). Complexes of this class reportedly serve as catalysts for oxidation/epoxidation, with chiral ruthenium complexes such as Ru-porphyrin, di-symmetric Ru-Schiff base, and Ru-bis amide complexes being employed in asymmetric epoxidation.

Ruthenium complexes contribute to sustainable processes through the conservation of fossil resources and energy. By a metathesis technology, palm oil and jojoba oil can be used to make high-performance olefins and other chemicals of specialty for a wide range of applications, including surfactants, detergents, lubricants, and biofuels (Butilkov et al., 2014). A representative example of metathesis starting from renewable raw materials is the self-metathesis of methyl oleate and cross-metathesis of methyl oleate with (Z)-2-butene-1,4-diol diacetate promoted by a series of indenylidene ruthenium catalysts (Kajetanowicz et al., 2014). Furthermore, a highly chemoselective procedure for obtaining valuable alcohols and higher-oxidation-state compounds by catalytic Z-selective cross-metathesis holds promise for a future application on the larger scale (Koh et al., 2014). In a patented procedure for valorization of biomass materials, cellulose was saccharidified using a ruthenium catalyst supported on carbonized cellulose containing SO_3H groups to give glucose in a 40% yield with 68% conversion of cellulose (Tanaka et al., 2012). Cross-metathesis of bio-sourced fatty nitriles with acrylonitrile provides an efficient way for the conservation of natural resources (Bidange et al., 2014). An important aspect in the industrial valorization of ruthenium complexes is the removal of ruthenium residues. It is mandatory in areas where the level of this heavy metal in the final products (pharmaceuticals, specialty polymers designed for food supplies, and biomedical, textile, or electronic applications) could be problematic for human health (Skowerski et al., 2014).

Overall, the high potential and versatility of the ruthenium complexes attest to the importance they assume in chemistry and other scientific and technical areas and undoubtedly ensure the emergence, in the near future, of further development in the service of the mankind.

References

Albani, B. A., Peña, B., Leed, N. A., de Paula, N. A. B. G., Pavani, C., Baptista, M. S., Dunbar, K. R., Turro, C. (2014) Marked improvement in photoinduced cell death by a new trisheteroleptic complex with dual action: Singlet oxygen sensitization and ligand dissociation. *J. Am. Chem. Soc.* **136**: 17095–17101.

Baraut, J., Massard, A., Chotard, F., Bodio, E., Picquet, M., Richard, P., Borguet, Y., Nicks, F., Demonceau, A., Le Gendre, P. (2015) Assessment of catalysis by arene-ruthenium complexes containing phosphane or NHC groups bearing pendant conjugated diene systems. *Eur. J. Inorg. Chem.* **2015**: 2671–2682.

Bergamo, A., Sava, G. (2015) Linking the future of anticancer metal-complexes to the therapy of tumour metastases. *Chem. Soc. Rev.* **44**(24): 8818–8835.

Bidange, J., Fischmeister, C., Bruneau, C., Dubois, J.-L., Couturier. J.-L. (2015) Cross metathesis of bio-sourced fatty nitriles with acrylonitrile. *Monatsh. Chem.* **146**, 1107–1113.

Butilkov, D., Lemcoff, N. G. (2014) Jojoba oil olefin metathesis: A valuable source for bio-renewable materials. *Green Chem.* **16**, 4728–4733.

Dickerson, M., Sun, Y., Howerton, B., Glazer, E. C. (2014) Modifying charge and hydrophilicity of simple Ru(II) polypyridyl complexes radically alters biological activities: Old complexes, surprising new tricks. *Inorg. Chem.* **53**: 10370–10377.

Di Giovanni, C., Vaquer, L., Sala, X., Benet-Buchholz, J., Llobet, A. (2013) New dinuclear ruthenium complexes: Structure and oxidative catalysis. *Inorg. Chem.* **52**: 4335–4345.

Ding, F., Sun, Y., Verpoort, F., Dragutan, V., Dragutan, I. (2014) Catalytic activity and selectivity of a range of ruthenium complexes tested in the styrene/EDA reaction system. *J. Mol. Catal. A: Chem.* **386**: 86–94.

Dragutan, V., Dragutan, I., Delaude, L., Demonceau, A. (2009) Exploring new achievements in olefin metathesis catalysts. Part 1. Highlights on *N*-heterocyclic carbene ruthenium complexes, *Chem. Today,* **27**: S9–S12.

Dragutan, I., Dragutan, V., Verpoort, F. (2014) Carbenoid transfer in competing reactions catalyzed by ruthenium complexes, *Appl. Organomet. Chem.* **28**: 211–215.

Duan, L., Wang, L., Li, F., Li, F., Sun, L. (2015) Highly efficient bioinspired molecular Ru water oxidation catalysts with negatively charged backbone ligands. *Acc. Chem. Res.* **48**, 2084–2096.

Gunanathan, C., Milstein, D. (2014) Bond activation and catalysis by ruthenium pincer complexes, *Chem. Rev.* **114**: 12024–12087

Huang, H., Zhang, P., Chen, H., Ji, L., Chao, H. (2015) Comparison between polypyridyl and cyclometallated ruthenium(II) complexes: Anticancer activities against 2D and 3D cancer models. *Chem. Eur. J.* **21**: 715–725.

Kajetanowicz, A., Sytniczuk, A., Grela, K. (2014) Metathesis of renewable raw materials Influence of ligands in the indenylidene type catalysts on self-metathesis of methyl oleate and cross-metathesis of methyl oleate with (*Z*)-2-butene-1,4-diol diacetate. *Green Chem.* **16**: 1579–1585.

Koh, M. J., Khan, R. K. M., Torker, S., Yu, M., Mikus, M. S., Hoveyda, A. H. (2015) High value alcohols and higher-oxidation-state compounds by catalytic Z-selective cross-metathesis. *Nature* **517**: 181–186.

Kothandaraman, J., Czaun, M., Goeppert, A., Haiges, R., Jones, J.-P., May, R. B., Surya Prakash, G. K., Olah, G. A. (2015) Amine-free reversible hydrogen storage in formate salts catalyzed by ruthenium pincer complex without pH control or solvent change. *ChemSusChem* **8**: 1442–1451.

Laine, T. M., Kärkäs, M. D., Liao, R.-Z., Åkermark, T., Lee, B.-L., Karlsson, E. A., Siegbahn, P. E. M., Åkermark, B. (2015) Efficient photochemical water oxidation by a dinuclear molecular ruthenium complex. *Chem. Commun.* **51**, 1862–1865.

Liu, J., Chen, Y., Li, G., Zhang, P., Jin, C., Zeng, L., Ji, L., Chao, H. (2015) Ruthenium(II) polypyridyl complexes as mitochondria-targeted two-photon photodynamic anticancer agents. *Biomaterials,* **56**: 140–153.

Malineni, J., Keul, H., Möller, M. (2015) An efficient *N*-heterocyclic carbene-ruthenium complex: Application towards the synthesis of polyesters and polyamides. *Macromol. Rapid Commun.* **36**, 547–552.

Marx, V. M., Sullivan, A. H., Melaimi, M., Virgil, S. C., Keitz, B. K., Weinberger, D. S., Bertrand, G., Grubbs, R. H. (2015) Cyclic alkyl amino carbene (CAAC) ruthenium complexes as remarkably active catalysts for ethenolysis. *Angew. Chem. Int. Ed.* **54**, 1919–1923.

Murray, B. S., Menin, L., Scopelliti, R., Dyson, P. (2014) Conformational control of anticancer activity: The application of arene-linked dinuclear ruthenium(II) organometallics. *Chem. Sci.* **5**: 2536–2545.

Nazarov, A. A., Meier, S. M., Zava, O., Nosova, Y. N., Milaeva, E. R., Hartinger, C. G., Dyson, P. J. (2015) Protein ruthenation and DNA alkylation: Chlorambucil-functionalized RAPTA complexes and their anticancer activity. *Dalton Trans.* **44**: 3614–3623.

Okamura, M., Masaoka, S. (2015) Design of mononuclear ruthenium catalysts for low-overpotential water oxidation. *Chem. Asian J.* **10**: 306–315.

Saraf, S. L., Fish, T. J., Benninghoff, A. D., Buelt, A. A., Smith, R. C., Berreau, L. M. (2014) Photochemical reactivity of Ru(II)(η6-*p*-cymene) flavonolato compounds. *Organometallics,* **33**: 6341–6351.

Skowerski, K., Gułajski, Ł. (2014) Purification strategies in olefin metathesis. In *Olefin Metathesis: Theory and Practice*, Grela, K., ed., John Wiley & Sons: Hoboken, NJ, USA, Chapter 25, pp. 559–571.

Storr, T. (2014) *Ligand Design in Medicinal Inorganic Chemistry*, John Wiley & Sons, Ltd.: Chichester, UK.

Tanaka Noble Metal Industrial Co, Ltd. (2012) Saccharification of biomass materials. *Jpn. Appl.* **873**: 2012.

Troian-Gautier, L., Moucheron, C. (2014) RutheniumII complexes bearing fused polycyclic ligands: From fundamental aspects to potential applications. *Molecules* **19**: 5028–5087.

Valente, A., Garcia, M. H. (2014) Syntheses of macromolecular ruthenium compounds: A new approach for the search of anticancer drugs. *Inorganics,* **2**: 96–114.

Wilson, J. J., Lippard, S. J. (2014) Synthetic methods for the preparation of platinum anticancer complexes, *Chem. Rev.* **114**: 4470–4495.

Xu, J., Boyer, C. (2015) Visible light photocatalytic thiol-ene reaction: An elegant approach for fast polymer post functionalization and step-growth polymerization. *Macromolecules* **48**, 520–529.

Chapter 2

Ruthenium in Medicinal Chemistry

Dinorah Gambino and Lucía Otero

Área de Química Inorgánica, Departamento Estrella Campos, Facultad de Química, Universidad de la República, Montevideo, Uruguay

dgambino@fq.edu.uy

Inorganic medicinal chemistry offers the development of bioactive metal compounds as a promising and attractive approach in the search for a pharmacological control of different diseases. After the discovery of its antineoplastic activity, cisplatin has dominated anticancer chemotherapy. Cisplatin and second-generation platinum derivatives are used worldwide, alone or combined with other drugs, for the treatment of a number of neoplasms despite their side effects and development of resistance mechanisms. Several attempts to develop metal-based drugs are currently in progress in order to overcome such limitations. In this sense, many other transition metals are being investigated. In particular, two ruthenium compounds, NAMI-A and KP1019, have undertaken phase I and phase II clinical evaluations representing a promising alternative to platinum

Dedicated to the memory of Professor Roberto A. Sánchez-Delgado.

Ruthenium Chemistry
Edited by Ajay Kumar Mishra and Lallan Mishra

ISBN 978-981-4774-39-0 (Hardcover), 978-1-315-11058-5 (eBook)
www.panstanford.com

drugs. In addition, many other ruthenium classical coordination compounds, either in oxidation state II or III, and organometallics have been widely explored in respect to their potential uses in medicine. This chapter focuses on the most relevant efforts carried out to develop ruthenium compounds as prospective agents for the treatment of not only cancer but also parasitic diseases of high prevalence. Those series of compounds for which especial efforts have been made to modify structures performing a sort of rational design and to identify their molecular targets and mode of action were emphasized.

2.1 Introduction

2.1.1 General Remarks

The serendipitous discovery of cisplatin antitumoral properties and its clinical introduction almost 40 years ago has determined the beginning of a modern Inorganic Medicinal Chemistry stage based on rational design. Cisplatin (Fig. 2.1) is still the lead compound in clinical use against various types of cancer, particularly ovarian, testicular, bladder, neck, and small cell lung cancer. Nevertheless, it has several clinical disadvantages, like systemic toxicity producing severe side effects, incidence of resistance to treatment and lack of response on many tumors, which severely limit its clinical value. Second and third generation of platinum-based drugs (carboplatin and oxaliplatin) (Fig. 2.1), rationally designed by chemical modification of the original cisplatin structure, are currently in clinical use but they did not lead to the expected improvements of biological behavior. The clinical success of platinum(II) drugs has expanded the field of cancer chemotherapy to other metal-based drugs. Although platinum drugs continue playing a central role in the treatment of cancer, being used worldwide in the chemotherapeutic regimes of around half of all cancer patients, extensive research is being currently performed for the development of prospective antitumoral drugs based on other metals (Galanski et al., 2003; Hall et al., 2007; Klein et al., 2009; Medici et al., 2015; Ronconi and Sadler, 2007).

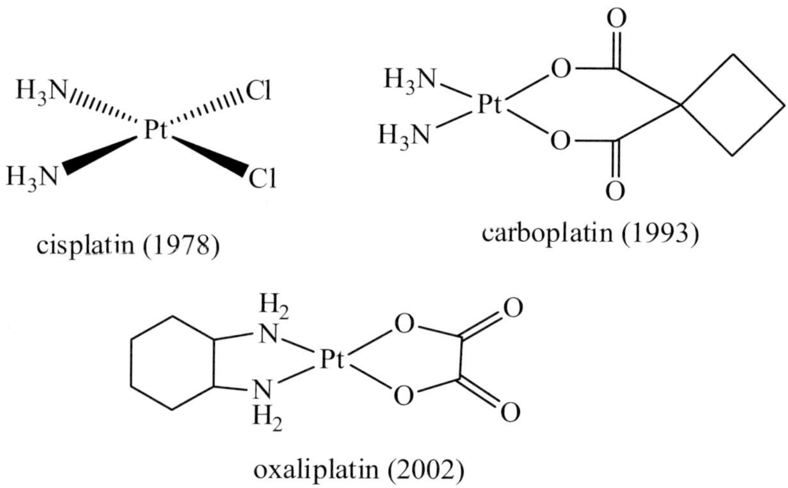

Figure 2.1 Structure and date of beginning of clinical use of Pt(II) antitumoral drugs: cisplatin, carboplatin, and oxaliplatin.

2.1.2 Ruthenium Compounds Potentiality as Prospective Drugs

During the last 25 years, ruthenium compounds either classical or organometallic have been widely explored in respect to their potential uses in medicine. In particular, they have attracted a considerable interest as the basis for new compounds to treat cancer diseases. This fact is mainly supported on some properties attributed to ruthenium-based compounds that suggested the possibility to get anticancer drugs endowed with a significant innovation versus the existing platinum drugs with their disadvantages of limited activity and toxicity. However, it should be noted that some of the statements that will be listed below, despite being supported by some experimental evidence, cannot be uncritically generalized (Alessio, 2016; Bergamo and Sava, 2011; Gambino and Otero, 2012). In physiological solutions, ruthenium is stable in two oxidation states, Ru(II) and Ru(III). Both states can form six-coordinated octahedral species which could allow the in vivo interconversion between both oxidation states without extra energy requirements for structural rearrangements. In addition, the steric and electronic properties of the complexes can be tuned by rationally selecting the two "extra" axial ligands (regarding square planar platinum compounds). The chemical relative "softness" of both ruthenium oxidation states makes them avid for ligands with sulfur and nitrogen donors which are ubiquitous in the major biomolecules, like DNA, serum and cellular proteins and enzymes. The rate of ligand exchange in ruthenium complexes is comparable to that for platinum ones, i.e., 10^{-2} to 10^{-3} s^{-1}. Reactivity of ruthenium compounds should be able to balance the inertness required for the compound to reach its target site (e.g., DNA) and minimize unnecessary side effects while allowing the sometimes required activation necessary for binding to the targets. However, no generalization is possible as both, relatively labile (e.g., NAMI-A) and relatively inert ruthenium compounds (e.g., KP1019), have anticancer activity (Alessio et al., 2016). As stated, the selection of ligands in classic coordination compounds as well as the possibility of generating organoruthenium compounds can intervene in the kinetics of the complexes to allow better control of their stability. Modification of the coordination environment around the central atom could also allow fine-tuning the redox properties of the ruthenium complexes. The reduction potential of each Ru(III) species has a biologically accessible value. Reducing agents like glutathion, ascorbate and some proteins are able to accomplish Ru(III) to Ru(II) reduction and, on the other hand, molecular oxygen and cytochrome oxidase can revert Ru(II) to Ru(III) (Allardyce et al., 2005). However, under biologically relevant conditions the parent compounds could undergo changes in their coordination producing metabolites of unknown $E°$ and binding to serum proteins (Alessio et al., 2016). In addition, redox potentials can be exploited to get ruthenium-based prodrugs. Inert Ru(III) compounds could undergo "activation by reduction" to be transformed, in some conditions, into the more reactive Ru(II) species. This activation would be produced in solid tumor tissues where fast-growing cells with insufficient vascularization generate a hypoxic (reducing) state and it would be prevented in healthy normally oxygenated cells allowing to achieve selectivity (Reisner et al., 2008; Yan et al., 2005).

Moreover, tumor cells often overexpress transferrin. The fact that ruthenium could be able to mimic iron in its interaction with this protein (as well as with albumin), could be used not only for transport but also to achieve more selective entry into cancer tissues (Ang and Dyson, 2006).

In addition, ruthenium could show further potentiality in Medicinal Chemistry for the design of "organic-like" small drugs due to its ability to organize organic ligands in a specific three-dimensional way forming structures with defined shapes that could for instance simulate organic bioactive molecules or biomolecules. Transition metals, and particularly ruthenium, provide an expanded set of coordination geometries for the generation of molecular diversity in respect to organic compounds. This approach could provide the opportunity of complementing the molecular diversity of organic chemistry for the discovery of new compounds with improved biological activities (Meggers, 2007).

2.2 Ruthenium Antitumoral Compounds

According to the World Health Organization (WHO), cancer is a group of diseases characterized by the uncontrolled growth of abnormal cells, which can invade and spread to distant organs or tissues of the body (metastasis). Metastases are the major cause of death from cancer. Cancer is a leading cause of death worldwide. Lung, prostate, colorectal, stomach, and liver cancer are the most common types of cancer in men, while breast, colorectal, lung, uterine cervix, and stomach cancer are the most common among women. Early detection and diagnosis and effective treatment are very important for increasing patient's survival. Depending on tumor stage and type the treatments include surgery, chemotherapy and/or radiotherapy (WHO—cancer, 2016).

Medicinal inorganic chemists are doing extensive research in order to explore new possibilities to overcome both major problems of current clinical antitumoral drugs: toxicity and drug resistance. In particular, different families of ruthenium compounds have been developed for potential use in the therapy of cancer (Barry and Sadler, 2013).

2.2.1 Classic Coordination Compounds

Among the classic coordination compounds, two ruthenium(III) compounds have entered clinical trials as prospective antitumoral drugs including clinical evaluations of phase I or preliminary phase II trials in patients: NAMI-A and KP1019 (Fig. 2.2). Both compounds, together with some relevant ruthenium organometallic compounds, were proposed as alternatives to platinum drugs for the therapy of cancer. NAMI-A was the first ruthenium antitumoral compound tested in human beings reaching a phase I/II therapeutic combination study with gemcitabine in patients with non-small cell lung cancer. Regrettably, the results showed that the combination is less active than gemcitabine treatment alone (Alessio, 2016; Bergamo et al., 2012; Bergamo and Sava, 2015).

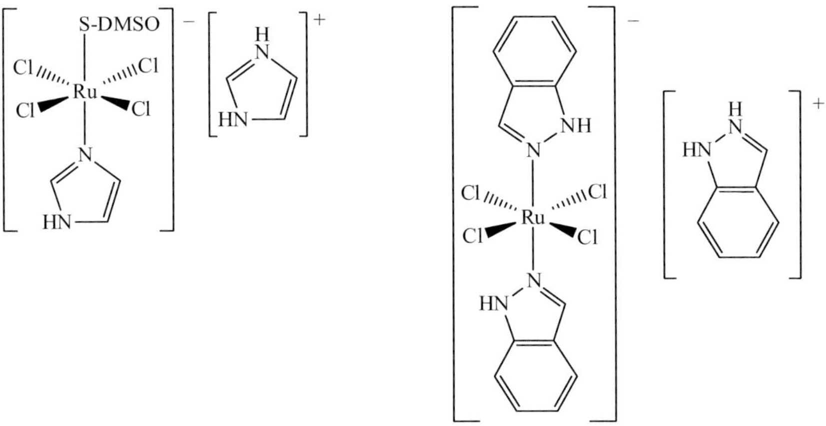

Figure 2.2 Chemical structure of NAMI-A (left) and KP1019 (right).

Although NAMI-A and KP1019 are structurally related they show a different cytotoxic profile. On one hand, at a first stage of studies NAMI-A showed almost no activity against primary tumors but it was highly active against secondary tumors or metastases. In particular, it was noncytotoxic against the 60 cell-line panel established by the National Cancer Institute for in vitro anticancer drug screening. Nevertheless, more recently, it showed to inhibit the proliferation of several leukemia cell lines (Alessio, 2016; Pillozi, 2014). On the other hand, KP1019 demonstrated high cytotoxicity against primary tumors, particularly against colorectal cancer resistant to cisplatin (Muhammad and Guo, 2014).

Studies performed up to the moment suggest that both compounds act through different mechanisms of those shown by platinum-based drugs. Their exact mechanism of action is not known at the molecular level. It is suggested that they interact with more than one target in the cancer cell. For instance, NAMI-A is more cytotoxic after reduction, binds to transferrin, causes apoptosis via the mitochondrial pathway and generates ROS. Cytoskeleton and transduction pathways associated to cell–cell and cell–matrix interaction were recognized as the main cell components altered after treatment in vitro and in vivo with this compound. NAMI-A undergoes rapid extracellular transformations and its metabolites interact predominantly with cell walls and components of the extracellular matrix and are not significantly internalized by cells. KP1019, undergoes DNA interactions similar to cisplatin but the induced conformational modifications are weaker for it than for cisplatin. The higher cytotoxicity of KP1019 compared to NAMI-A has been attributed to a higher cell uptake for the former, involving both passive diffusion and transferrin-dependent uptake. Phase I clinical studies performed to this compound and NKP-1339, the sodium salt analogue of KP-1019, revealed promising anticancer activity and low level of side-effects (Trondl et al., 2014; Bergamo et al., 2012; Allardyce and Dyson, 2016).

Although the final clinical trials of NAMI-A are somewhat disappointing this compound has greatly contributed to increasing the interest in the rational development of non-platinum anticancer metal compounds, particularly based on ruthenium. Taking in mind the commented results, several research groups are currently working on the rational design of new families of ruthenium compounds searching for improved biological properties, like solubility, lipophilicity, absorption, or tumor targeting.

Ru(II) polypyridyl complexes constitute another relevant series of compounds that have been developed as prospective antitumoral drugs. These compounds have been designed pointing out to DNA as molecular target. They bind DNA with high affinity and specificity, so that they would have potential to interfere with cellular processes involving this relevant biomolecule. In these complexes, the metal centre acts as a central scaffold to which active polypyridyl ligands are conjugated together with ancillary ligands. The structure of the former has been modified to improve affinity toward DNA. The latter were selected to favorably modulate cellular uptake and/or binding leading to improved cytotoxicity.

This approach allowed to generate and evaluate several closely related ruthenium compounds which allowed to perform structure–activity studies and get interesting correlations (Gill and Thomas, 2012).

In this context, Schatzschneider et al. developed a series of $[Ru(bpy)_2(NN)]^{2+}$ complexes, where bpy = bypyridine, by varying the size of the NN polypyridyl ligands (bpy, phen, dppz, dppn). The studies performed showed that cellular uptake and cytotoxicity was dependent of the surface area of the NN ligand. From this series the compound with the greatest surface area, $[Ru(bpy)_2(dppn)]^{2+}$ (Fig. 2.3), displayed a cytotoxicity toward HT-29 human colon adenocarcinoma and MCF-7 cells similar to that of cisplatin (Schatzschneider et al., 2008).

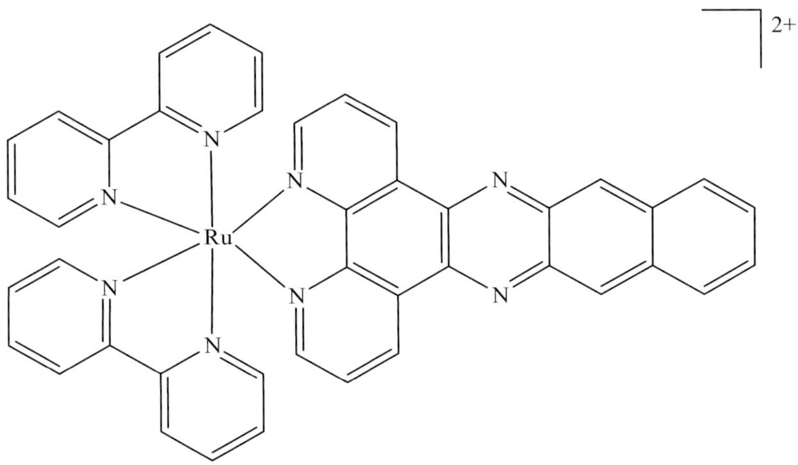

Figure 2.3 Structure of $[Ru(bpy)_2(dppn)]^{2+}$.

Further studies of this compound showed that it could also be useful in photodynamic chemotherapy and that it is a very efficient singlet oxygen sensitizer capable of cleaving duplex DNA with high efficiencies by photoexcitation (Foxon et al., 2009; Sun et al., 2010). The DNA-binding of ruthenium(II) polypyridyl complexes has been extensively studied and it has been shown that these compounds display high DNA-binding affinities. But although in the past decades DNA has been considered their main target, recent studies revealed that mitochondria is another important target (Gorzoni et al., 2016; Zeng et al., 2016).

Ruthenium nitrosyl complexes of nonheme tetraazamacrocycles ligands, with particular emphasis on cyclam and its derivatives, have been studied in detail as antitumor compounds but also as vasodilators by Tfouni et al. These compounds are particularly interesting due to the enhanced thermodynamic stability and slow substitution kinetics imparted to the complexes by the tetradentate tetraazamacrocyclic ligand. NO is an endogenous signaling molecule involved in several physiological and pathological processes. This type of complexes have shown interesting biological activities emerging from the controlled release of NO in biological medium upon reduction or upon irradiation. For example, *trans*-$[Ru^{II}(NO)Cl(cyclam)](PF_6)_2$ (Fig. 2.4) showed a low in vitro cytotoxic activity on B16–F10 melanoma cancer cells in the dark (IC_{50} 6.8 mM) but, under light irradiation (366 nm) it lead at a dose of 1×10^{-4} M to about 50% of cell death on this cell line. In addition, this compound showed low cytotoxicity on a mammalian cell model (Rodrigues et al., 2016).

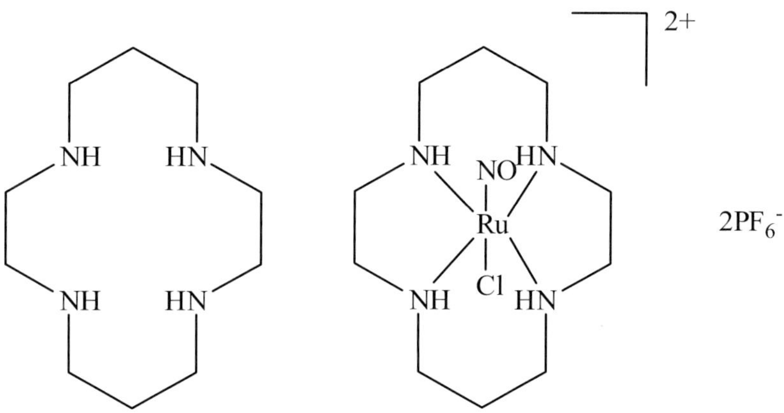

Figure 2.4 Structure of cyclam and *trans*-$[Ru^{II}(NO)Cl(cyclam)](PF_6)_2$.

Although NO-releasing ruthenium complexes have been explored as anticancer agents due to their sensitivity toward light and concomitant formation of oxygen reactive species, the therapeutic efficacy of the released NO depends on the concentration reached in the target (Wink et al., 2008). The cytotoxic effect depends on the stability of the compound in the way to the cell target, the intracellular amount of the ruthenium complex, the amount of the NO delivered and the time and intensity of the light used for irradiation. It has been demonstrated for *cis*-$[Ru(NO)(bpy)_2(4\text{-pic})](PF_6)_3$, where bpy = 2,2′-bipyridine and 4-pic = 4-picoline, that encapsulation within liposomes can improve the biological profile of the compound.

2.2.2 Organometallic Compounds

Organometallic compounds are defined as metal compounds showing at least one covalent metal–carbon bond. This group of compounds, classically developed for catalytic purposes, currently offer shows huge possibilities in the design of novel metal-based drugs, due to the high structural variety they offer in respect to organic compounds:

- different coordination polyhedra ranging at least from linear to octahedral
- higher diversity of variation in stereochemistry
- kinetic stability and possible control of kinetic properties
- suitable lipophilicity to trespass membranes

They also offer wider opportunities than classical coordination compounds for the design of novel metal-based compounds with new modes of action. In this context, different classes of organometallic compounds have shown to be promising candidates for antitumoral therapy, among them metallocenes, half-sandwich, and carbonyl compounds (Gasser et al., 2011).

In particular, Ru^{II}-arene half-sandwich compounds were designed as metal-based drugs acting on DNA, but none of them has entered clinical trials to date. The Ru-arene system offers wide possibilities for incorporation of different ligands (X, Y, Z) or functional groups on the arene moiety (Fig. 2.5).

These "piano stool" shaped compounds have the ability of interacting covalently with guanine residue in DNA by substitution of a labile chloride ligand and, in addition, by intercalating through selected expanded bicycle arene moieties or minor groove binding. In this context, two main rational design efforts were performed by the groups of Sadler and Dyson. Sadler et al. have developed a family of related Ru-arene compounds $[(\eta^6\text{-arene})Ru(en)(Cl)]$, where en = ethylenediamine, including different arene moieties. For these compounds, it has been established that their mechanism of action has many analogies to that of cisplatin. Detailed kinetic studies showed that the hydrolysis of the Ru–Cl bond is strongly influenced by the nature of the coligands. As in the case of cisplatin, the hydrolysis is suppressed in the blood due to the high chloride content of this fluid and after entering the cell the hydrolysis takes place because of the huge lowering of chloride concentration. Similarly to cisplatin the aqua cation $[(\eta^6\text{-arene})Ru(en)(H_2O)]_2^+$ binds to nuclear DNA with a high affinity for the N7 position of guanine bases. Unlike cisplatin that mainly forms single strand bifunctional adducts, these compounds are only able to form monofunctional adducts with DNA (Gasser et al., 2011; Peacock, 2008).

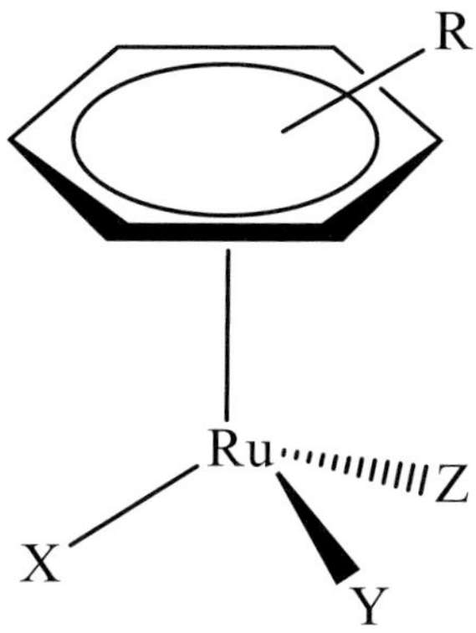

Figure 2.5 General structure of half sandwich Ru(II)-arene compounds.

This family of compounds developed by Sadler's group demonstrated cytotoxicity to a wide range of cancer cell lines, including cisplatin-resistant ones (Bruijnincx and Sadler, 2009; Nofke, 2012). In particular, RM175 (ONCO4417) (Fig. 2.6) has demonstrated in vitro and in vivo antitumoral activity on mammary carcinoma and antimetastatic properties (Bergamo et al., 2010).

RAPTA series of complexes, developed by Dyson et al., are Ru(II) half sandwich compounds that include different arene moieties together with monodentate PTA (1,3,5-triaza-7-phosphoadamantane) or methylated PTA as coligand. This coligand would confer these compounds interesting biological properties, in particular improved water solubility. Structural changes were correlated with biological activity searching for

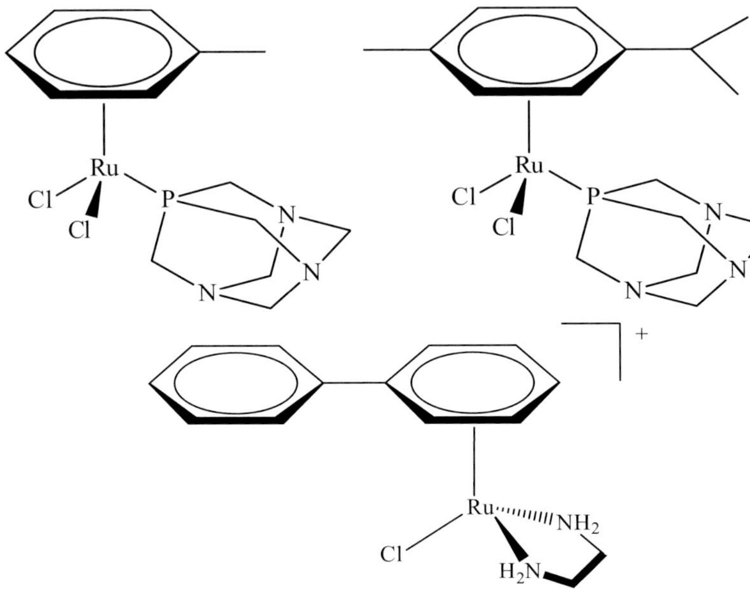

Figure 2.6 Structures of RAPTA-T (upper structure left) and RAPTA-C (upper structure right) by Dyson et al. and ONCO4417 (RM175) by Sadler et al. (lower structure).

more efficient compounds. Interestingly, RAPTA compounds have shown effect on metastatic tissues. From the whole initial series of structurally related compounds, RAPTA-T (Fig. 2.6) showed the highest selectivity toward tumor cells over normal cells. It has been demonstrated that it modulates cell cytoskeleton and lowers cell ability to migrate and metastasize. Although the mechanism of action is yet not clear, the interaction with DNA seems not to be determinant of the bioactivity. In general, the in vitro DNA-damaging ability of RAPTA compounds does not correlate with the observed activity, suggesting a non-DNA-based mechanism of cytotoxicity. Proteins have been proposed as molecular targets of RAPTA compounds. In particular, the inhibitory effect of RAPTA-T on cathepsin B has been demonstrated. Ruthenium was detected in the cytoskeleton and membranes by ICP-MS, which could be expected as cytoskeletal proteins are crucial for metastatic processes. The activity of RAPTA-C (Fig. 2.6) on preclinical models together with its fast clearance from organs and bloodstream and low toxicity make this compound highly promising for further development. RAPTA-C is more effective in inhibiting tumor growth than doxorubicin. In addition, it is reasonably water soluble and tolerates low pH. Therefore, currently in vivo studies based on oral administration are being performed by the Dyson's group (Allardyce and Dyson, 2016; Bergamo et al., 2012; Gasser and Metzler-Nolte, 2011; Murray et al., 2016).

Another completely different approach in the trend to develop ruthenium-based antitumoral compounds has been established by Meggers et al. (Bergamo, 2012; Meggers, 2007). He postulates to explore the opportunities of transition metal inorganic compounds to build small molecules with defined three-dimensional structures that could mimic biologically relevant organic molecules. These compounds are inert scaffolds that could show activity but not due to the presence of a metal center of pharmacological interest but by mimicking certain biological molecules. In particular, he developed ruthenium(II) compounds that are analogous to inhibitors of staurosporin that show selectivity and affinity for certain enzymes. These compounds act mimetizing staurosporin due to the shape of the molecule and this shape results from the coordination of the ruthenium center to suitable ligands. Figure 2.7 shows one of the compounds developed by Meggers'group using this design approach.

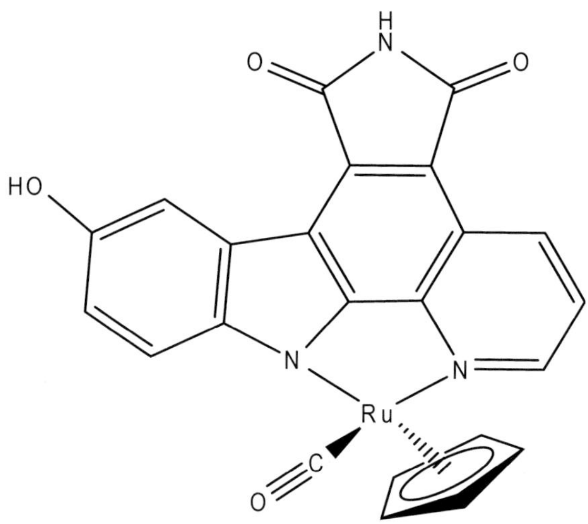

Figure 2.7 Chemical structure of one of the enantiomers of DW1/2.

2.3 Ruthenium Antiparasitic Compounds

The development of new chemotherapeutic agents for the treatment of parasitic diseases is urgently needed since most of the current drugs in clinical use are decades old and show disadvantages like high toxicity, limited efficacy, development of resistance and need of prolonged treatments.

As previously stated, metal based-drugs and, in particular, ruthenium compounds have been deeply explored in the search of potential drug candidates for cancer therapy and they have demonstrated high potentiality in this field. In addition, based on the proposed resemblances in the metabolic pathways of highly proliferative cells, like parasites and tumor cells, a correlation between antiparasitic and antitumor activities has been proposed. In this sense, both organic and metal-based antitumoral drugs have also shown antiparasitic activities and vice versa (Gambino and Otero, 2012; Farrell et al., 1984; Kinnamon et al., 1979; Sánchez-Delgado and Anzellotti, 2004). Thus, the development of bioactive metal compounds, and, in particular, the development of ruthenium-based compounds, has appeared as a promising and attractive approach in the search for a pharmacological control of these diseases (Gambino and Otero, 2012; Navarro et al., 2010).

The selection of ruthenium as central atom is also based on the parasites' need of iron and on the iron-ruthenium chemical resemblances. Alike to highly proliferative tumor cells, parasites show higher demands of iron than normal host cells to properly carry out essential biological functions. In addition, iron uptake is very important for virulence. As a result, parasites have improved mechanisms for internalizing or sequestering host iron (Francisco et al., 2008; Sutak et al., 2008). Thus, by mimicking iron, the uptake of ruthenium compounds into the parasite could be favored. In particular, the binding of ruthenium antiparasitic compounds to transferrin could be a suitable "trojan horse" way of entrance of ruthenium-based antiparasitic drugs into the parasite. In the same way, the masked entrance of antiparasitic organic ligands coordinated to ruthenium could be a way to circumvent parasite resistance to these organic drugs (Costa Pessoa and Tomaz, 2010).

Different strategies have been developed in order to obtain metal compounds bearing antiparasitic activity. As it will be detailed below, the coordination of a ligand bearing antiparasitic activity in order to improve its bioavailability or to achieve a metal-drug synergism through a dual or multiple mechanism of action has been the most deeply explored. In addition, metal coordination of DNA intercalators, intending this biomolecule to be a target in the parasite or the metal coordination of non bioactive ligands in order to obtain metal inhibitors of parasite specific enzymes has also been studied (Gambino, 2011; Gambino and Otero, 2012; Navarro et al., 2010; Sánchez-Delgado et al., 2004). Many examples of all these different approaches have been described in the literature. However, only those that show a rational based design and a systematic development will be described in the following sections.

2.3.1 Ruthenium Compounds against Trypanosomatid Parasites

World Health Organization (WHO) currently categorizes a diverse group of seventeen communicable ailments as Neglected tropical diseases (NTDs). Although they represent a tremendous burden, some of these diseases have historically received low attention by the pharmaceutical industry due to the fact that they mainly affect populations living in poverty in tropical and subtropical regions, without adequate sanitation and in close contact with infectious vectors and domestic animals and livestock. They affect more than one billion people in 149 countries and cost developing countries huge amounts of resources every year (WHO—neglected diseases, 2016).

Among these NTDs, American trypanosomiasis (Chagas disease), Human African trypanosomiasis and leishmaniasis are infections produced by genomically related trypanosomatid parasites and are mainly transmitted through the bite of specific insects. Chagas disease is a potentially life-threatening illness caused by the protozoan kinetoplastid parasite, *Trypanosoma cruzi* (*T. cruzi*). Originally, it was an ancient and endemic disease confined to Latin America but in the last decade it has spread to non endemic countries in other continents due to immigration of unknowingly infected people and lack of controls of the different parasite transmission ways. At least 10000 people die every year from Chagas disease, and about 25 million people are in risk of acquiring the disease (WHO—neglected diseases, 2016).

Human African trypanosomiasis (HAT), also known as sleeping sickness, is a parasitic disease caused by two subspecies of *Trypanosoma brucei brucei*, namely *T. brucei gambiense* or *T. brucei rhodesiense*. Wild and domestic animals can host these parasites and may represent under particular conditions an important reservoir of infection. The disease mostly affects poor populations living in remote rural areas of Sub-Saharan Africa, although cases have been reported in suburban areas of big cities. In the last 20 years, the incidence of the disease has been decreasing according to WHO initiatives. In particular, in 2014 the number of new cases reported dropped below 4000 and the estimated number of actual cases was 15000 (WHO—neglected diseases, 2016).

Leishmaniases comprise a group of diseases caused by protozoan parasites from more than 20 *Leishmania* species. There are three main forms of leishmaniasis: cutaneous, visceral, and mucocutaneous. The disease affects worldwide but mainly countries of Latin America, Africa and Asia. (WHO—neglected diseases, 2016).

The leading investigation on ruthenium complexes bearing anti *T. cruzi* activity was initiated in 1993 by Sanchez-Delgado et al. (Sánchez-Delgado et al., 1993). This work involved the development of metal complexes, including ruthenium ones, with clotrimazole (CTZ) and ketoconazole (KTZ). CTZ and KTZ are well-established antifungals that act as sterol biosynthesis inhibitors. Considering that the sterol biosynthesis of pathogenic fungi is similar to that of trypanomatid parasites, these antifungals can potentially be antiparasitic agents. The complexation of these drugs to a transition metal, ruthenium in particular, is based on the paradigm of a metal-drug synergy, in which the activity of a metal-drug complex would be higher than the sum of its individual components not only through an improved pharmacological profile but also acting on multiple targets (Sánchez-Delgado et al., 2004; Sánchez-Delgado and Anzellotti, 2004). In fact, the inorganic complex $[RuCl_2(CTZ)_2]$ (Fig. 2.8a) displayed a very high anti *T. cruzi* activity (IC_{50} = 5 nM), higher than the free clotrimazole (IC_{50} = 5.8 μM) with no toxicity to mammalian cells. The metal-drug synergy was demonstrated as clotrimazole fragment appeared to dissociate and exert its known sterol biosynthesis inhibition while the remaining ruthenium fragment would covalently bind to parasitic DNA to cause cell damage (Sánchez-Delgado et al., 1993; Sánchez-Delgado et al., 1998; Navarro et al., 2000). Unfortunately, the low water solubility of the complex prevented its effects in vivo.

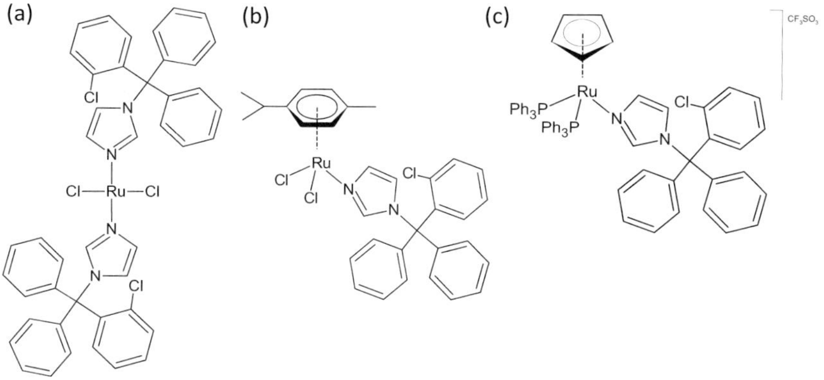

Figure 2.8 Ru-CTZ complexes (a) $[RuCl_2(CTZ)_2]$ and (b) $[Ru^{II}(\eta^6\text{-}p\text{-cymene})Cl_2(CTZ)]$ and (c) $[RuCp(PPh_3)_2(CTZ)](CF_3SO_3)$.

Based on these promising results, Sánchez Delgado's group developed eight new Ru-CTZ complexes belonging to two series: four octahedral Ru^{II} and Ru^{III} coordination compounds (*cis,fac*-$[Ru^{II}Cl_2(DMSO)_3(CTZ)]$, *cis,cis,trans*-$[Ru^{II}Cl_2(DMSO)_2(CTZ)_2]$, $Na[Ru^{III}Cl_4(DMSO)(CTZ)]$, Na[*trans*-$Ru^{III}Cl_4(CTZ)_2$]) and four organometallic Ru-*p*-cymene compounds ($[Ru^{II}(\eta^6\text{-}p\text{-cymene})Cl_2(CTZ)]$, $[Ru^{II}(\eta^6\text{-}p\text{-cymene})(bipy)(CTZ)](BF_4)_2$, $[Ru^{II}(\eta^6\text{-}p\text{-cymene})(en)(CTZ)](BF_4)_2$, $[Ru^{II}(\eta^6\text{-}p\text{-cymene})(acac)(CTZ)](BF_4)$, where bipy = bipyridine, en = ethlylenediamine, acac = acetylacetonate (Martínez et al., 2012). Ancillary ligands were included in order to modulate complexes' physicochemical properties and in particular to improve their water solubility. All obtained compounds were tested in vitro against both *T. cruzi* epimastigotes (Y strain) and *L. major* promastigotes and their cytotoxic activity on human osteoblasts was also established. The most promising compound was the organometallic piano stool complex, $[Ru^{II}(\eta^6\text{-}p\text{-cymene})Cl_2(CTZ)]$, (Fig. 2.8b) that resulted 110 and 58 times more active than free CTZ against *L. major* and *T. cruzi*, showing IC_{50} values in the submicromolar range (0.0015 and 0.1 μM, respectively). The inclusion of the organometallic fragment leads to a 6-fold increase in the anti *T. cruzi* activity respect to the original $[RuCl_2(CTZ)_2]$ complex. In addition, this organometallic compound showed excellent selectivity indexes, greater than 500 and 75 for *L. major* and for *T. cruzi*, respectively, making it a good candidate for further in vivo and mechanistic studies. In fact, recently published work showed that $[Ru^{II}(\eta^6\text{-}p\text{-cymene})Cl_2(CTZ)]$ as well as $[Ru^{II}(\eta^6\text{-}p\text{-cymene})(acac)(CTZ)](BF_4)$ significantly reduced the lesion size in mice exposed to *L. major* cutaneous infection. A mitochondrial dependent apoptotic-like death in the extracellular form of the parasite was observed for these compounds implicating phospholipids externalization, mitochondrial depolarization and DNA fragmentation pathway (Iniguez et al., 2016). Recently, a new Ru-CTZ organometallic compound $[RuCp(PPh_3)_2(CTZ)](CF_3SO_3)$, where *Cp* = cyclopentadienyl fragment was described by our group. The compound was evaluated in vitro on *T. cruzi* (Y strain) and on the infective form of *T. brucei brucei* strain 427 (cell line 449). Results showed that complexation of the bioactive CTZ to the $\{RuCp(PPh_3)\}$ moiety lead to a significant increase of the antiparasitic activity showing a 6-fold and 40-fold higher activity than free CTZ on *T. cruzi* and on the infective form of African trypanosomes, respectively. Related to the mechanism of action, this compound affects sterol membrane biosynthesis at the same step as terbinafine, inhibiting the squalene-2,3-epoxidase enzyme (Rodríguez-Arce et al., 2015).

Based on a similar approach, our group has been working since 2004 on the development of metal complexes bearing potential antiparasitic activity. Selected bioactive ligands were 5-nitrofuryl containing thiosemicarbazones (TSC) (Fig. 2.9a). These ligands, containing the 5-nitrofurane pharmacophore, were synthesized as nifurtimox analogues showing very good activities in vitro against *T. cruzi* (Aguirre et al., 2004). Various series of ruthenium complexes with these bioactive ligands were rationally designed, including in the ruthenium coordination sphere co-ligands that could modulate some of their physicochemical properties. In this sense, four series of complexes of the formula $[Ru^{II}Cl_2(TSC)_2]$, $[Ru^{III}Cl_3(DMSO)(TSC)]$, $[Ru^{II}Cl(PPh_3)(TSC\text{-}H)_2]$,

$[Ru^{II}Cl_2(TSC)(HPTA)_2]Cl_2$, with TSC-H = deprotonated form of the thiosemicarbazone ligand, HPTA = protonated form of 1,3,5-triaza-7-phosphaadamantane, PPh_3 = triphenylphosphine, were developed and their activity in vitro against *T. cruzi* was evaluated (Pagano et al., 2009; Sarniguet et al., 2014). In addition, a series of organometallic compounds including the *core* Ru-*p*-cymene were also obtained. Dimeric compounds of the formula $[Ru^{II}_2(\eta^6\text{-}p\text{-cymene})_2(TSC\text{-}H)_2]^{2+}$ with hexafluorophosphate or chloride as counterions were developed and evaluated against both *T. cruzi* and *T. brucei brucei* (Demoro et al., 2012; Demoro et al., 2013). In general, all obtained classical coordination compounds showed poor to moderate activity against *T. cruzi* being, most of them, less active than the corresponding thiosemicarbazone ligand and the reference drug nifurtimox. However, in order to get new bases for the rational design of new ruthenium based complexes, correlations between activity-lipophilicity and water solubility could be established (Pagano et al., 2009; Sarniguet et al., 2014). On the other hand, the hypothesis of a synergistic effect caused by ruthenium complexation with a bioactive ligand was demonstrated by $[Ru^{II}_2(\eta^6\text{-}p\text{-cymene})_2(TS4\text{-}H)_2](PF_6)_2$ compound (Fig. 2.9b). This ruthenium organometallic compound showed higher activity against trypomastigotes of *T. cruzi* than the reference drug nifurtimox (IC_{50} = 20.0 μM on Y strain and 24.7 μM on Dm28c clone trypomastigotes) and a very significant increase in growth inhibitory activity on *T. brucei brucei* with respect to the free TS4 ligand (IC_{50} > 100 μM). Furthermore, selectivity was also evidenced for this compound as it was about 50-fold more potent on this parasite than on the studied mammalian cell line (Demoro et al., 2012; Demoro et al., 2013).

The potential mechanism of action of the Ru-TSC compounds was studied. The hypothesis of a dual mechanism was demonstrated for most of them: the production of oxidative stress showed that the mechanism of the 5-nitrofuryl thiosemicarbazone ligands remains in the obtained complexes and, on the other hand, the complexes' interaction with DNA was also evidenced (Sarniguet et al., 2014; Demoro et al., 2013).

Another example of the metal-ligand synergistic effect approach was developed by Franco et al. (Nogueira Silva et al., 2008). A complex of the formulae *trans*-$(Ru(Bz)(NH_3)_4SO_2)(CF_3HSO_3)_2$ with Bz = benznidazole (*N*-benzyl-2-(2-nitro-1*H*-imidazol-1-yl) acetamide) as bioactive ligand was obtained and evaluated in vitro and in vivo against *T. cruzi*. The obtained complex was much more active against both epimastigote (IC_{50} = 127 μM) and trypomastigote (IC_{50} = 79 μM) forms of *T. cruzi* (Y strain) than the free Bz (IC_{50} > 1 mM for both forms). In the in vivo studies, Ru-Bz complex, at a thousand-fold smaller concentration than the considered optimal dose for Bz (385 μmol/kg/day), was able to eliminate the nests of amastigotes in myocardium tissue and skeletal muscles and to decrease myocarditis in treated mice without showing acute toxicity. The electron withdrawing ability of the Ru(II) center that would favor the nitro group reduction as well as the increased water solubility promoted by the coordination of Bz to the hydrophilic *trans*-$[Ru(NH_3)_4L]^{2+/+/0}$ moiety could explain the increased activity.

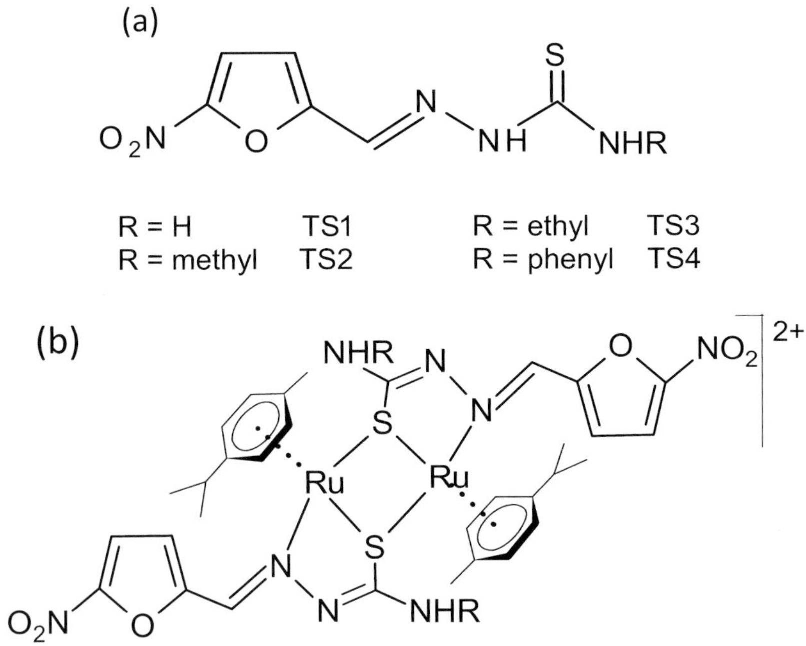

Figure 2.9 (a) Bioactive 5-nitrofuryl containing thiosemicarbazones (b) $[Ru^{II}_2(\eta^6\text{-}p\text{-cymene})_2(TSC\text{-}H)_2]^{2+}$ compounds.

On the other hand, the same group has developed ruthenium-nitrosyl complexes bearing antitrypanosomal activity. This approach is based on the fact that nitric oxide has an important role in the modulation of the hosts's immune response upon parasitic invasion. Ruthenium-nitrosyl complexes, which could release NO after reduction of NO^+ in the biological media, would be useful for developing potential chemotherapeutic agents for parasitic diseases (Tfouni et al., 2012). In this sense, two series of compounds of the formula *trans*-$[Ru(NO)(NH_3)_4(L1)](X)_n$, and $[Ru(NO)(bpy)_2(L2)](X)_n$ (Fig. 2.10) with L1 = pirazine (pz), L-histidine (L-hist), N-bonded imidazole (imN), isonicotinamide (isn), pyridine (py), nicotinamide (nic), isonicotinic acid (isa), 4-picoline (4-pic), C-bonded imidazole (imC), triethylphosphite ($POEt_3$), sulfite (SO_3) and L2 = N-bonded imidazole (imN), N-bonded methyl imidazole and sulfite (SO_3) were evaluated in vitro and in vivo against *T. cruzi* and/or *Leishmania* parasites (Tfouni et al., 2012; Silva et al., 2007; Silva et al., 2009; Silva et al., 2010; Guedes et al., 2010; Melo et al., 2010).

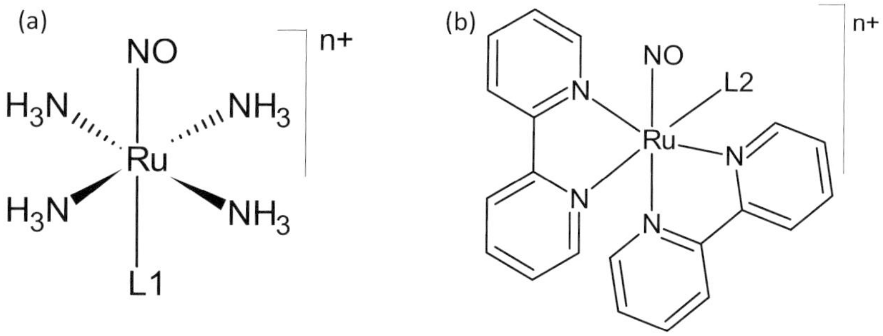

Figure 2.10 Ruthenium nitrosyl complexes: (a) *trans*-$[Ru(NO)(NH_3)_4(L1)]^{n+}$, (b) *cis*-$[Ru(NO)(bpy)_2(L2)]^{n+}$. See text for L1 and L2.

Most of the evaluated Ru-NO complexes resulted more active in vitro against the epimastigote form of *T. cruzi* (Y strain) than the reference drug benznidazole (IC_{50} = 3180 μM) but equal or less active than this drug against the trypomastigote form (IC_{50} = 53 μM). Compounds with isn (*trans* series), imN (both series) and SO_3 (*cis* series) were evaluated in vivo. All of them, highly increased the time of survival of the infected mice as well as a reduction in the myocardium inflammation and a decrease in the number of amastigote nests were observed. On the other hand, most complexes were more active in vitro against leishmanial promastigotes than the reference NO donor sodium nitroprusside ($Na_2[Fe(CN)_5NO)]$). In particular, *trans*-

$[Ru(NO)(NH_3)_4(N\text{-}im)]^{n+}$, having the best selective index (SI = 18), was evaluated in vivo inhibiting the parasite growth similarly to the pentavalent antimony drug glucantime but at a concentration 66-fold smaller. Results strongly suggested that not only the liberation of NO but the inhibitory effect against *T. cruzi* GAPDH (glyceraldehyde-3-phosphate dehydrogenase) would be responsible for the antiparasitic activity of these Ru-NO compounds (Tfouni et al., 2012; Silva et al., 2007; Silva et al., 2009; Silva et al., 2010; Guedes et al., 2010; Melo et al., 2010).

2.3.2 Ruthenium Compounds against Malaria Parasites

Malaria is the most widespread, and one of the most severe, tropical parasitic diseases and it continues to be a major cause of mortality and morbidity in poor regions in the developing world. The disease is caused by Plasmodium parasites. There are 5 parasite species that cause malaria in humans, but *P. falciparum* and *P. vivax* constitute the greatest threat. The parasites are spread to people through the bite of an infected female *Anopheles* mosquito (who.int/mediacentre/factsheets). Mortality from malaria has increased in the last 20 years mainly due to the development of resistance of the parasites to antimalarial drugs, particularly to the first-line drug chloroquine.

In order to find an answer to the problem of resistance of some *Plasmodium* strains to chloroquine (CQ), this drug has been selected for the development of ruthenium classical and organometallic compounds by Sanchez Delgado et al. This approach is based on the fact that ruthenium coordination would modify CQ electronic distribution leading to changes in basicity and lipophilicity and making it less recognizable to the parasite's defense mechanism. In addition, the metal-drug synergism would lead to an improvement in the antiparasitic activity. The first developed compound, $[RuCl_2(CQ)]_2$, (Fig. 2.11a) showed a 5-fold enhancement of activity on the Colombian FCB1 resistant parasite strain in respect to CQDP (chloroquine diphosphate). In addition, the in vivo assays showed a reduction of 94% of the parasitemia for the ruthenium complex while CQDP causes a 55% reduction. No acute toxicity effects on mice during prolonged treatment were observed (Gambino and Otero, 2012; Sánchez-

Delgado et al., 1996; Sánchez-Delgado and Anzellotti, 2004; Sánchez-Delgado et al., 2004).

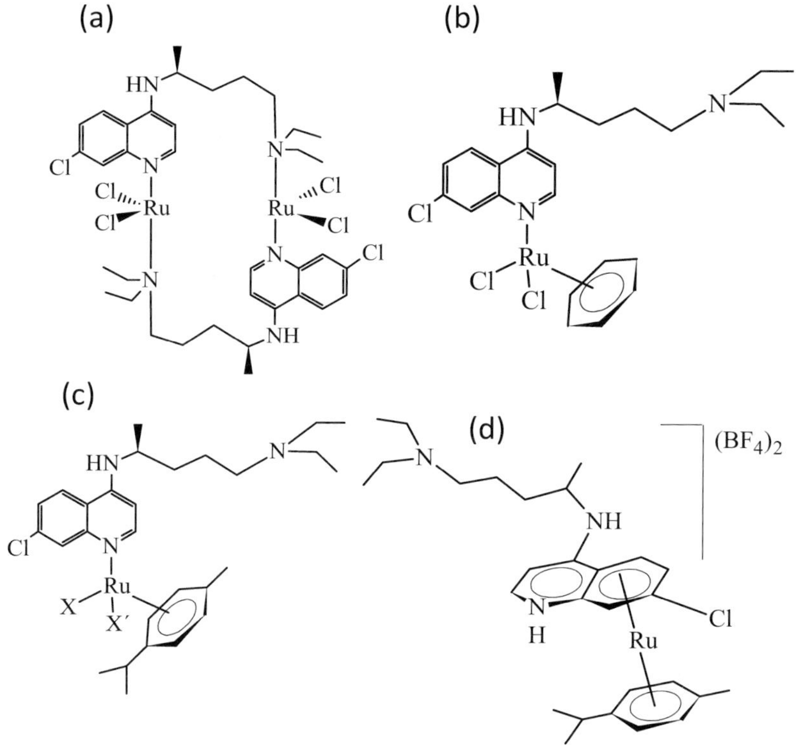

Figure 2.11 Ru-CQ compounds. (a) $[RuCl_2(CQ)]_2$, (b) Ru(II) half-sandwich arene CQ compounds: $[Ru(\eta^6\text{-}p\text{-cymene})Cl_2(CQ)]$, $[Ru(\eta^6\text{-}p\text{-cymene})(H_2O)_2(CQ)](BF_4)_2$ and $[Ru(\eta^6\text{-}p\text{-cymene})(en)(CQ)](PF_6)_2$, (c) $[Ru(\eta^6\text{-}p\text{-benzene})Cl_2(CQ)]$ and (d) $[Ru(\eta^6\text{-}p\text{-cymene})(\eta^6\text{-CQDP})](BF_4)_2$.

The mechanism of antimalarial action of $[RuCl_2(CQ)]_2$ was investigated by studying DNA (due to the presence of ruthenium and CQ) and the inhibition of hemozoin generation (the main mechanism of action of CQ) as potential parasite targets (Egan and Helder, 1999; Martínez et al., 2008). The complex is rapidly hydrolyzed in aqueous solution and binds to DNA but heme aggregation inhibition seems to be the main mechanism of action of the ruthenium complex as it is for free CQ. The enhancement in lipophilicity due to complexation of CQ to ruthenium was related to a better accumulation in the parasite leading to a more efficient inhibition of heme aggregation that would explain the activity of this ruthenium compound in CQ resistant *Plasmodium* strains (van Schalkwyk and Egan, 2006).

Based on these promising results, Sánchez-Delgado et al. developed five new Ru-arene-CQ compounds (Fig. 2.11b–d) (Rajapakse et al., 2009; Navarro et al., 2012). Four of them were regular Ru(II) half-sandwich compounds but [Ru(η^6-*p*-cymene)(η^6-CQDP)][BF_4]$_2$ showed an unprecedented η^6 bonding through the carbocyclic ring of CQ (Fig. 2.11d).

Most of the new organoruthenium compounds were more active than CQ against three CQ-resistant and four CQ-sensitive *P. falciparum* strains. In addition, binding of CQ to ruthenium in these particular organometallic scaffolds lowered or at least maintained the cytotoxicity of CQ on normal mammalian cells (normal mouse splenocytes C57B6 and human foreskin fibroblast Hs27 cells) (Martínez et al., 2010). As previously described for [$RuCl_2$(CQ)]$_2$ complex, both DNA and inhibition of heme aggregation were established as targets for these Ru-arene compounds (Navarro et al., 2012; Martínez et al., 2009; Martínez et al., 2010). In addition, the interaction of the ruthenium compounds with proteins (human serum albumin (HSA) and transferrin) was studied. Results showed that compounds interact covalently with HSA and apotransferrin through the exchangeable ligands when available. However, [Ru(η^6-*p*-cymene)(η^6-CQDP)](BF_4)$_2$ although covalently saturated also interacted with both serum proteins. HSA interaction turned to be reversible which would favor the role of this protein as a carrier. As expected, all studied compounds showed high binding affinities for apotransferrin (1.77×10^3–1.18×10^5 M^{-1}) but lower than that of iron (Martínez et al., 2011).

Recently, another series of organoruthenium complexes with general formula [Ru(*p*-cymene)(CQ)(N-N)]$^{2+}$ where N-N = 2′-bipyridine, 5,5′-dimethyl-2,2′-bipyridine, 1,10-phenanthroline, 4,7-diphenyl-1,10-phenanthroline were evaluated against CQ-sensitive 3D7 and CQ-resistant W2 strains of *Plasmodium falciparum*, with potency and selectivity indexes similar to those of CQ. In addition, the parasitemia of *P. berghei*-infected mice was reduced by treatment with the complexes (Macedo et al., 2016).

Based on the same approach half-sandwich ruthenium *p*-cymene compounds were obtained but instead of CQ, CQ-analogue compounds were selected as bioactive ligands (Glans et al., 2012; Ekengard et al., 2015). These ligands were designed with the aim of including further coordination atoms and/or modifying the electronic properties of the molecules trying to obtain compounds with an improved antimalarial activity of both the free ligands and their ruthenium *p*-cymene complexes. In these sense, a pyridine, an imidazole, a salicylaldimine moiety or an amine-(N-methyl) imidazolyl chelating unit were included at the N7 position of the CQ framework and the corresponding ruthenium *p*-cymene complexes were synthesized. However, neither the ligands, nor the complexes resulted more active than CQ on both sensitive and resistant *Plasmodium* strains.

Another interesting approach for the development of novel metal-based antimalarials has been the inclusion of a metallocene center into the scaffold of CQ to enhance its pharmacological properties and circumvent *P. falciparum* resistance. In particular, the ferrocene compound ferroquine (FQ) is being developed by Sanofi-Aventis under phase II clinical trials as the first bioorganometallic drug candidate (Fig. 2.12) (Dive and Biot, 2008; Biot, 2004; Dubar et al., 2008).

Ruthenoquine, its ruthenium analogue, showed a similar activity to that of FQ (Fig. 2.12) (Beagley et al., 2002; Beagley et al., 2003; Blackie et al., 2007).

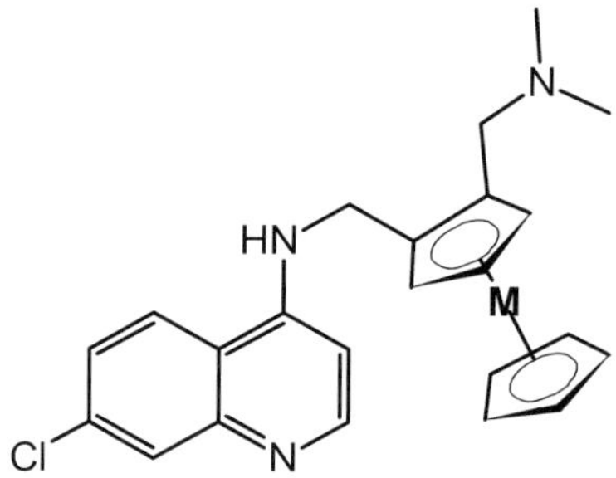

Figure 2.12 Ferroquine (M = Fe) and ruthenoquine (M = Ru) structures.

In addition, Moss and Chibale synthesized and evaluated other CQ analogues containing the ruthenocenyl moiety by changing the substituents on the aliphatic side chain of CQ obtaining similar activities. As previously stated for other CQ complexes, the activity of these metallocenes was related to an increase in lipophilicity as a consequence of the inclusion of the organometallic moiety even though an inherent antiplasmodial activity associated with the metallocene moiety itself was not discarded (Blackie et al., 2007).

2.4 Concluding Remarks and Perspectives

Ruthenium is undoubtedly the star metal in the present search for therapeutic agents. Even though most of the published literature is mainly related to the treatment of cancer, a significant amount of research has been conducted to obtain therapies for many other disorders.

Related to cancer therapy, ruthenium compounds have shown broad diversity, in terms of activity, toxicity, and mechanisms of action. Even compounds that are structurally very similar behave differently and have surprisingly different biological properties. Their in vivo anticancer activities also differ as some are active against primary tumors, but others only against metastases. In addition, some compounds are believed to interact with DNA, whereas this is questionable or unlikely for others. Thus, there are still no general guidelines for the synthesis of new active species and this diversity makes the research in this field an appealing area of work.

Related to the antiparasitic activity, previously discussed data clearly show that the strategy of combining ruthenium and a bioactive ligand in a single molecule could produce an enhancement of the activity of the ligand, either by affecting the parasite's mechanism of resistance or by bringing in new targets due to the presence of the ruthenium ion that lead to additive effects or a metal-drug synergism. However, systematic research is still needed to provide ruthenium-based leads, or at least hits, for further drug development for the therapy of neglected parasitic diseases. In this sense, the establishment of structure-activity relationships and the detailed knowledge on molecular mechanism of action instead of the work through isolated serendipitous efforts, will surely aid to achieve the goals of antiparasitic ruthenium drugs development.

References

Aguirre G, Boiani L, Cerecetto H, González M, Denicola A, Otero L, Gambino D, Rigol C, Olea-Azar C, Faundez M, Aguirre G, Boiani L, Cerecetto H, González M, Denicola A, Otero L, Gambino D, Rigol C, Olea-Azar C, Faundez M (2004). In vitro activity and mechanism of action against the protozoan parasite *Trypanosoma cruzi* of 5-nitrofuryl containing thiosemicarbazones. *Bioorg Med Chem*, **12**: 4885–4893.

Alessio E (2017). Ru anticancer complexes. Thirty years of the drug candidate NAMI-A and the myths in the field of ruthenium anticancer compounds: A personal perspective. *Eur J Inorg Chem*, **12**: 1549–1560.

Allardyce CS, Dorcier A, Scolaro C, Dyson PJ (2005). Development of organometallic (organo-transition metal) pharmaceuticals. *Appl Organomet Chem*, **19**: 1–10.

Allardyce CS, Dyson PJ (2016). Metal-based drugs that break the rules. *Dalton Trans*, **45**: 3201–3209.

Ang WH, Dyson PJ (2006). Classical and non-classical ruthenium-based anticancer drugs: Towards targeted chemotherapy. *Eur J Inorg Chem*, **20**: 4003–4018.

Barry NPE, Sadler PJ (2013). Exploration of the medical periodic table: Towards new targets. *Chem Commun*, **49**: 5106–5131.

Beagley P, Blackie MAL, Chibale K, Clarkson C, Meijboom R, Moss JR, Smith PJ, Su H (2003). Synthesis and antiplasmodial activity in vitro of new ferrocene–chloroquine analogues. *Dalton Trans*, 3046–3051.

Beagley P, Blackie MAL, Chibale K, Clarkson C, Moss JR, Smith PJ (2002). Synthesis and antimalarial activity in vitro of new ruthenocene–chloroquine analogues. *J Chem Soc Dalton Trans*, 4426–4433.

Bergamo A, Gaiddon C, Schellens JHM, Beijnen JH, Sava G (2012). Approaching tumour therapy beyond platinum drugs. Status of the art and perspectives of ruthenium drug candidates. *J Inorg Biochem*, **106**: 90–99.

Bergamo A, Masi A, Peacock AFA, Habtemariam A, Sadler PJ, Sava G (2010). In vivo tumour and metastasis reduction and in vitro effects on invasion assays of the ruthenium RM175 and osmium AFAP51 organometallics in the mammary cancer model. *J Inorg Biochem*, **104**: 79–86.

Bergamo A, Sava G (2011). Ruthenium anticancer compounds: Myths and realities of the emerging metal-based drugs. *Dalton Trans*, **40**: 7817–7823.

Bergamo A, Sava G (2015). Linking the future of anticancer metal-complexes to the therapy of tumour metastases. *Chem Soc Rev*, **44**: 8818–8835.

Biot C (2004). Ferroquine: A new weapon in the fight against malaria. *Curr Med Chem Anti-Infective Agents*, **3**: 135–147.

Blackie MAL, Beagley P, Croft SL, Kendrick H, Moss JR, Chibale K (2007). Metallocene-based antimalarials: An exploration into the influence of the ferrocenyl moiety on in vitro antimalarial activity in chloroquine-sensitive and chloroquine-resistant strains of Plasmodium falciparum. *Bioorg Med Chem*, **15**: 6510–6516.

Bruijnincx PCA, Sadler PJ (2009). Controlling platinum, ruthenium and osmium reactivity for anticancer drug design. *Adv Inorg Chem*, **61**: 1–62.

Costa Pessoa J, Tomaz I (2010). Transport of therapeutic vanadium and ruthenium complexes by blood plasma components. *Curr Med Chem*, **17**: 3701–3738.

Demoro B, Rossi M, Caruso F, Liebowitz D, Olea-Azar C, Kemmerling U, Maya JD, Guiset H, Moreno V, Pizzo C, Mahler G, Otero L, Gambino D (2013). Potential mechanism of the anti-trypanosomal activity of organoruthenium complexes with bioactive thiosemicarbazones. *Biol Trace Elem Res*, **153**: 371–381.

Demoro B, Sarniguet C, Sánchez-Delgado R, Rossi M, Liebowitz D, Caruso F, Olea-Azar C, Moreno V, Medeiros A, Comini MA, Otero L, Gambino D (2012). New organoruthenium complexes with bioactive thiosemicarbazones as co-ligands: Potential anti-trypanosomal agents. *Dalton Trans*, **41**: 1534–1543.

Dive D, Biot C (2008). Ferrocene conjugates of chloroquine and other antimalarials: The development of ferroquine, a new antimalarial. *ChemMedChem*, **3**: 383–391.

Dubar F, Khalife J, Brocard J, Dive D, Biot C (2008). Ferroquine, an ingenious antimalarial drug–thoughts on the mechanism of action. *Molecules*, **13**: 2900–2907.

Egan TJ, Helder MMM (1999). The role of haem in the activity of chloroquine and related antimalarial drugs. *Coord Chem Rev*, **190–192**: 493–517.

Ekengard E, Glans L, Cassells I, Fogeron T, Govender P, Stringer T, Chellan P, Lisensky G, Hersh WH, Doverbratt I, Lidin S, de Kock C, Smith PJ, Smith GS, Nordlander E (2015). Antimalarial activity of ruthenium(ii) and osmium(ii) arene complexes with mono- and bidentate chloroquine analogue ligands. *Dalton Trans,* **44**: 19314–19329.

Farrell NP, Williamson J, McLaren DJM (1984). Trypanocidal and antitumour activity of platinum-metal and platinum-metal-drug dual-function complexes. *Biochem Pharmacol,* **33**: 961–968.

Foxon SP, Alamiry MAH, Walker MG, Meijer AJHM, Sazanovich IV, Weinstein JA, Thomas JA (2009). Photophysical properties and singlet oxygen production by ruthenium(II) complexes of benzo[i]dipyrido[3,2-a:2′,3′-c]phenazine: A spectroscopic and TD DFT study. *J Phys Chem A,* **113**: 12754–12762.

Francisco AF, de Abreu Vieira PM, Arantes JM, Pedrosa ML, Martins HR, Silva M, Veloso VM, de Lana M, Bahia MT, Tafuri WL, Carneiro CM (2008). Trypanosoma cruzi: Effect of benznidazole therapy combined with the iron chelator desferrioxamine in infected mice. *Exp Parasitol,* **120**: 314–319.

Galanski M, Arion VB, Jakupec MA, Keppler BK (2003). Recent developments in the field of tumor-inhibiting metal complexes. *Curr Pharm Des,* **9**: 2078–2089.

Gambino D (2011). Potentiality of vanadium compounds as anti-parasitic agents. *Coord Chem Rev,* **255**: 2193–2203.

Gambino D, Otero L (2012). Perspectives on what ruthenium-based compounds could offer in the development of potential antiparasitic drugs. *Inorg Chim Acta,* **393**: 103–114.

Gasser G, Ott I, Metzler-Nolte N (2011). Organometallic anticancer compounds. *J Med Chem,* **54**: 3–25.

Gill MR, Thomas JA (2012). Ruthenium(II) polypyridyl complexes and DNA—from structural probes to cellular imaging and therapeutics. *Chem Soc Rev,* **41**: 3179–3192.

Glans L, Ehnbom A, de Kock C, Martínez A, Estrada J, Smith PJ, Haukka M, Sánchez-Delgado RA, Nordlander E (2012). Ruthenium(II) arene complexes with chelating chloroquine analogue ligands: Synthesis, characterization and in vitro antimalarial activity. *Dalton Trans,* **41**: 2764–2773.

Gorzoni Doro F, Queiroz Ferreira K, Novais da Rocha Z, Caramori G, Gomes AJ, Tfouni E (2016). The versatile ruthenium(II/III) tetraazamacrocycle complexes and their nitrosyl derivatives. *Coord Chem Rev*, **306**: 652–677.

Guedes PMM, Oliveira FS, Gutierrez FRS, da Silva GK, Rodrigues GJ, Bendhack LM, Franco DW, Do Valle Matta MA, Zamboni DS, Santana da Silva R, Santana Silva J (2010). Nitric oxide donor *trans*-[RuCl([15]aneN)NO] as a possible therapeutic approach for Chagas' disease. *Brit J Pharmacol*, **160**: 270–282.

Hall MD, Mellor HR, Callaghan R, Hambley TW (2007). Basis for design and development of platinum(IV) anticancer complexes. *J Med Chem*, **50**: 3403–3411.

Iniguez E, Varela-Ramirez A, Martínez A, Torres CL, Sánchez-Delgado RA, Maldonado RA (2016). Ruthenium-Clotrimazole complex has significant efficacy in the murine model of cutaneous leishmaniasis. *Acta Trop*, **164**: 402–410.

Kinnamon K, Steck EA, Rane ES (1979). Activity of antitumor drugs against African Trypanosomes. *Antimicrob Agents Chemother*, **15**: 157–160.

Klein AV, Hambley TW (2009). Platinum drug distribution in cancer cells and tumors. *Chem Rev*, **109**: 4911–4920.

Macedo TS, Colina-Vegas L, Da Paixão M, Navarro M, Barreto BC, Oliveira PCM, Macambira SG, Machado M, Prudêncio M, D'alessandro S, Basilico N, Moreira DRM, Batista AA, Soares MBP (2016). Chloroquine-containing organoruthenium complexes are fast-acting multistage antimalarial agents. *Parasitology*, **143**: 1543–1556.

Martínez A, Carreon T, Iniguez E, Anzellotti A, Sánchez A, Tyan M, Sattler A, Herrera L, Maldonado RA, Sánchez-Delgado RA (2012). Searching for new chemotherapies for tropical diseases: Ruthenium–clotrimazole complexes display high in vitro activity against *Leishmania major* and *Trypanosoma cruzi* and low toxicity toward normal mammalian cells. *J Med Chem*, **55:** 3867–3877.

Martínez A, Rajapakse CSK, Jalloh D, Dautriche C, Sánchez-Delgado RA (2009). The antimalarial activity of Ru–chloroquine complexes against resistant *Plasmodium falciparum* is related to lipophilicity, basicity, and heme aggregation inhibition ability near water/*n*-octanol interfaces. *J Biol Inorg Chem*, **14**: 863–871.

Martínez A, Rajapakse CSK, Naoulou B, Kopkalli Y, Davenport L, Sánchez-Delgado RA (2008). The mechanism of antimalarial action of the ruthenium(II)–chloroquine complex $[RuCl_2(CQ)]_2$. *J Biol Inorg Chem*, **13**: 703–712.

Martínez A, Rajapakse CSK, Sánchez-Delgado RA, Varela-Ramirez A, Lema C, Aguilera RJ (2010). Arene–Ru(II)–chloroquine complexes interact with DNA, induce apoptosis on human lymphoid cell lines and display low toxicity to normal mammalian cells. *J Inorg Biochem,* **104**: 967–977.

Martínez A, Suárez J, Shand T, Magliozzo RS, Sánchez-Delgado RA (2011). Interactions of arene-Ru(II)-chloroquine complexes of known antimalarial and antitumor activity with human serum albumin (HSA) and transferrin. *J Inorg Biochem,* **105**: 39–45.

Medici S, Pean M, Nurchi VM, Lachowicz JI, Crisponi G, Zoroddu MA (2015). Noble metals in medicine: Latest advances. *Coord Chem Rev,* **284**: 329–350.

Meggers E (2007). Exploring biologically relevant chemical space with metal complexes. *Curr Opt Chem Biol,* **11**: 287–292.

Melo Pereira JC, Carregaro V, Costa DL, Santana da Silva J, Cunha FQ, Franco DW (2010). Antileishmanial activity of ruthenium(II)tetraammine nitrosyl complexes. *Eur J Med Chem,* **45**: 4180–4187.

Muhammad N, Guo Z (2014). Metal-based anticancer chemotherapeutic agents. *Cur Opt Chem Biol,* **19**:144–153.

Murray BS, Babak MV, Hartinger CG, Dyson PJ (2016). The development of RAPTA compounds for the treatment of tumors. *Coord Chem Rev,* **306**: 86–114.

Navarro M, Castro W, Biot C (2012). Bioorganometallic compounds with antimalarial targets: Inhibiting hemozoin formation, *Organometallics,* **31**: 5715–5727.

Navarro M, Gabbiani G, Messori L, Gambino D (2010). Metal-based drugs for malaria, trypanosomiasis and leishmaniasis: Recent achievements and perspectives. *Drug Discov Today,* **15**: 1070–1077.

Navarro M, Lehmann T, Cisneros-Fajardo EJ, Fuentes A, Sánchez-Delgado RA, Silva P, Urbina JA (2000). Toward a novel metal-based chemotherapy against tropical diseases: Part 5. Synthesis and characterization of new Ru(II) and Ru(III) clotrimazole and ketoconazole complexes and evaluation of their activity against *Trypanosoma cruzi. Polyhedron,* **19**: 2319–2325.

Noffke AL, Habtemariam A, Pizarro AM, Sadler PJ (2012). Designing organometallic compounds for catalysis and therapy. *Chem Commun,* **48**: 5219–5246.

Nogueira Silva JJ, Pavanelli WR, Salazar Gutierrez FR, Chagas Alves Lima F, Borges Ferreira da Silva A, Santana Silva J, Franco DW (2008).

Complexation of the Anti-Trypanosoma cruzi drug benznidazole improves solubility and efficacy. *J Med Chem,* **51**: 4104–4114.

Pagano M, Demoro B, Toloza J, Boiani L, González M, Cerecetto H, Olea-Azar C, Norambuena E, Gambino D, Otero L (2009). Effect of ruthenium complexation on trypanocidal activity of 5-nitrofuryl containing thiosemicarbazones. *Eur J Med Chem,* **44**: 4937–4943.

Peacock AFA, Sadler PJ (2008). Medicinal organometallic chemistry: Designing metal arene complexes as anticancer agents. *Chem Asian J,* **3**: 1890–1899.

Pillozzi S, Gasparoli L, Stefanini M, Ristori M, D'Amico M, Alessio E, Scaletti F, Becchetti A, Arcangeli A, Messori L (2014). NAMI-A is highly cytotoxic toward leukaemia cell lines: Evidence of inhibition of KCa 3.1 channels. *Dalton Trans,* **43**: 12150–12155.

Rajapakse CSK, Martínez A, Naoulou B, Jarzecki AA, Suárez L, Deregnaucourt C, Sinou V, Schrével J, Musi E, Ambrosini G, Schwartz GK, Sánchez-Delgado RA (2009). Synthesis, characterization, and in vitro antimalarial and antitumor activity of new ruthenium(II) complexes of chloroquine. *Inorg Chem,* **48**: 1122–1131.

Reisner E, Arion VB, Keppler BK, Pombeiro AJL (2008). Electron-transfer activated metal-based anticancer drugs. *Inorg Chim Acta,* **361**: 1569–1583.

Rodrigues FP, Carneiro ZA, Mascharak P, Curtia C, S. da Silva R (2016). Incorporation of a ruthenium nitrosyl complex into liposomes, the nitric oxide released from these liposomes and HepG2 cell death mechanism. *Coord Chem Rev,* **306**: 701–707.

Rodríguez Arce E, Sarniguet C, Moraes TS, Vieites M, Tomaz AI, Medeiros A, Comini MA, Varela J, Cerecetto H, González M, Marques F, García MH, Otero L, Gambino D (2015). A new ruthenium cyclopentadienyl azole compound with activity on tumor cell lines and trypanosomatid parasites. *J Coord Chem,* **68**: 2923–2937.

Ronconi L, Sadler PJ (2007). Using coordination chemistry to design new medicines. *Coord Chem Rev,* **251**: 1633–1648.

Sánchez-Delgado RA, Anzellotti A (2004). Metal complexes as chemotherapeutic agents against tropical diseases: Trypanosomiasis, malaria and leishmaniasis. *Mini Rev Med Chem,* **4**: 23–30.

Sánchez-Delgado RA, Anzellotti A, Suárez L (2004). In: Sigel H., Sigel A. (eds.), *Metal Ions in Biological Systems 41: Metal Ions and Their Complexes in Medication,* Marcel Dekker, New York, p. 379.

Sánchez-Delgado RA, Lazardi K, Rincon L, Urbina JA, Hubert AJ, Noels AN (1993). Toward a novel metal-based chemotherapy against tropical diseases. 1. Enhancement of the efficacy of clotrimazole against Trypanosoma cruzi by complexation to ruthenium in $RuCl_2(clotrimazole)_2$. *J Med Chem*, **36**: 2041–2043.

Sánchez-Delgado RA, Navarro M, Lazardi K, Atencio R, Capparelli M, Vargas F, Urbina JA, Bouillez A, Noels AF, Masi D (1998). Toward a novel metal based chemotherapy against tropical diseases 4. Synthesis and characterization of new metal-clotrimazole complexes and evaluation of their activity against. *Trypanosoma cruzi. Inorg Chim Acta*, **275–276**: 528–540.

Sánchez-Delgado RA, Navarro M, Pérez H, Urbina JA (1996). Toward a novel metal-based chemotherapy against tropical diseases. 2. Synthesis and antimalarial activity in vitro and in vivo of new ruthenium- and rhodium-chloroquine complexes. *J Med Chem*, **39**: 1095–1099.

Sarniguet M, Vieites M, Cipriani M, Lapier J, Toloza JD, Maya C, Olea V, Moreno D, Gambino D, Otero L (2014). Water soluble ruthenium complexes bearing activity against protozoan parasites. *Biol Trace Elem Res*, **159**: 379–392.

Schatzschneider U, Niesel J, Ott I, Gust R, Alborzinia H, Wolfl S (2008). Cellular uptake, cytotoxicity, and metabolic profiling of human cancer cells treated with ruthenium(II) polypyridyl complexes $[Ru(bpy)_2(N\text{–}N)]Cl_2$ with N–N = bpy, phen, dpq, dppz, and dppn. *ChemMedChem*, **3**: 1104–1109.

Silva JJN, Guedes PMM, Zottis A, Balliano TL, Silva FON, Lopes LGF, Ellena J, Oliva G, Andricopulo AD, Franco DW, Santana Silva J (2010). Novel ruthenium complexes as potential drugs for Chagas's disease: Enzyme inhibition and in vitro/in vivo trypanocidal activity. *Brit J Pharmacol*, **160**: 260–269.

Silva JJN, Osakabe AL, Pavanelli WR, Silva JS, Franco DW (2007). In vitro and in vivo antiproliferative and trypanocidal activities of ruthenium NO donors. *Brit J Pharmacol*, **152**: 112–121.

Silva JJN, Pavanelli WR, Pereira JCM, Silva JS, Franco DW (2009). Experimental chemotherapy against *Trypanosoma cruzi* infection using ruthenium nitric oxide donors. *Antimicrob Agents Chemother*, **53**: 4414–4421.

Sun Y, Joyce LE, Dickson NM, Turro C (2010). Efficient DNA photocleavage by $[Ru(bpy)_2(dppn)]^{2+}$ with visible light. *Chem Commun*, **46**: 2426–2428.

Sutak R, Lesuisse E, Tachezy J, Richardson DR (2008). Crusade for iron uptake in unicellular eukaryotes and its significance for virulence. *Trends Microbiol*, **16**: 261–268.

Tfouni E, Ramos Truzzi D, Tavares A, Gomes AJ, Figueiredo LE, Franco DW (2012). Biological activity of ruthenium nitrosyl complexes. *Nitric Oxide*, **26**: 38–53.

Trondl R, Heffeter P, Kowol CR, Jakupec MA, Berger W, Keppler BK (2014). NKP-1339, the first ruthenium-based anticancer drug on the edge to clinical application. *Chem Sci*, **5**: 2925–2932.

van Schalkwyk DA, Egan TJ (2006). Quinoline-resistance reversing agents for the malaria parasite *Plasmodium falciparum. Drug Resist. Update*, **9**: 211–226.

Wink DA, Ridnour LA, Hussain SP, Harris CC (2008). The reemergence of nitric oxide and cancer. *Nitric Oxide*, **19**: 65–67.

http://www.who.int/cancer/en/ visited October 2016.

http://www.who.int/neglected_diseases/diseases/en/, accessed 30 September 2016.

http://www.who.int/mediacentre/factsheets/fs094/en/, accessed 30 September 2016.

Yan YK, Melchart M, Habtemariam A, Sadler PJ (2005). Organometallic chemistry, biology and medicine: Ruthenium arene anticancer complexes. *Chem Commun*, **14**: 4764–4776.

Zeng C, Zhang C, Lai S, Yinb H, Tang B, Wana D, Liu Y (2016). Anticancer activity studies of ruthenium(II) polypyridyl complexes against human gastric carcinoma SGC-7901 cell. *Inorg Chem Commun*, **70**: 210–218.

Chapter 3

An Overview of Ruthenium Complexes as a Potential Sensing Agent

Goutam Kumar Patra and Anupam Ghorai

Department of Chemistry, Guru Ghasidas Vishwavidyalaya, Koni, Bilaspur, Chhattisgarh, India

patra29in@yahoo.co.in

Recently, luminescent transition metal complexes have appeared as attractive candidates for the development of responsive chemosensors and cell imaging probes due to their abundant photophysical, photochemical, and electrochemical properties. Among these complexes, ruthenium(II) complexes with polypyridyl ligands, such as 2,2′-bipyridine (bpy), 1,10-phenanthroline (phen), and/or *ortho* phenanthroline derivatives, have been constantly reported as luminescent chemosensors/probes for oxygen, bioactive molecules, metal cations, and anions in the last few years. As one of useful luminescent reporters for biosensing and bioimaging, ruthenium(II) complexes offer several unique advantages including metal-to-ligand charge transfer (MLCT) based visible-light excitation and emission, large Stokes shifts (typically greater than 150 nm), high chemical and photochemical stabilities, low cytotoxicity, good water solubility, and high response efficiency. In addition, as the result of MLCT-

Ruthenium Chemistry
Edited by Ajay Kumar Mishra and Lallan Mishra

ISBN 978-981-4774-39-0 (Hardcover), 978-1-315-11058-5 (eBook)
www.panstanford.com

based emission, ruthenium(II) complexes have also been widely investigated as an important material for use in electro-generated chemi-luminescence analysis with advantages of high stability, sensitivity, and lower environmental sensitivity. The versatile nature of Ru complexes in sensing is found as amperometric sensor, electrochemical sensor, solid-state sensor and simple chemosensor. It has the potentiality to sense different analytes in different forms such as—(1) being immobilised with a solid support like alumina, polytyramine film, graphene modified electrode, etc. (2) aptamer–Ru complex system (3) Ru-nano shell (4) complex of different bioactive compounds (5) single-colour quantum dots–ruthenium complex (QDs–Ru) assembling dyads (6) sub-micron ruthenium oxide thin-film, etc. The Ru(II) polypyridyl complexes containing imidazole and/or free metal ion binding sites have received growing attention as optical sensors for pH. Dipyrazinylpyridine-appended homoleptic ruthenium complex was also found to exhibit pH sensing properties. Hetero bimetallic Ru(II)–Ru(I) complex was also exploited as a potential chemosensor for anions and pH sensing. Unlike other metal complexes, the versatility and uniqueness of Ru complexes as chemosensor is observed in sensing of uncommon analytes like organo tin halides, hypochlorous acid, peroxy nitrite, etc., which are rarely detected according to the literature. Ru complexes being covalently bonded with photo conducting polymer can also act as photosensitizers. Ru complexes are also used as pH and temperature sensor.

3.1 Introduction

The development of molecular probes capable of detecting cations, anions and neutral species has attracted considerable attention due to their important roles in biological, industrial and environmental processes. In this endeavour extensive work has been devoted to transition metal based fluorescent and/or colorimetric systems, because of their good water solubility, high chemical and photo stability, intense polarised luminescence, red emission, large Stoke shifts and long lifetimes compared to purely organic luminophores (Hupp et al., 2000; Roger and Wolf, 2002). In contrast to conventional singlet-emitter organic fluorophores, transition metal complexes display triplet emission due to spin–orbit coupling imparted by the heavy atom effect, which leads to efficient singlet–triplet state mixing and enhancement of phosphorescence quantum efficiency. The phosphorescence behaviour of metal complexes has found potential for the construction of organic light-emitting diodes (OLEDs) to display or lighting applications (Schubert et al., 2012; Williams et al., 2011). Out of the transition metal complexes, ruthenium complexes are one of the most investigated chemical systems due to their remarkable chemical stability and intriguing ground- and excited-state photophysical and redox properties. Taking advantage of these favourable properties, optical and electrochemical sensors based on the ruthenium complexes have been rapidly developed for the detection of versatile analytes (Boal and Rosenzweig, 2009).

In comparison to the organic fluorophores, the excited state properties of ruthenium complexes are complicated and can include metal-to-ligand charge-transfer (MLCT), ligand-to-ligand charge transfer (LLCT), intra-ligand charge-transfer (ILCT), ligand-to-metal charge transfer (LMCT), metal–metal-to-ligand charge-transfer (MMLCT), ligand-to-metal–metal charge transfer (LMMCT) and metal-to-ligand–ligand charge-transfer (MLLCT) states. The properties of the excited states are highly sensitive to the metal centre, type of the ligands and nature of the local environments, allowing the photophysical properties (e.g. emission wavelength, lifetime and intensity) of metal complexes to be tailored for specific applications (Park et al., 2009). Indeed the charge separation in the MLCT excited state would result in the complexes being usually sensitive to their environment and with appropriate incorporation of receptors, they may act as reporters of the environment. Thus, various sensing applications by making use of Ru complexes can be achieved by the incorporation and design of suitable ligands to the complexes.

3.2 Sensing Property of Ruthenium Complexes

Though there are several review articles on the use of transition metal complexes as chemosensors, the detailed description of recent luminescent Ru complexes as sensing agent is scanty in the literature. This chapter describes the ion binding properties of the Ru complexes containing host receptor(s) R onto their organic moiety (where R denotes a receptor for versatile analytes) and their optical responses, with particular attention to the changes in their luminescence properties upon addition of these analytes in solution. Interaction of cationic and neutral complexes of Ru with gaseous, bioactive and hazardous molecules has also been discussed.

3.3 Ruthenium Complexes as Cation Sensor

Metal cations play an important role in both biological and environmental systems. For example, numerous d-block transition metal ions are involved in normal human physiology and pathology. However, certain heavy metal ions such as Hg^{2+} and Cd^{2+} are toxic to living organisms. The Na^+/K^+ pump maintains the electrochemical gradient in animal cells. Therefore, it is not surprising that continuous effort has also been invested into the development of chemosensors for recognition of cations. Ru complexes serve well for this purpose in detecting cations (Li and Zamble, 2009).

Similar to the other transition metal chemosensors, the Ru complexes, extensively used as chemosensors for the detection of cations, commonly incorporate a receptor motif on their ancillary ligands for the recognition of cations. Another way of cation sensing is the chemodosimeter approach, in which the probe irreversibly reacts with cations showing some visible colour change and significant variation in its luminescence property.

Zheng et al. reported a Ru(II) complex, $[Ru(bpy)_2(Htppip)](ClO_4)_2 \cdot H_2O \cdot DMF$ (**RuL**) where, bpy = 2,2′-bipyridine and Htppip = 2-(4-(2,6-di(pyridin-2-yl)pyridin-4-yl)phenyl)-1Himidazo[4,5-f][1,10]phenanthroline (Fig. 3.1) and thoroughly investigated its cation-sensing properties in both neat CH_3CN and aqueous 4-(2-hydroxyethyl)-1-piperazineethanesulphonic acid buffer (pH = 7.2)/CH_3CN (71/1, v/v) solutions by UV–visible absorption, emission, ^{1}H NMR spectra and DFT studies (Zheng et al., 2013).

These analyses revealed that **RuL** acts as an efficient colorimetric sensor for Fe^{2+}, as evidenced by an obvious colour change from pale yellow to light red-purple. This chemosensor exhibits high selectivity towards Fe^{2+} over the other cations studied (Na^+, Mg^{2+}, Ba^{2+}, Mn^{2+}, Fe^{3+}, Co^{2+}, Ni^{2+}, Cu^{2+}, Zn^{2+}, Cd^{2+}, Hg^{2+}, and Ag^+) through the formation of 1:2 Fe^{2+}–**RuL** complex (Fig. 3.2). Interestingly the colorimetric limit of detection (LOD) for Fe^{2+} in the CH_3CN aqueous solution was determined to be ~4.58×10^{-8} M which is slightly higher than the LOD value of ~4.46×10^{-8} M determined in neat CH_3CN.

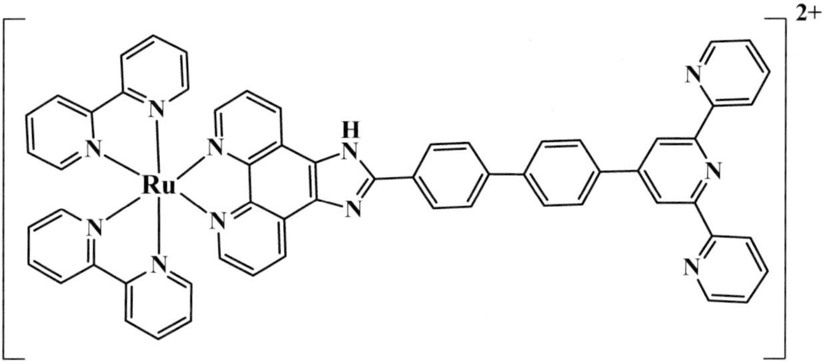

Figure 3.1 $[Ru(bpy)_2(Htppip)]^{2+}$.

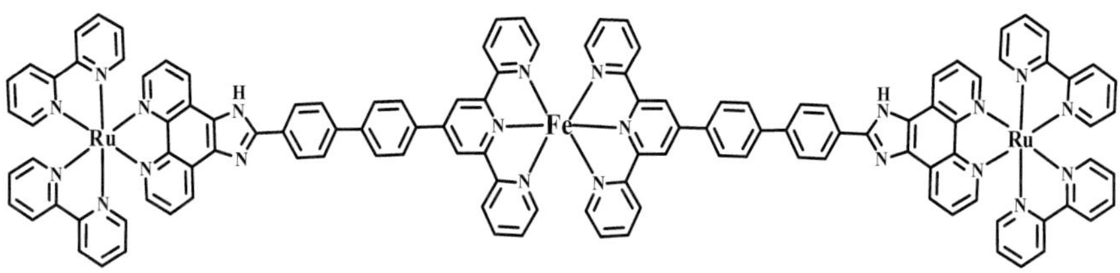

Figure 3.2 $\{Fe^{2+} + 2[Ru(bpy)_2(Htppip)]^{2+}\}$.

Baitalik et al. also reported a monometallic Ru(II) complex of composition $[(bipy)_2Ru(tpy\text{-}Hbzim\text{-}dipy)](ClO_4)_2$, (Fig. 3.3) (where tpy-Hbzim-dipy = 4′-[4-(4,5-di-pyridin-2-yl-1H-imidazol-2-yl)-phenyl]-[2,2′;6′,2′]terpyridine), which can act as a colorimetric sensor for Fe^{2+} ion and showed colour change from yellow orange to deep red-violet in CH_3CN solution with a detection limit of as low as 6.68×10^{-9} M (Baitalik et al., 2014).

Mercury in particular is a toxic metal, which when accumulates in the vital organs of human beings and animals causes poisonous effects that results in serious haematological destruction, such as kidney malfunctioning and brain damage. Therefore, monitoring and precise determination of mercuric ion concentration in water and thus in relevant biological matrices are extremely beneficial for the environmental and toxicological monitoring.

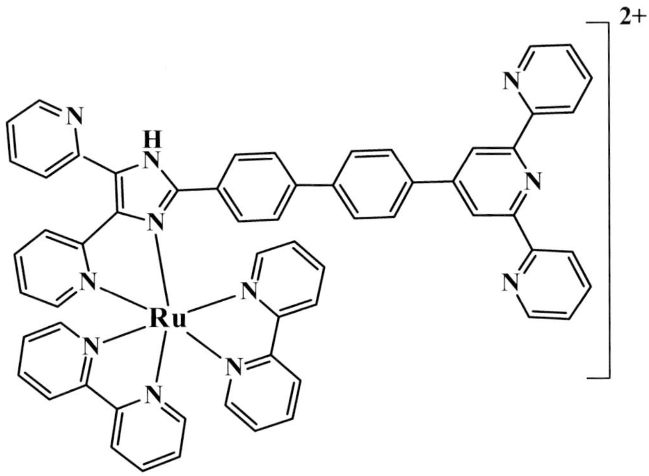

Figure 3.3 [(bipy)$_2$Ru(tpy-Hbzim-dipy)]$^{2+}$.

A new Ru(II)complex, (*cis*-Ru *bis*[2,2-bipyridine]-*bis*[4-aminothiophenol]-*bis*[hexafluorophosphate]) was found to have the potentiality to detect Hg^{2+} in aqueous solution up to 0.4 ppm (Hamid et al., 2011). As shown in (Fig. 3.4), Ru(II)-bis(bipyridine) was selected to act as the carrier for the sulphur-containing 4-aminothiophenol moiety, which acted as the receptor for obtaining a new chemosensor for Hg^{2+} ions. The incorporation of the Hg^{2+} ions into thiols of the coordinated 4-aminothiophenol ligand would be expected to influence the absorption properties of the Ru(II)-bipyridine core and thus allowing access to a new potential chemosensor based on the chromophore-spacer-receptor concept.

Another novel colorimetric probe based on a mesoporous nanocrystalline TiO_2 film sensitized with a ruthenium dye (Fig. 3.5a) has been synthesized by Palomares, and co-workers (Palomares et al., 2004). They efficiently exploited it for the detection of Hg^{2+} ions with easily discernible colour change. Moreover, by adsorbing the sensor molecules onto high-surface-area mesoporous metal oxide films, an easy-to-use system for the dip sensing of mercury (II) in aqueous solution has been developed. These films can be reused for sensing if they are washed with an aqueous solution of KI which presumably removes the mercury from the surface by forming its stable iodide complex.

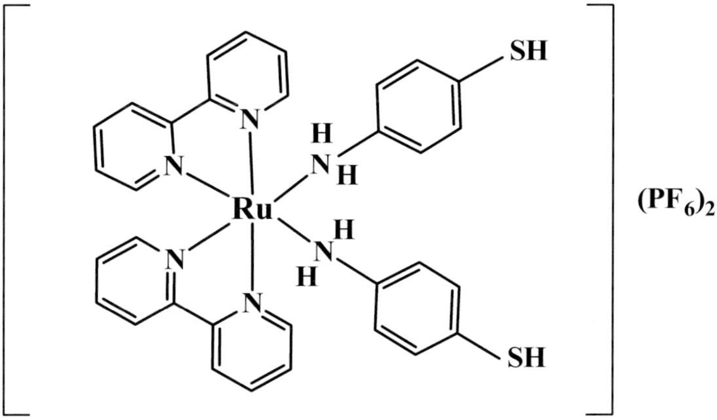

Figure 3.4 *cis*-Ru *bis*[2,2-bipyridine]-*bis*[4-aminothiophenol]-*bis*[hexafluorophosphate].

The same researchers later designed and synthesized another novel Ru-polypyridyl complex (Fig. 3.5b) for detection of Hg^{2+} ions. There, following an entirely different approach, Hg^{2+} was sensed by using a coordination complex of Ru(III) with substituted bipyridine and thiocyanate ions. Hg^{2+} coordinated reversibly to the sulphur atom of the dye's NCS groups and a colour change was observed even at sub-micromolar level concentration of mercury (Palomares et al., 2005).

The bis(2,2′-bipyridine)(4-methyl-2,2′-bipyridine-4′-carboxylic acid)ruthenium(II). $2PF_6$ complex was used by Chun-Lian et al. as a fluorescent chemosensor (FCS) to recognize Cu(II) in EtOH/H_2O (1:1, v/v) solution (Chun-Lian et al., 2006). The chemosensor, consisting of Ru(II) ion, was complexed with the bipyridyl units having the carboxyl groups (Fig. 3.6).

This carboxyl group interacted with the metal ions and the response mechanism of the sensor is based on the fluorescence quenching of complex by binding with Cu(II). The sensor was applied to the quantification of Cu(II) with a linear range covering from 5.0×10^{-8} to 1.0×10^{-4} M and a detection limit of 4.2×10^{-8} M. It is worth mentioning that the response behaviour of the complex to Cu(II) is pH independent in medium condition (pH 4.0–8.0),

However, a too much higher pH value would lead to formation of the precipitation of $Cu(OH)_2$, which in turn, would reduce its complexation with sensor, so fluorescence intensity would increase. With pH lower than 4.0, the fluorescence intensity of the probe decreases obviously with the decrease of the pH value. The possible mechanism for the pH-dependent luminescence of these complexes was proposed to be proton-induced quenching. From the view of sensitivity and response speed of sensor, it was found that pH 6.0 was optimum experimental condition for this probe. The important features of this sensor are high selectivity, neutral medium and large Stokes shift.

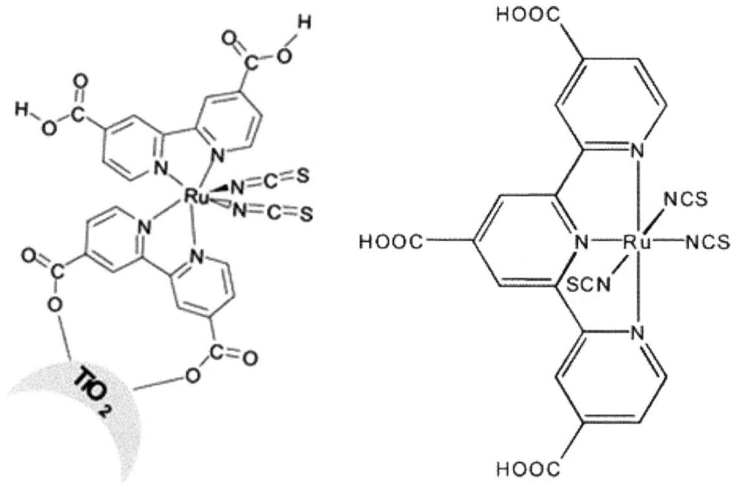

(a) (b)

Figure 3.5 Ru-polypyridyl complexes for detection of Hg^{2+}.

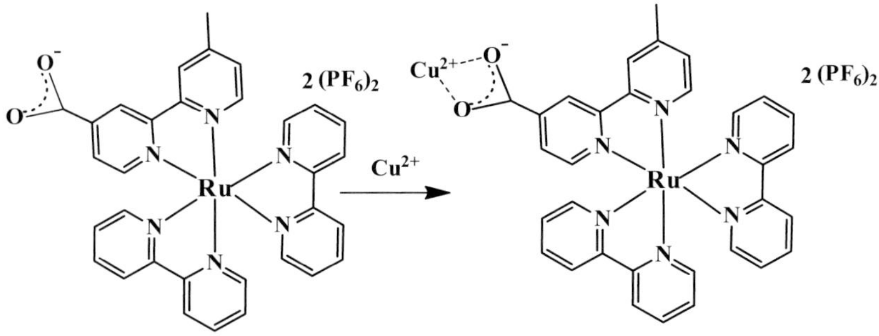

Figure 3.6 Proposed binding mode of the Ru(II) complex with Cu(II) ion.

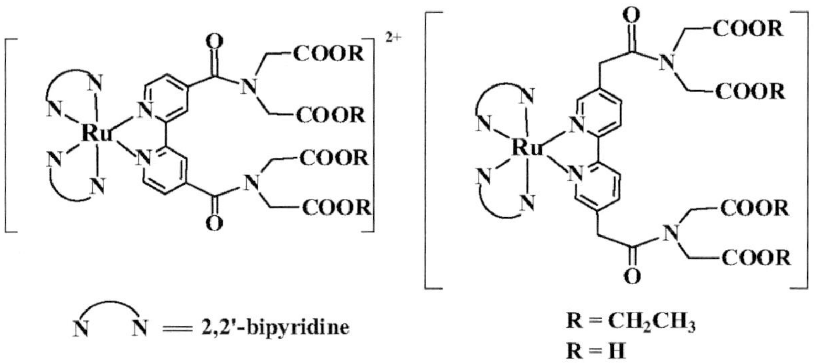

Figure 3.7 Ru-polypyridyl complexes for detection of Ln^{3+} and Eu^{3+}.

Yam et al. showed that the ruthenium(II) polypyridine complexes with iminodiacetic acid containing ligands (Fig. 3.7) can be used for the detection of Ln(III) and Eu(III) exploiting their photo-physical properties (Yam et al., 2000). Fluorescent chemo-dosimeters have become a powerful tool for sensing trace amount of metal ions owing to their simplicity, short response time and high sensitivity. Recently the use of upconversion luminescence (UCL) as the detection signal has become an interesting strategy for sensing and bioimaging (Wang et al., 2005; Liu et al., 2010). This unique UCL mechanism allows rare-earth nanophosphors to display some special advantages as photoluminescent probes in bioimaging, such as a large anti-Stokes shift of several hundred nm, an absence of auto fluorescence from biological samples, remarkable light penetration depth in tissue and no photo bleaching. Taking these advantages Qian et al. synthesized a novel chromophoric Ru complex-assembled nanophosphor (N719-UCNPs) (Scheme 3.1) and exploited it as a highly selective water-soluble probe for upconversion luminescence sensing and bio imaging of intracellular mercury ions (Qian Liu et al., 2011).

Further application of N719-UCNPs in sensing Hg^{2+} was confirmed by optical titration experiment and upconversion luminescence live cell imaging. Using the ratiometric upconversion luminescence as a detection signal, the detection limit of Hg^{2+} for this nanoprobe in water was down to 1.95 ppb, lower than the maximum level (2 ppb) of Hg^{2+} in drinking water set by the US Environmental Protection Agency. Importantly, the nanoprobe

N719-UCNPs has been shown to be capable of monitoring changes in the distribution of Hg^{2+} in living cells by upconversion luminescence bioimaging.

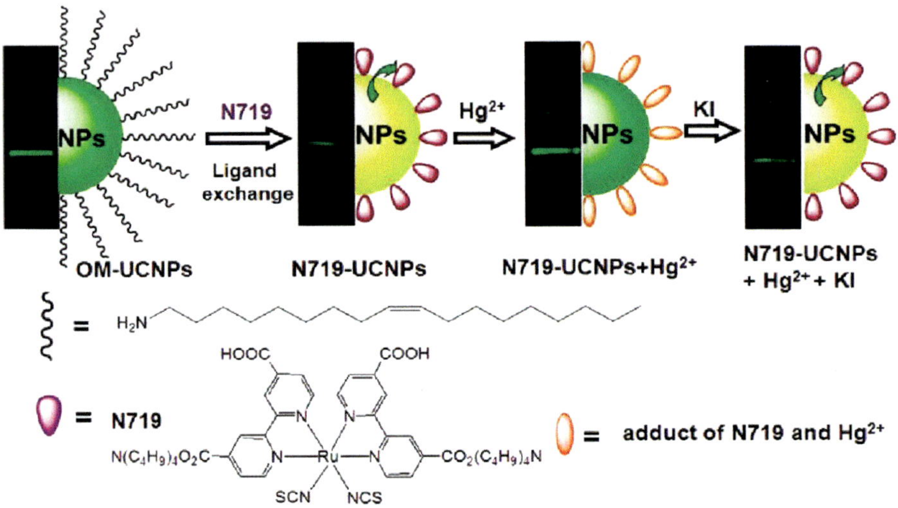

Scheme 3.1 Synthesis of the ruthenium complex N719-modified upconversion nanophosphor (N719-UCNPs) and their UCL response to Hg^{2+}.

A novel water-soluble ruthenium(II) complex as a turn-on copper(II) ion luminescent sensor based on *o*-(phenylazo)aniline (Fig. 3.8) was designed and synthesized by Zhang and co-workers (Zhang et al., 2015). In this probe, the azo group would undergo a specific oxidative cyclization reaction with copper(II) ions and turn into high luminescent benzotriazole, triggering significant luminescent increments which were linear to the concentrations of copper(II) ions. This chemodosimeter is distinguished by its high sensitivity (over 80-fold luminescent *switch-on* response), good selectivity and low detection limit (4.42 nM) in water. The changes of the emission intensity of the probe in the presence of other metal ions or amino acids were negligible. Moreover, the copper(II) luminescent sensor exhibited good photo stability under light irradiation. Furthermore, the proposed chemosensor is successfully utilised in biological samples assay and imaged copper(II) ions in living pea aphids.

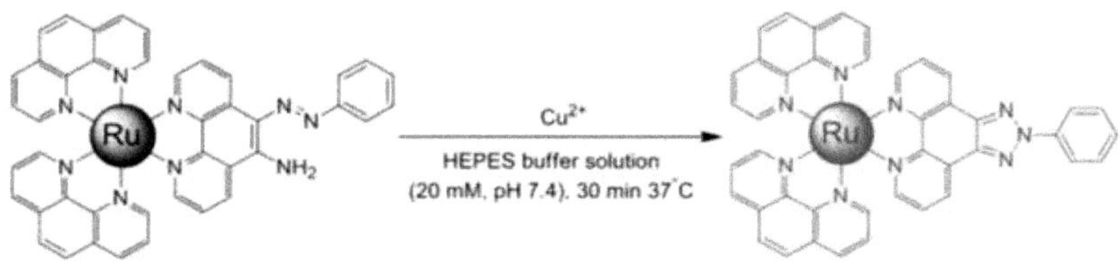

RuMAZO
off

RuTAZO
on

Figure 3.8 Water soluble Ru-complex for detection of Cu^{2+}.

3.4 Ruthenium Complexes an Anion Sensor

Anions are ubiquitous throughout biological systems. Among the range of biologically important anions, the fluoride anion has attracted growing attention due to its established role in preventing dental caries. Fluoride anion is also being explored extensively as a treatment for osteoporosis, a type of fluoride toxicity that generally manifests itself clinically in terms of increasing bone density. Acetate is a critical component of numerous metabolic processes. Acetate production and oxidation rate have been frequently used as an indicator of organic decomposition in marine sediments. Phosphate anions are very important anionic species in living organisms. Naturally occurring phosphate-binding protein (PBP) selectively and strongly bind hydrogen phosphate. The nitrite (NO_2^-) ion also has been used as a vasodilator, bronchodilator, intestinal relaxant or laxative in medicine and investigated for the treatment of heart diseases (Gale et al., 2012). However, high levels of nitrite are toxic to aquatic life. Consequently, the design of new phosphorescent metal complex sensors for anions is of high importance, and has received intense attention in the general fields of both luminescent sensing and supramolecular chemistry (Sancenon et al., 2011). Compared with organic molecules, transition metal complexes can offer unique geometries for the correct positioning of the ligands in the right conformations so that their receptor "host" groups can effectively converge toward the external anion "guest". Among all metal complexes, Ru complexes are extensively used for anion sensing (Hayes et al., 2003).

There are three main approaches in designing of various Ru complexes for anions. The first is to introduce an acidic hydrogen group into the ligands of the complexes. Weakly basic anions can form hydrogen bonds with the acidic hydrogen, while strongly basic anions can induce deprotonation. The second strategy is to employ the specific Lewis acid–base interactions between the anions and Lewis acids such as boron or tin atoms. The third is the chemodosimeter approach that utilizes a specific chemical reaction between the anion and a reactive functional group on the metal complex. A series of artificial receptors, based on Ru complex system, have been designed and synthesized for detection of anions. In this section, the sensing of anions by Ru complexes is illustrated on the basis of these three main strategies.

3.4.1 Anion Sensing through Acidic Hydrogen

Gunnlaugsson and co-workers reported the use of $[Ru(bpy)_3]^{2+}$ derived from 1,8-naphthalimide conjugate for the detection of anions (Fig. 3.9a) (Gunnlaugsson et al., 2010). In the absence of anions, excitation of the complex at 432 nm produced MLCT emission at 615 nm along with a weak shoulder at 505 nm assigned to the naphthalimide emission, suggesting that efficient energy transfer takes place from the naphthalimide singlet excited state to the MLCT. The addition of anions results in quenching of luminescence, with the strongest effect observed for F^-, and then followed by AcO^- and Cl^-. An interesting feature of this design is that the anion recognition event at the naphthalimide site interferes with the energy transfer mechanism from the naphthalimide moiety to the Ru(II) MLCT excited state, causing both the naphthalimide and Ru-centred emissions to be effectively quenched. Emission titration spectroscopy revealed a 1:1 binding ratio of the metal complex to the anion, while NMR studies suggested that the anions interacted with the 4-amino moiety of the naphthalimide motif through hydrogen bonding.

Later, the same group appended an aryl urea motif to one of the ligands in $[Ru(bpy)_3]^{2+}$ to generate complex (b) for anion sensing (Fig. 3.9b). Interestingly, the MLCT emission of complex (b) was sensitive to acetate, phosphate and pyrophosphate ions, but not fluoride ion (Gunnlaugsson et al., 2012). Phosphate enhanced the luminescence intensity of the Ru(II) complex, whereas the addition of pyrophosphate and acetate resulted in a reduced emission response, with nearly complete quenching (>90%) of luminescence observed for pyrophosphate. The lack of response of (b) to F^- ions is noteworthy considering the presence of acidic hydrogen atoms in its urea functionality. NMR experiments showed that all of the anions directly interacted with the urea protons.

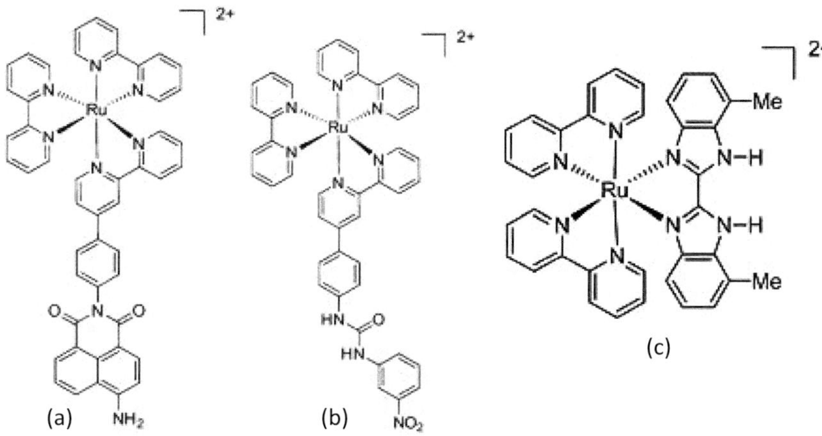

Figure 3.9 Ru-bipyridyl complexes for detection of anions.

$[Ru(bpy)_2(DMBbimH_2)](PF_6)_2$ ($DMBbimH_2$ = 7,7′-dimethyl-2,2′-bibenzimidazole) has been investigated as a luminescent sensor for anions (Fig. 3.9c) (Ye et al., 2011). The imidazole scaffold is a bifunctional unit which possesses both electron lone pairs for coordination with the central metal ion, as well as an NH group for hydrogen bonding to anions in the second coordination sphere. $[Ru(bpy)_2(DMBbimH_2)]^{2+}$ displayed an intermediate pKa value of 6.2, compared with 7.2 and 5.7 for previously reported analogues with 2,2′-biimidazole and 2,2′-bibenzimidazole motifs, respectively. This moderate acidity was significant for sensing weakly basic anions such as Cl^-, Br^-, I^-, or NO_3^-, as it conferred sufficient acidity for hydrogen bonding to those anions without promoting unwarranted deprotonation. Upon binding of weakly basic anions, the emission intensity at 617 nm was enhanced by about 35–40%, with a small red-shift. The enhancement of emission intensity was attributed to the rigidifying effect of hydrogen bonding on the NH bond, which reduces non-radiative decay of the MLCT state through the NH pathway. On the other hand, strongly basic anions such as F^- and AcO^- deprotonated the metal complex, reducing the luminescence intensity through intramolecular quenching of the emissive MLCT state by electron-transfer from the anionic deprotonated amide moiety. This study demonstrated that the hydrogen bonding affinity of these complexes toward anions can be fine-tuned by small changes to the peripheral groups of the anion binding site in order to achieve the desired acidity for anion detection and discrimination.

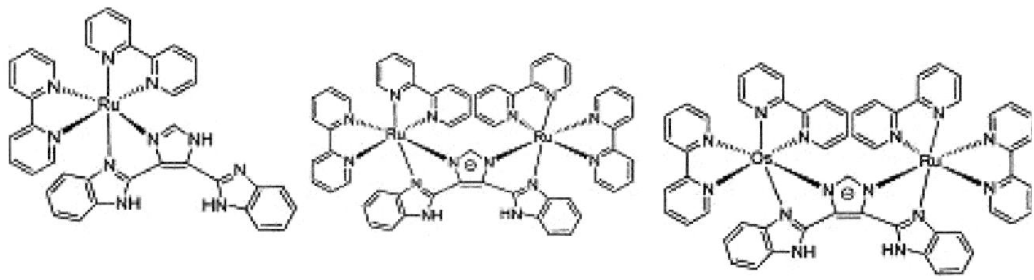

(a) (b) (c)

Figure 3.10 Ru-bimetallic complexes for anion sensing.

The 4,5-bis(benzimidazole-2-yl)imidazole (H_3Imbzim) scaffold was utilized by Baitalik and co-workers to construct monometallic Ru^{II} (Fig. 3.10a), or homobimetallic Ru^{II}-Ru^{II} (1.10b) and heterometallic Ru^{II}-Os^{II} complexes (Fig. 3.10c) for anion sensing (Baitalik et al., 2010). In the bimetallic complexes, the central anionic imidazole moiety bridges the two metal centres, while the flanking benzimidazole subunits provides both coordinative bond formation to the metal and NH groups for anion sensing in the second coordination sphere. The chelating structure of the 4,5-bis(benzimidazole-2-yl)imidazole bidentate ligand enforces a *syn* conformation for the two NH groups of the benzimidazole subunits. Additionally, in the heterometallic ruthenium(II)–osmium(II) complex, photo-induced energy transfer from the ruthenium unit to the osmium unit can also take place. These complexes were quenched by F^- and AcO^- ions, while minimal changes in luminescence were observed upon addition of other anions. A 1:1 hydrogen-bonded adduct was formed at a lower concentrations of anions, whereas stepwise deprotonation of the two benzimidazole NH groups occurred at higher concentrations of anions.

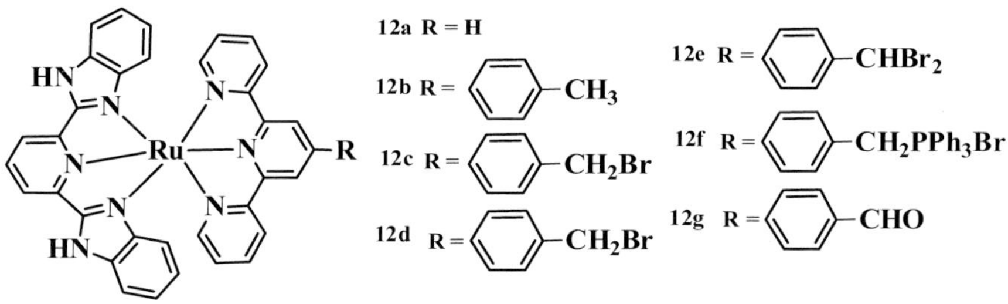

Figure 3.11 Heteroleptic Ru-terpyridyl complexes for anion sensing.

The same group have also utilized the terpyridine (tpy) and the tridentate 2,6-bis(benzimidazole-2-yl)pyridine (H_2pbbzim) motif and constructed a series of luminescent heteroleptic bis-tridentate [Ru(tpy)(H_2pbbzim)]$^{2+}$ complexes (Fig. 3.11). To stabilize the MLCT state, electron-withdrawing groups were introduced to the *tpy* ligand to fine-tune the electronic properties of these complexes. Compared with the parent complexes [Ru(tpy)$_2$]$^{2+}$ or [Ru(H_2pbbzim)$_2$], which are practically non-luminescent at room temperature with very short excited state lifetimes, the heteroleptic complexes exhibited extended room-temperature lifetimes that increase with electron-withdrawing substitution at the *tpy* ligand. The addition of F^- or AcO^- ions quenches the luminescence of the complexes with a concomitant red-shift from 690 to 740 nm, while weakly basic anions such as Cl^-, Br^-, I^-, NO_3^- and ClO_4^- had minimal effect. The selectivity of the complex for F^- and AcO^- was generally mirrored in the colorimetric, electrochemical and NMR titration results. The mechanism of luminescence was attributed to deprotonation of the NH proton from the benzimidazole moiety, aided by the electron-withdrawing character of the metal centre. In a later work, a series of hetero- and homoleptic tridentate Ru(II) were generated containing the phenanthrene-imidazole (HImzphen) motif as an anion recognition site (Fig. 3.12) (Baitalik et al., 2011). Similar to previous studies, basic anions such as F^- and AcO^- quenched the luminescence emission of the complexes.

Interestingly, a blue-shift was also observed for [Ru(tpy)$_2$]$^{2+}$ derivatives (Fig. 3.12) on addition of F^- ion, whereas a red-shift was observed in the corresponding MLCT absorption band. This was attributed to different degrees of destabilization of the *tpy*-HImzphen π^* orbital and the $Ru^{II}(d\pi)^6$ metal-centred orbital in the ground state and excited state complexes. The anion sensing properties of the complexes were paralleled in the absorption, NMR spectroscopic and cyclic voltammetric results.

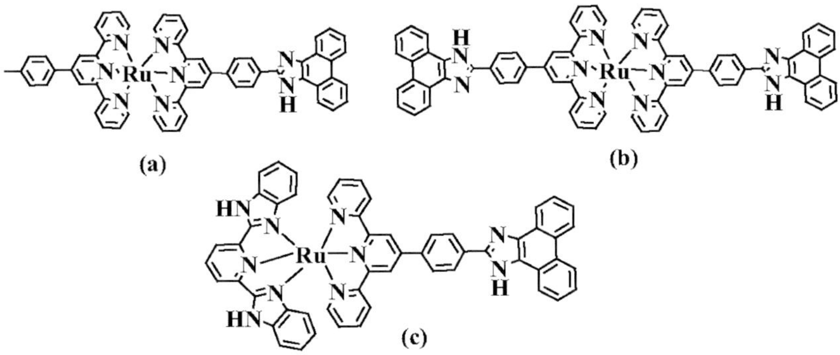

Figure 3.12 Ru-complexes based on phenanthrene-imidazole (HImzphen) motif.

Shang et al. reported a series of artificial receptors, based on Ru-sulphonamido system (Fig. 3.13). The interaction of these receptors with biologically important anions was determined by UV-vis, ^1H NMR titration and electrochemical experiments (Shang et al., 2009). Results indicate that these receptors show high recognition abilities for fluoride (F^-) or acetate (AcO^-), moderate affinities for dihydrogen phosphate ($H_2PO_4^-$) or hydroxyl (OH^-) and almost no affinities for chloride (Cl^-), bromide (Br^-) and iodide (I^-). ^1H NMR titration shows that the interaction between the receptors and anions depends on the hydrogen-bond formation. The two –NH of the sulphonamido system in these complexes are in *ortho*-configuration. According to the optimized geometry, the space for anionic coordination of these systems is better fit for the size of fluoride and acetate anion. Therefore, sulphonamido receptors with –NH located in *ortho* positions have a strong affinity to fluoride and acetate anion. The advantages of using receptors (Fig. 3.13) for the detection of various anions are as follows: (1) the receptors have strong binding ability for F^- and AcO^-; (2) The receptors accompany visible colour changes when they interact with F^- and AcO^-. Therefore, the sulphonamido complexes may be used as convenient naked-eye detection tool, such as the detection of F^- in toothpaste.

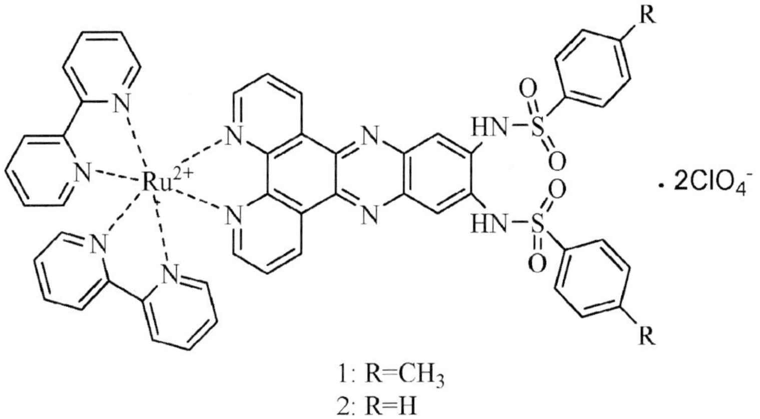

Figure 3.13 Ru-sulphonamido system for detection of anions.

3.4.2 Anion Sensing through Lewis Acidic Sites

Ligands containing Lewis acidic groups capable for coordination with anions can be utilized as part of a metal complex for anion sensing. For example, three-coordinate boron compounds are known to bind to fluoride and cyanide ions, resulting in disruption of the π–π conjugation around the boron centre. Depending on the nature of the linker region, binding of the anion at the Lewis acidic site can influence the photophysical properties of the metal centre, resulting in a change in emission.

A series of $[Ru(bpy)_3]^{2+}$-based complexes containing tri arylboron groups has been synthesized for anion detection by Wang et al. (Wang et al., 2011) (Fig. 3.14). Compared with the parent molecule $[Ru(bpy)_3]^{2+}$, the boryl-functionalized complexes display red-shifted MLCT emission, which was attributed to the lowering of the LUMO energy through π-conjugation of the *bpy* moiety with the electron withdrawing *tri* arylboron group. Meanwhile, the *t-butyl-bpy* complexes exhibited lower phosphorescence energies compared with their *bpy* counterparts, which could be explained by the raising of the HOMO ($\mathbf{t_{2g}}$) energy level by the electron-donating *t-butyl* groups. The addition of fluoride or cyanide ions caused a 10–40 nm blue-shift in the emission of the complexes to 585–600 nm, consistent with the change of the boryl group from electron-withdrawing to electron-donating upon anion binding. The shift in phosphorescence wavelength upon analyte binding could be potentially exploited for the ratiometric detection of anions. This study was the first to demonstrate the tuning of MLCT phosphorescence in Ru(II) complexes by fluoride and cyanide anions.

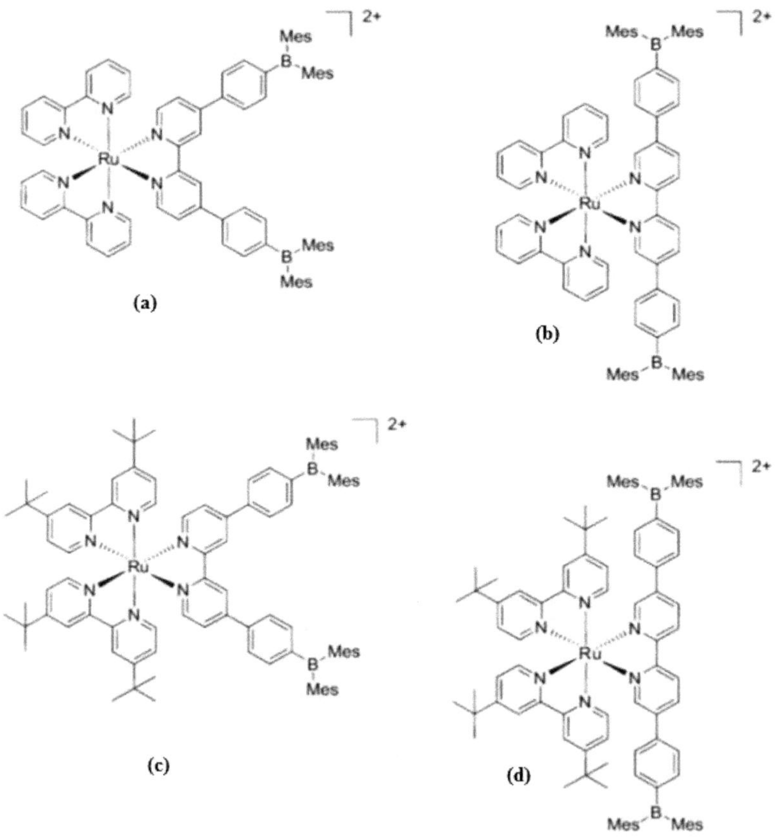

Figure 3.14 Ru-complexes containing tri arylboron groups.

3.4.3 Anion Sensing through Chemodosimeter Approach

Chemodosimeters achieve analyte recognition via chemical transformations involving bond breakage and bond formation

(Li et al., 2012). Chemodosimeters usually comprise two units: (1) a reactive functional group that is capable of carrying out a chemical transformation in the presence of a specific analyte and (2) functionalities that recognize the chemical reaction with an altered spectroscopic signal. The chemical transformation involved in the design of chemodosimeters is usually highly specific, thus providing enhanced selectivity for the analytes.

In 2011, Das and co-workers have developed a Ru(II) complex luminescent chemodosimeter for nitrite anion detection (Fig. 3.15) (Das et al., 2011). Their design was based on the reaction of a vicinal diamine group situated on the phenanthroline ligand with N_2O_3, which is produced from NO_2^- under acidic conditions. These results in the formation of a triazole derivative which shows a small red-shift and an approximate doubling of luminescence intensity compared with the parent diamine complex. Furthermore, the system was selective for NO_2^- over other reactive oxygen and nitrogen species (RONS) such as NO_3^-, $H_2O_2^-$, O_2^- and ClO^-.

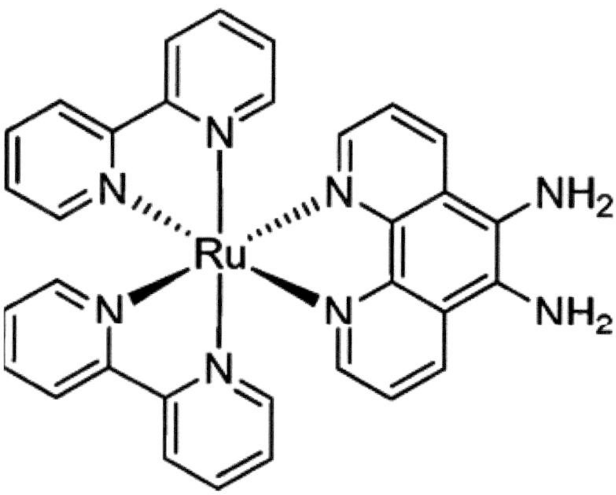

Figure 3.15 Ru-complexes for nitrite anion detection.

A trinuclear heterobimetallic Ru(II)-Cu(II) donor-acceptor complex, $\{RuII(tbubpy)(CN)_4\text{-}[CuII(dien)]_2\}(ClO_4)_2$ (*tbubpy*)4,4′-di-tert-butyl-2,2′-bipyridine; dien (diethylenetriamine) (Fig. 3.16), has been synthesized and successfully used as an chemodosimetric ensemble for the specific detection of cyanide ion in aqueous DMF (Wong et al., 2004). The MLCT transition and luminescence properties of the solvatochromic $[Ru^{II}(tbubpy)(CN)_4]^{2-}$ donor are perturbed by the coordination of two Cu(II) acceptors but restored in the presence of CN^-. Spectroscopic and mass spectrometric studies confirmed the cleavage of the cyano bridge between Ru(II) and Cu(II) of the chemodosimetric ensemble after the binding of cyanide to the Cu(II) centres. The overall binding constant, K_B, between the complex and CN^- is measured to be $(7.39 \pm 0.23) \times 10^6\ M^{-1}$. A detection limit of 1.2 μM (0.03 ppm) of CN^- in aqueous DMF (pH 7.4) was achieved. Thermodynamic evaluation shows that the analyte specificity of this chemodosimeter is attributable to the relative stability of the donor-acceptor complex to that of adducts formed between the acceptor metal centre and the analytes.

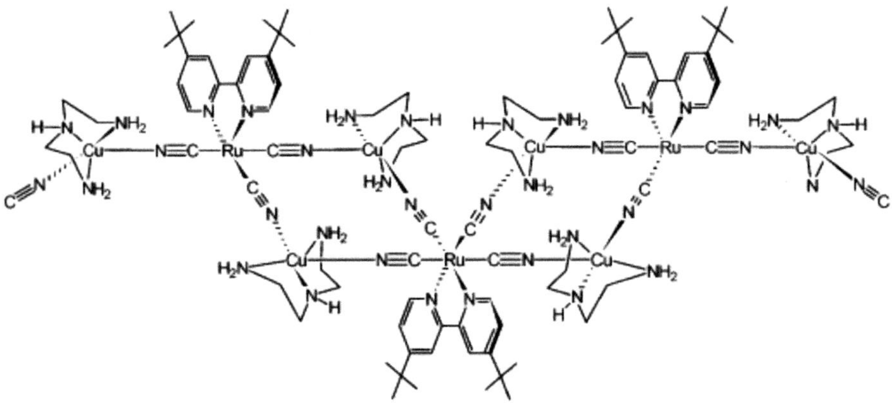

Figure 3.16 Ru(II)-Cu(II) donor-acceptor complex for CN^- detection.

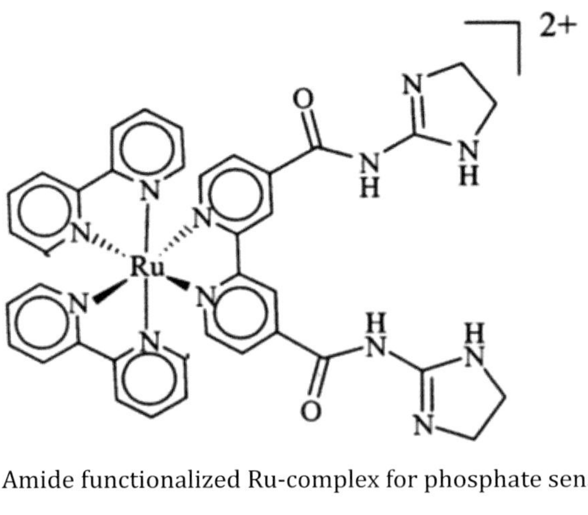

Figure 3.17 Amide functionalized Ru-complex for phosphate sensing.

Another $[Ru(bpy)_3]^{2+}$ based assembly with amide functionalities (Fig. 3.17), designed by Watanabe and co-workers, binds anionic phosphor-diesters (Watanabe et al., 1998). For example, addition of 10 equivalents of anionic tetraethylammonium diphenyl phosphate to a 0.02 mM acetone solution of the complex yielded a factor of 1.3-fold increase in the luminescence intensity. The authors suggested that the increase is a consequence of binding-induced rigidification and inhibition of activity in vibrational modes responsible for nonradiative decay. Again, an alternative interpretation is that anion binding electrostatically destabilizes the coordinated *bpy* ligand created in the emissive MLCT state. Destabilization, in turn, should increase the excited state-ground state energy gap, thereby decreasing the nonradiative decay rate and increasing the emission quantum yield.

3.5 Ruthenium Complexes as Gas Sensor

Certain gaseous molecules, such as oxygen (O_2) and nitric oxide (NO), play pivotal roles in biological and environmental science. Oxygen analysis is important in determining the suitability of water for living organisms (biological oxygen demand), blood gas analysis, pressure-sensitive paints, in vivo analysis, and in combustion monitoring. Besides, aberrant intracellular O_2 levels, is a sign of cancer and can impair mitochondrial respiration (Vaupel et al., 2001).

The emission of luminescent metal complexes can be quenched by interaction with particular chemical species. O_2 is a well-documented quencher of the triplet excited state of phosphorescent metal complexes, which takes place mainly via energy transfer from the triplet ground state of O_2 to form singlet oxygen. Therefore, it is not surprising that the most oxygen phosphorescent chemosensors are designed based on the phosphorescence quenching mechanism, thus resulting in a "switch-off" response of the metal complex in the presence of O_2. In order to examine the relative efficiency of the O_2 sensor, a Stern–Volmer plot is commonly employed to determine the quenching constant, which is an indirect measure of the O_2 sensing capacity of sensor.

Since the excited states of many species that can be quenched by oxygen are also sensitive to environment and respond to metal ions, oxidants, reductants, surfactants, DNAs, proteins, etc., the key to practical systems is separating the sensor molecule from all these interferents while still allowing access to oxygen. The simplest and most widely used approach is to put the sensor into a gas permeable, solvent impermeable membrane. Some of the most successful sensors have been based on silicone elastomers since the basic polydimethyl siloxane structure can have excellent oxygen diffusion and quenching with the added benefit of good bio compatibility. Besides, various polymers (polymethylmethacrylate, polyvinyl chloride, polystyrene, derivatized cellulose), plasticizers in polymer (to improve permeability and hence response) have also been used as common supports. Another versatile anchoring scheme uses sol–gel inclusion. Because of their long excited state lifetimes, large shifts between excitation and emission, and efficient emissions, metal complexes immobilized within a solid supporting material have become important oxygen sensors (Huang et al., 2012). Ru complexes in particular are found to be very efficient in sensing of oxygen. Complex crystal structures containing void spaces have also been employed for this purpose as O_2 molecules can freely move inside the crystal and quench the luminescence.

A series of $[Ru(bpy)_2(phen)]^{2+}$ complexes derivatized with ethynylated phenyl, pyrenylethynylene and pyrene units appended to the phen ligand (Fig. 3.18) have been investigated for their abilities to detect O_2 (Zhao et al., 2010, 2012). The emission lifetime and intensity were tuneable by modification of the pendant group. Among these, the complex (Fig. 3.18c) was found to possess sufficiently long lifetime (τ = 58.4 μs) and excellent solubility in organic solvents and polymers, making it a suitable candidate to be developed as an O_2 sensor. This complex displayed high sensitivity and selectivity for O_2 over air or N_2, as demonstrated by a significant decrease in phosphorescence intensity in solution or IMPES-C sensing film.

Later on, Wang and co-workers reported two closely related polypyridyl Ru(II) analogues appended with aryl ethynylene ligands (Fig. 3.18e,f) (Wang et al., 2010). These complexes could also function as chemosensors for O_2 in various polymer films, although their luminescence lifetimes were lower than for complexes 3.18a–d. Zeolites, with their novel cage structure also provide an alternative support. For example, zeolite Y consists of 13 Å super cages with 7 Å windows, ensuring that a sensitizer like $[Ru(bpy)_3]^{2+}$, with a diameter of 12 Å is securely entrapped, thus preventing leaching of the dye (Balzani et al., 2001). Oxygen is an efficient quencher of photoexcited $[Ru(bpy)_3]^{2+}$ by an energy transfer process, and the resulting deactivation of $[Ru(bpy)_3]^{2+}$ allows for spectroscopic monitoring by emission intensity (McBride et al., 1977). Meier et al. have exploited this encapsulation technique for making optical oxygen sensors (Meier et al., 1995). They used dealuminated zeolite Y as host with Si/Al ratio of 6 and reported the quenching of the chemosensor upon exposure to gas phase O_2. However, this method of oxygen sensing was implicated with a problem that the accessibility of the entrapped $[Ru(bpy)_3]^{2+}$ inside the zeolite towards O_2 was hindered. In a recent study, it was determined that this was primarily due to intrazeolitic water, which impeded the diffusion of O_2 through the zeolite (Dutta et al., 2003). Hence, the dissolved oxygen in water cannot be detected using the aluminous zeolites, since water blocks the diffusion of O_2.

The same researchers later reported a highly siliceous (Si/Al > 100) zeolite that proposed a sensor, designed for dissolved oxygen alleviating the problems of intrazeolitic O_2 diffusion. The conventional procedure using ion-exchange of hexammine ruthenium could not be used here because of the lack of ion-exchange sites in the siliceous zeolites. The emission quenching of intrazeolitic $[Ru(bpy)_3]^{2+}$ by oxygen dissolved in water was studied. Dealumination of the zeolitic framework leads to a hydrophobic environment that promotes oxygen transfer from water to the zeolite. PDMS films with the siliceous zeolites were fabricated and the quenching of $[Ru(bpy)_3]^{2+}$ with dissolved oxygen was tested. Stern–Volmer plots were obtained and Non-linearity in the Stern–Volmer plots was explained as arising from nonhomogeneous distribution of the $[Ru(bpy)_3]^{2+}$ within the zeolite (Fig. 3.19), with more dye nearer the zeolite surface. The long-term stability of the $[Ru(bpy)_3]^{2+}$-siliceous zeolite is expected to be high, since there will be no dye leaching and photodecomposition is also minimized. A dissolved oxygen sensor based on the quenching of fluorescence from a ruthenium dye complex entrapped in a porous sol-gel film is reported by McEvoy et al. (McEvoy et al., 1996). Sol-gel-derived silica films were fabricated by dip-coating onto planar and optical fibre substrates. The films were pre-doped with the oxygen-sensitive ruthenium complex [Ru-*tris*(4,7-diphenyl-l, l0-phenanthroline)], the fluorescence of which is quenched in the presence of oxygen. The structure and behaviour of sol-gel films are dependent on the fabrication parameters. In particular, enhancement of the surface hydrophobicity increases the quenching response in water. This is achieved by using suitable proportions of modified precursors of silica of the form $R(OEt)_3Si$, where R is an alkyl group, in the standard fabrication procedure. On increasing the ratio of modified precursor, the quenching response in the aqueous phase increases. A very low limit of detection, 6 ppb, was achieved by the authors for these modified films. Using a high-brightness blue LED, combined with a miniature photodiode detection system, this oxygen-sensitive ruthenium complex immobilized in sol-gel-derived porous silica coatings may act as a low-cost, high-performance, portable dissolved oxygen sensor in many varied situations from aeration control to on-line river pollution monitoring.

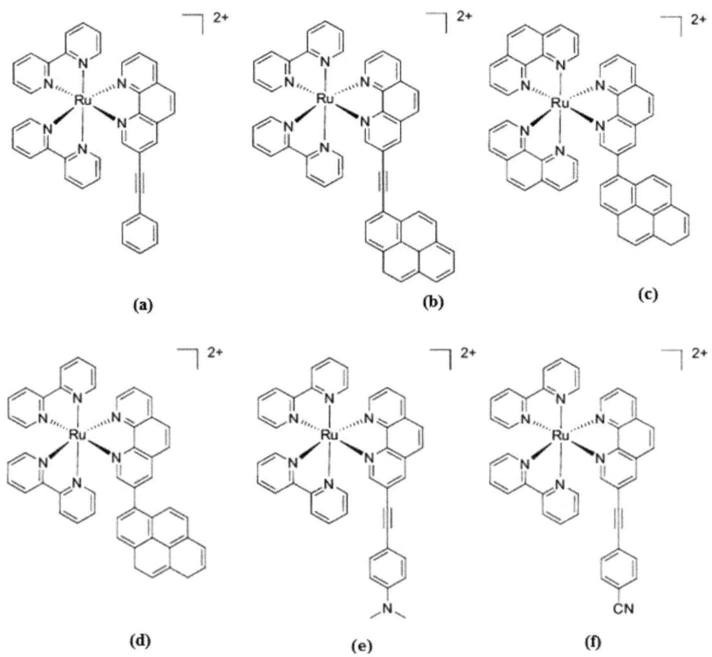

Figure 3.18 Ru complexes used for O_2 sensing.

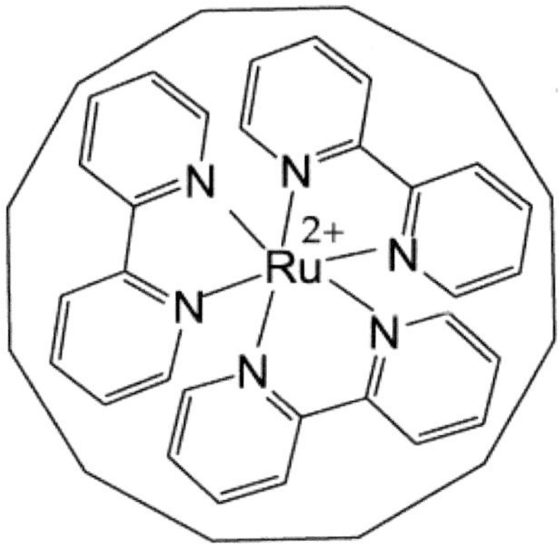

Figure 3.19 Ru complex in zeolite cage structure.

3.6 Ruthenium Complexes as Sensing Agent of Neutral Species

Ruthenium complexes are playing a sharply increasing role in inorganic and medicinal chemistry not only due to its abundant uses as a potential sensing agent for metal cations, anions and gases but also it has promising sensing capability towards different neutral molecules including amino acids, DNA, drug molecules and hazardous compounds.

3.6.1 Detection of Amino Acids

Amino acids are a class of molecules that play a fundamental role in the human body as both the basic building blocks of proteins and mediators in various biochemical pathways. For instance, tryptophan and tyrosine are precursors of the important neurotransmitters serotonin and dopamine, respectively, aspartate and glycine can be used for the biosynthesis of nucleotides (Roth et al., 1997), while histidine is involved in tissue generation and repair pathway (Hong et al., 1999). To this end, increasing attention has been paid to the development of selective and sensitive detection methods for amino acids. Most metal complex chemosensors for the detection of cysteine (Cys) or homocysteine (Hcy) usually contain appended aldehyde functionality on the ancillary ligand(s). The aldehyde reacts readily with Cys (or Hcy) to form a thiazinane (or thiazolidine) group, which leads to a change in luminescence signal. An alternative strategy is to employ an addition reaction between the thiol group of the amino acid with an unsaturated double bond (such as 1,4-addition to an enone) in order to disrupt the π-conjugation of the metal complex, thus resulting in luminescent response. Both approaches can be considered as part of the chemodosimeter strategy.

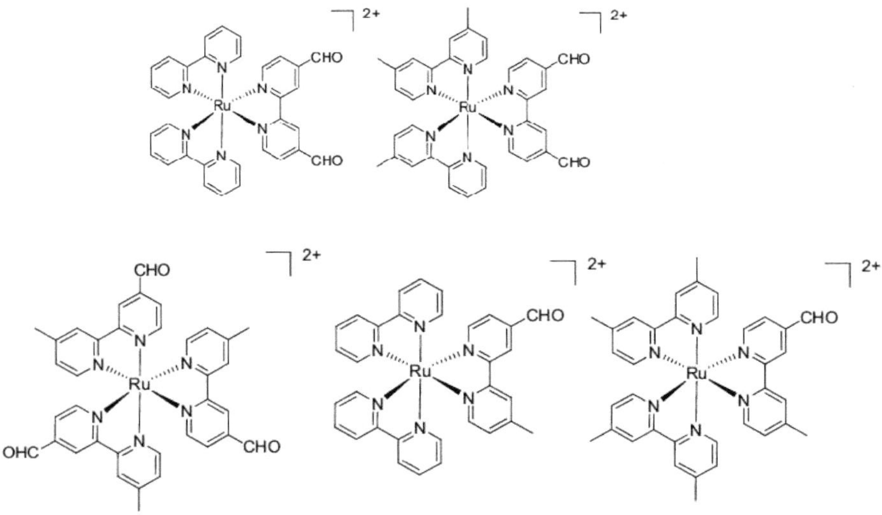

Figure 3.20 $[Ru(bpy)_3]^{2+}$ complexes functionalized with aldehyde group(s) for sensing Cys and Hcy.

A Ru(II) polypyridyl complex (Fig. 3.20) functionalized with aldehyde groups for the selective detection of Cys and Hcy has been reported (Yuan et al., 2010). In this design, the aldehyde groups have a dual function as both selective receptors for amino acids and as strong electron-withdrawing groups that quench the background luminescence of the complex. The addition of Cys and Hcy triggers the formation of the thiazinane (Cys) or thiazolidine (Hcy) motif, which results in a significant luminescence enhancement and blue-shifted emission of the complex, which was attributed to the impairment of electron-withdrawing ability of the ligand upon reaction. The detection limits of the complexes for Cys and Hcy were 1.4 and 1.2 μM, respectively. The probe was highly selective for Cys and Hcy even in the presence of interfering biological substances such as other amino acids, DNA and proteins. Later, ruthenium(II) polypyridyl complexes (mono- or di-substituted with aldehyde groups at various positions) were synthesized for the detection of Cys and Hcy using a similar mechanism. An approximately 10-fold enhancement in luminescence emission was reported at saturating concentrations of Hcy, and detection limits of 0.3 and 1 M were reported for Hcy and Cys, respectively.

3.6.2 Detection of DNA Molecule

Over the past decades, the design and synthesis of metal complexes that bind to DNA with specific site in order to develop novel chemotherapeutics, chemical probes and highly sensitive diagnostic agents for DNA have received considerable attention because DNA is not only the central carrier of genetic information in living organisms but also the data encoded within DNA can be transcripted and translated into a variety of functional and structural proteins that are essential for life. Although much work has been done on planar Pt(II) complexes, some of the most interesting recent work has used octahedral Ru(II) complexes with α-diimine ligands pioneered by Barton et al. (Barton et al., 1986, 1989). For sufficiently extended π systems, the α-diimine may bind in the major groove of the DNA helix by intercalation of one ligand into the base pair stacking. The remaining two ligands sterically interfere with the phosphate backbone and block full insertion.

Since the metal complex is trigonal, it exists in Λ and Δ enantiomeric forms (shown in Fig. 3.21 for $[Ru(Ph_2phen)_3]^{2+}$). The Λ isomer experiences less severe steric crowding than the Δ isomer on intercalation into the helical optically active DNA duplex, which causes tighter binding for the Λ isomer and a higher affinity. This feature is used to probe different DNA conformations. To date, considerable work has been done on these Ru(II) systems with a variety of α-diimine ligands. Some important features that govern the binding of Ru-complexes with DNA molecule are (1) Ru(II) complexes make nearly ideal probes because of the long τ values (0.1–10 ms), bright luminescences, and variable excited state redox properties for DNA nicking. Their emissions allow use of a wealth of luminescence techniques (2) Large hydrophobic ligands are required to obtain specific intercalative binding versus a non-specific ionic binding. Intercalative binding is required for chiral recognition. Selectivity goes as follows: $Ru(bpy)_3^{2+}$ < $Ru(phen)_3^{2+}$ < $Ru(Ph_2phen)_3^{2+}$. For smaller ligands such as phen, there was controversy over the precise binding mode, although for the phen, binding is probably not intercalative (Bottomley et al., 1997) (3) Chiral recognition occurs with B-DNA but not Z-DNA. Barton suggested using L isomers to identify Z-DNA regions; the L isomer does not bind to B-DNA but does to Z-DNA (4) Single stranded DNA binds electrostatically and shows little chiral recognition (5) Highly reducing excited states photocleave DNA at guanine. O_2 quenching is less specific. Gill and Thomas have recently reviewed the use of inert Ru(II) polypyridyl complexes as structural probes for DNA, and have highlighted the inter-disciplinary nature of this research in cellular imaging and therapeutic applications (Gill and Thomas, 2012). Metal complexes that display a change in luminescent behaviour upon binding may be utilized as probes for DNA. The archetypical examples are the "molecular light switch" ruthenium(II) complexes $[Ru(bpy)_2(dppz)]^{2+}$ and $[Ru(phen)_2(dppz)]^{2+}$ developed by the groups of Barton and Sauvage in the early 1990s (Barton et al., 1992). These complexes are non-emissive in aqueous solution due to non-radiative decay mechanisms of the excited state by solvent interactions. However, protection of the *dppz* moiety within the hydrophobic environment of DNA results in intense ^{3}MLCT luminescence. This emission is generated from the ^{1}MLCT excited state ligand. The *dppz* ligand and its derivatives have thus been commonly utilized for the construction of "molecular light switch" metal complexes for DNA (Tuite et al., 2011).

The first transition-metal complex-based two photon absorbing luminescence lifetime probes for cellular DNA was presented by Baggaley et al. (Baggaley et al., 2014). This allowed cell imaging of DNA free from endogenous fluorophores and potentially facilitates deep tissue imaging. In this study, ruthenium(II) luminophores (Fig. 3.22) are used as phosphorescent lifetime imaging microscopy (PLIM) probes for nuclear DNA in both live and fixed cells. The DNA-bound probes display characteristic emission lifetimes of more than 160 ns, while shorter-lived cytoplasmic emission is also observed. These timescales are orders of magnitude longer than conventional FLIM, leading to previously unattainable levels of sensitivity, and auto fluorescence free imaging.

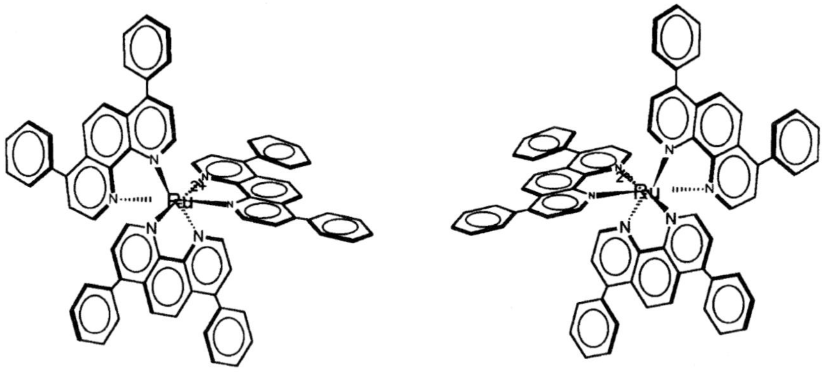

Λ Δ

Figure 3.21 A typical structure for a Ru(II) sensor–probe for DNA.

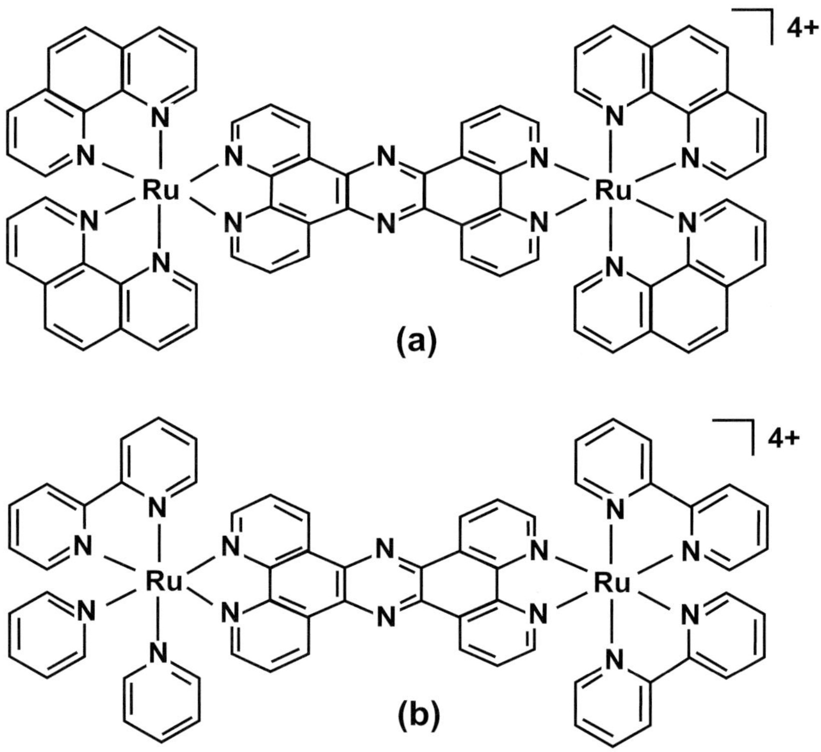

Figure 3.22 Structures of the Ru complexes used in DNA sensing.

3.6.3 Detection of Other Biomolecules

Amyloid-β, an extracellular protein fragment, is self-assembled to form aggregates, which is implicated in many neurodegenerative disorders such as Alzheimer's disease, Parkinson's disease and type II diabetes (Dobson et al., 2003). Alzheimer's disease is a common amyloidosis disease, characterized by severe cognitive dysfunction and memory impairment leads to complete loss of independent living and self-carrying capacity (Skovronsky et al., 2009). Therefore, the detection of β-amyloid demands rapt attention of researchers in the field of sensors.

Metal complexes represent viable alternatives to organic small molecules for the development of structural probes for biomolecules. Compared with the mostly tetrahedral or trigonal planar conformations usually exhibited by organic compounds, metal complexes possess distinct geometries that can be harnessed to create structural complexity and to access novel chemical space. The ligands of the metal complex can be carefully arranged for selective interactions with biomolecules (Meggers E. (2009)). Furthermore, the long phosphorescent lifetime of metal complexes allows their phosphorescence to be readily distinguished in the presence of endogenous fluorophores likely to be present in biological environments by use of time-resolved spectroscopy or fluorescence lifetime imaging microscopy. Finally, metal complexes generally benefit from a modular synthesis which allows analogues of the hit structures to be synthesized rapidly and conveniently, compared with the somewhat lengthy, linear and protecting group-laden syntheses of complex organic molecules.

Taking these advantages of a metal complexes over its organic counterpart, a classical "molecular light switch" complex $[Ru(bpy)_2dppz]^{2+}$ (*dppz* = dipyridophenazine) has been utilized as a luminescent probe for Aβ-fibrils (Fig. 3.23) (Marti et al., 2011). The authors suggested that the sensing mechanism of the Aβ-fibrils by $[Ru(bpy)_2dppz]^{2+}$ would be analogous to that offered by intercalation of the metal complex into DNA. The protection of the complex within the hydrophobic interior of the peptide may suppress non-radiative decay of the excited state, thereby enhancing luminescence emission. Using time-resolved emission spectroscopy, the phosphorescence of the complex in response to Aβ-fibrils could be readily distinguished in the presence of a strongly fluorescent background. Furthermore, the aggregation and fibrillation of monomeric Aβ could be monitored in real-time using the Ru(II) complex as a probe.

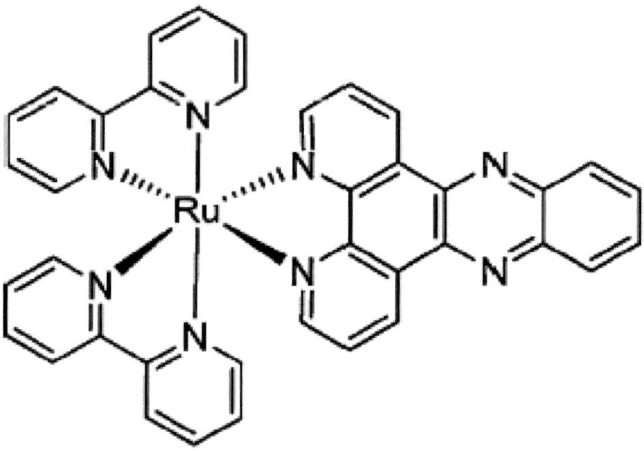

Figure 3.23 Chemical structure of molecular light switch $[Ru(bpy)_2dppz]^{2+}$.

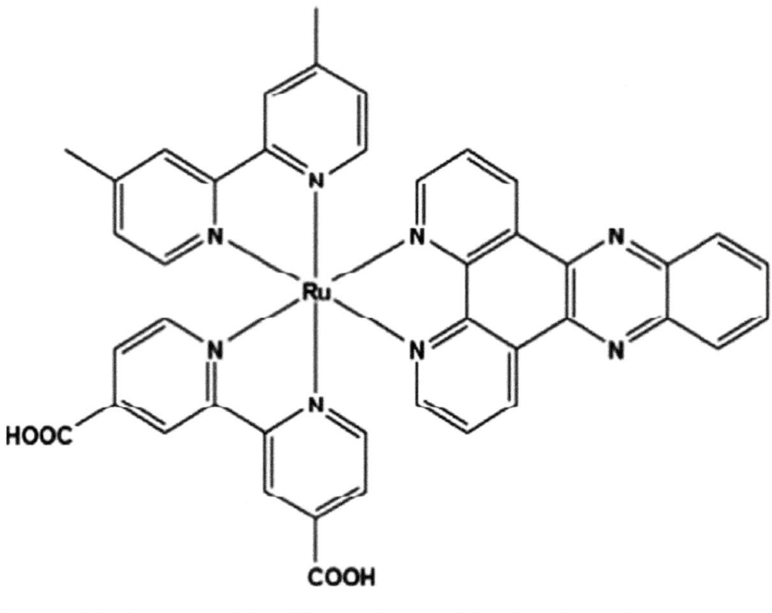

Figure 3.24 Structure of Ru(II) complex used for detection of biomolecules.

Seenivasan and co-workers recently reported a new, simple, label free sensor [Ru(dmbpy)(dcbpy)dppz)] complex (dmbpy; 4,40-

dimethyl-2,20-bipyridine,dcbpy;4,40-dicorboxy-2,20-bipyridine, dppz; dipyridophenazine) (Fig. 3.24) intercalated aptamer based recognition of amyloid-β (Seenivasan et al., 2015). For the first time, they elucidate the application of the RNA aptamer–Ru(II) complex system for specific recognition of amyloid monomer and inhibit the oligomer/fibril formation. Aptamers are DNA or RNA molecules that can specifically bind to a wide range of targets from small molecules to whole cells that can be selected through an in vitro selection method known as systematic evolution of ligands by exponential enrichment (SELEX). Interestingly, afore mentioned Ru(II) complex shows weak luminescence intensity in the aqueous medium but it shows strong luminescence intensity in the presence of RNA aptamer.

3.6.4 Detection of Hazardous Compounds

Ru complexes are also known to be used as a potential sensing agent for detection of hazardous compounds. Many important environmental pollutants including polychlorinated biphenyls, dioxins and some toxic organometallics are also halogen-abundant. Organotin compounds (OTCs) represented by the formula $RnSnX_{(4-n)}$, where Sn is the tin atom, R is an alkyl or aryl group, X is usually a halide compounds or hydroxide anion, and n ranges from 1 to 4. Organotin compounds are known for their wide distribution and strongly toxic effect on marine organisms. The degradation products of OTCs usually exist as di- or mono-organotin complexes, both with high biological activities. XB-driven molecular recognition has been used to develop the first FCS for organotin halides. Chromophore-bridged binuclear ruthenium complexes (Scheme 3.2) containing multiple isocyanide ligands were synthesized by Niu et al. and were exploited as XB-active receptors with AIE characteristics (Niu et al., 2013). Halogen bonding (XB) is the interaction based on donation of electron density from a Lewis base to an electron-deficient halogen atom and it is a frequently occurring non-covalent interaction parallel to hydrogen bonding in molecular self-assembly processes. Interaction with organotin halides leads to aggregation of the receptor molecules in solution and the resulting off/on fluorescence responses in the near-infrared region. Visual detection of organotin halides at the micromolar concentration level can be easily carried out by the developed FCS. This study confirms that cooperation of multiple XBs in solution is able to support high-affinity binding between small molecules. Furthermore, the monomer-to-aggregate conversion of chromophores, which usually results in high-contrast spectral changes, has proven to be a suitable signalling strategy for XB-motivated molecular sensing. Both the recognition and signalling mechanisms discussed by the authors are highly constructive for further designs of FCS targeting challenging species such as dioxins and dioxin like polychlorinated biphenyls.

3.7 pH Sensing

The measurement of pH is important for areas ranging from cardiac critical care to acid rain pollution of streams. Absorption spectroscopy has long been used for measuring pH. Changes in the absorption spectra for species having pH dependent ground state chemistry can be used as both quantitative pH determinations and as titration end point detectors. However, absorption spectra are not so sensitive, which hence increases the difficulty of using them with very small sensors. Because of its central importance, pH measurement via luminescent sensors has been an area of considerable study. A number of intensity-based systems using primarily organic dyes with pH-sensitive functionality have been suggested (Wolfbeis et al., 1998; Maier et al., 2002). Rather fewer systems employing metal complexes have been reported. A variety of metal ions have been used in both single and multicomponent systems (Grigg et al., 1992; Wolfbeis et al., 1998). The general approach for the design of pH-sensitive luminescent metal complexes is to modify a core ligand for pH sensitivity. Typically, either a 2,2′-bipyridine or 1,10-phenanthroline structure is used and an alcohol, acid, or amine function attached to provide the pH-sensitive element. Complexes based on the platinum metal ions Ru^{2+}, Os^{2+}, and Re^{+} are among the most commonly used. Complexes with these metals are normally luminescent, and photochemically robust, and exhibit either near UV or visible absorption, and can be "tuned" by a variety of synthetic methods (Demas et al., 2001). Herein, we discuss Ru complexes as pH sensor in details. An easy scheme for measuring pH is excited state acid–base chemistry. The excited state molecule has substantially more energy (190 kJ mol^{-1} or greater for a visible emitting species), and the excited state will generally have a completely different electron distribution from the ground state. Both of these factors can cause pK* to differ from pK. In some cases, the pK values can differ by 5–10 pK units. This would be equivalent to converting an acid of the strength of acetic acid into a non-acid or a super acid. The first example reported for an excited state acid base reaction of a platinum group metal diimine complex was for $RuL_2(CN)_2$. The two cyanides can be protonated at the N end, which dramatically blue shifts the MLCT excited states. The two pK values are on the order of 0–1. However, on excitation the complex becomes a super acid with a highly negative pK. While fundamentally very interesting, these complexes are not interesting for monitoring pH values in the range of greatest interest ($5 < pH < 9$), although they may have some utility in the extreme acidity range of much less than 0. However, one does not necessarily expect the usable range for an indicator in the ground state to be the same as if the ground state species were used as an indicator. For example, with $[Ru(3,4,7,8\text{-}Me_4phen)_2(4,7\text{-}(HO)_2\text{-}phen)_2]^-$, the pH titration curve using ground state absorption has a break at 5 while the excited state emission titration curve has a break at pH 2 (Demas et al., 1976). There is an alternative approach to emission pH sensing. The technique involves building a composite molecule that has a luminescent piece, which is covalently attached to a fragment that is pH sensitive and can quench the excited portion differently depending on its state of protonation. The most common version of this type of sensor uses electron transfer quenching since the redox potentials of many species are dependent on their state of protonation. Thus, depending on whether a species is protonated, it can act as a quencher in one form and a non-quencher in the other. This is shown schematically in Fig. 3.25. Examples of such built-in quenching systems involve α-diimine ligands with CH_2 coupled pendant phenol groups where the phenolate ion is a much more efficient electron transfer quencher than the phenol. The usable pH range of this system is about 6–10. Another example includes $-CH_2NR_2$ groups since amines can be good reductive excited state quenchers of RuL complexes. A variation on this mechanism is where the Ru(II) emission shows a pronounced variation in emission characteristics with the charge on the pendant groups. These systems are usable in the 0–8 pH range. Figure 3.26 (X) shows a titration curve for a diamino Ru(II) complex supported on a polymer. These data reveal two important issues. The double hump clearly indicates the presence of more than one pK*, which is not surprising given the two basic nitrogens. However, in solution, one of the two pK* values is at 7.5, which would suggest that it would be perfect for physiological measurements. Inspection of the polymer supported titration curve reveals that the pK* has shifted substantially to more basic regions, and the polymer supported sensor is not suitable for physiological measurements. This result once again demonstrates that the transfer of solution data to polymer systems is not always smooth and many surprises will occur.

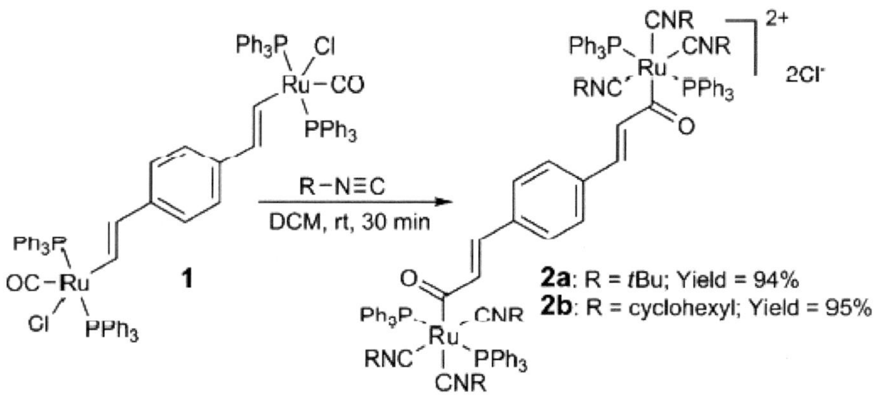

Scheme 3.2 Synthesis of the binuclear ruthenium complexes.

A good variation on pH sensing utilizes the Forster mechanism of Fig. 3.26 (Y). Protonation of a pH-sensitive dye changes its absorption spectrum. By proper selection of the dye, the absorption emission overlap of the donor and acceptor vary and the degree of quenching of the donor changes.

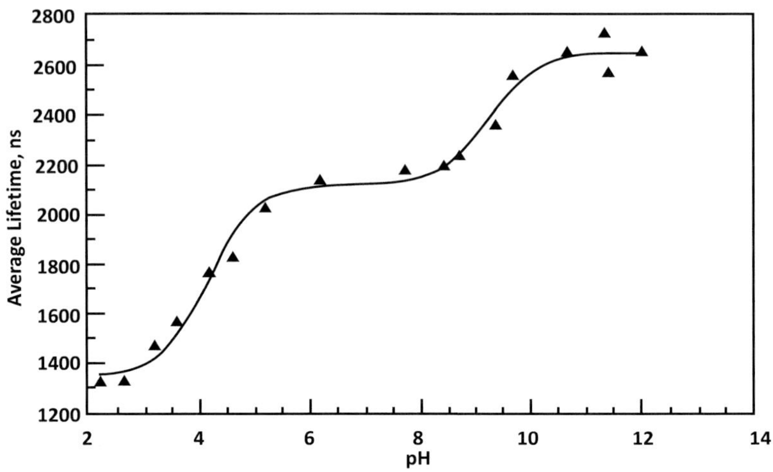

Figure 3.25 Various strategies used to measure pH.

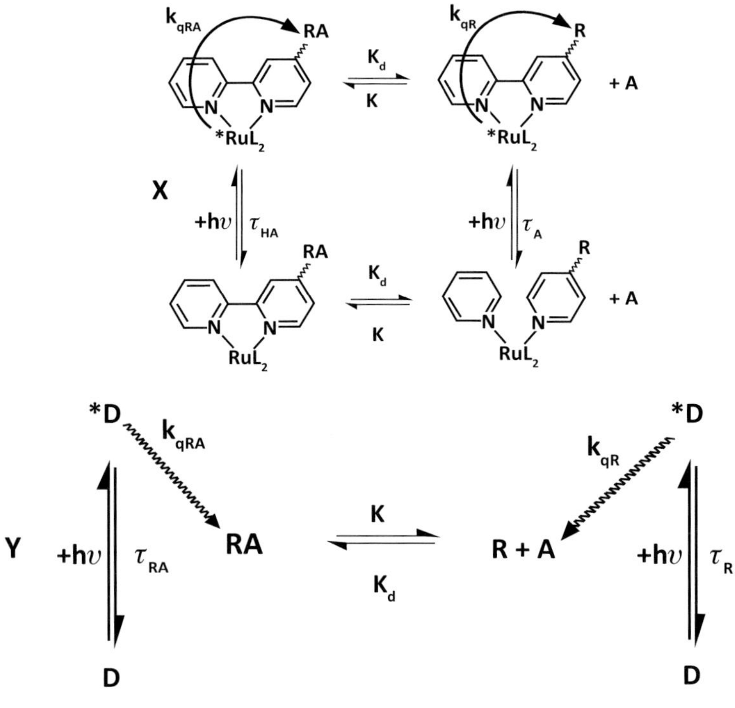

Figure 3.26 Lifetime pH titration curve for polymer immobilized [Ru(Ph$_2$ phen)$_2$(DEAM)$_2$bpy]$^{2-}$ ((DEAM)$_2$bpy-4,4′-bis(diethylamino)methyl-2′-bipyridine) in air-saturated water. The solid line is the two pK* fit to the data.

The problem with this approach is how to organize it so that the pH sensor energy acceptor is always maintained at a suitable distance from the donor. This problem can be largely overcome by making the donor and the acceptor, an ion pair. This energy transfer approach has also been modified to produce a Ru complex sensor by Fig. 3.26 (Y) (Huber et al., 1998). Here $[Ru(Ph_2bpy)_3]^{2-}$ donor is used in conjunction with a bromothymol blue anion as the indicator. Since the dye has a pK of 7.5, it works in the physiological range. The sensor ion pair is supported in a hydrophobic membrane. Co-extraction of a proton from the aqueous phase along with the chloride counter ion into the membrane provides a proton to change the more highly quenching blue form of the dye into the less quenching yellow protonated form. The luminescence of the ruthenium complex is monitored by changes in the apparent lifetime using a phase shift measurement or by intensity methods. Because of the long lifetimes of the ruthenium complex, phase measurements are inexpensive and low cost blue LEDs can be used as the excitation source. Another Ru complex using the pH-sensitive ligand 5-carboxy-1,10-phenanthroline was synthesized and characterized by Higgins et al. (Higgins et al., 2005). The complex showed monotonic changes in both luminescent intensity and lifetime with pH values over the range $2 < pH < 9$. Here the impact of various structural features on both the range of pH sensitivity and dynamic response was studied using both intensity and lifetime measurements (Higgins et al., 2005). It was possible to predictably tune the pH sensitivity range over about 1.5 pKa units. While significant variation in the dynamic response range was observed, the correlation with structural features needs further study. Recently, a new heterobimetallic ruthenium(II)–rhenium(I) complex of $[Ru(bpy)_2(HL)Re(CO)_3Cl](ClO_4)_2 \cdot 6H_2O$ (**RuHLRe**) {bpy = 2,2′-bipyridine and HL=2-(4-(2,6-di(pyridin-2-yl)pyridin-4-yl)phenyl)-^{1}H-imidazo[4,5-f]-[1,10]phenanthroline} was synthesised by Li et al. (2014). The ground-and excited-state acid–base properties of **RuHLRe** were studied using UV–Vis absorption spectrophotometric and spectrofluorimetric titrations in a 100:1 (v/v) Britton–Robinson buffer–CH_3CN solution combined with luminescence lifetime measurements. The complex exhibited two-step separate protonation–deprotonation processes in both the ground and excited states. The complex acted as pH-induced "*off–on–off*" luminescence switches (I_{on}/I_{off} = 31.0 and 14.6), with one of the switching actions being driven by pH variations over the physiological pH range (5.3–8.0).

3.8 Temperature Sensing

The significant and predictable temperature dependence of the emission of metal complexes suggests their use as temperature probes, although to our knowledge no one has actually used them in such applications. In particular, Crosby suggested Ru-diimine complexes as cryogenic sensors. These complexes have very large temperature coefficients below liquid nitrogen temperatures and huge coefficients at liquid helium temperatures. For example, the luminescence lifetime of $Ru(bpy)_3^{2-}$ changes from 5 to 200 ms ongoing from 77 to 4 K (Crosby et al., 1973). Their multi-microsecond lifetimes and ease of excitation with blue LEDs should make them superb remote cryogenic sensors.

In view of the great difficulty of controlling sample temperature at cryogenic temperatures, particularly in cold finger arrangements, the Crosby sensors would appear to be attractive for measurement probes. For example, it would be very easy to paint a cold finger probe with the complex in a polymer. Then, by interrogating the lifetime at different positions, the temperature profile across the device could be easily measured. By using a focused laser beam, a resolution of a 100 mm or less should be possible. Alternatively, using an optical imaging system, the entire temperature profile of a device could be measured at once. Of course, this approach is also possible at room or elevated temperature.

Especially in the cryogenic supersonic wind tunnel where both pressure and temperature are important parameters, metal complexes are attractive temperature sensors. $[Ru(terpy)_2]_2$ is not luminescent at room temperature but glows brightly at low temperatures. This feature makes it the basis of temperature-sensitive luminescence paints. By suitable modification of the complex, the temperature range over which it can respond can be altered. However, even at room temperature, the temperature coefficient is large enough to be very reasonable. Based on the thermal quenching via only an upper d–d level, we have carried out a theoretical analysis on the molecular parameters of a sensor molecule for optimizing sensor response for different temperatures (Demas et al., 1992). The model is very attractive for predicting behaviour and designing systems for use in the liquid nitrogen (77 K) to above ambient (ca. 350–450 K) temperature range. This model is inappropriate for the low temperatures of Crosby's measurements where two or three closely spaced sublevels of the emitting state manifold are involved.

3.9 Conclusion and Perspective

Ruthenium complexes offer a range of desirable advantages that make them viable alternatives to organic molecules as luminescent chemosensors. First, they possess high luminescence quantum yields and long phosphorescence lifetimes that can allow them to function effectively in the presence of a high fluorescent background by use of time-resolved emission spectroscopy. Secondly, their photophysical properties are highly sensitive to changes in structure or in the environment, such as upon coordination or reaction of the analyte with a recognition motif conjugated to an ancillary ligand. Finally, their modular synthesis allows facile synthesis of a range of analogues with bearing different functional groups for the fine-tuning of their chemical and/or photophysical properties. In this chapter, recent examples of Ru complex chemosensors for cations, anions, gas molecules and biomolecules reported in the last three years are mostly highlighted. Not only these analytes but also the successful application of Ru complex as pH sensor and a temperature-sensitive agent are also vividly discussed. Moving forward, one potential avenue for improvement is in the development of phosphorescent chemosensors with higher selectivity. For example, while most of the recently reported anion chemosensors containing an acidic hydrogen group have achieved excellent discrimination between strongly and weakly basic anions, the selectivity exhibited between two strongly basic anions such as fluoride and acetate have been relatively modest. As an interesting exceptional example, $[Ru(bpy)_3]^{2+}$-

derived conjugate was reported by Gunnlaugsson and co-workers, which can distinguish fluoride from other strongly basic anions such as acetate and pyrophosphate, although the mechanism for this selectivity has not yet been fully established. Another long-term goal is for the development of metal complex chemosensors capable of functioning in aqueous solution with minimal organic solvent, such as the ruthenium(II) nitrite ion chemodosimeter reported by Das and co-workers which required 10% acetonitrile solution. Finally, we believe that the application of luminescent metal complexes to detect pathologically relevant peptides such as beta-amyloid deserves further attention in the literature.

References

Baggaley E, Martin RG, Nicola HG, David T, Igor VS, Stanley WB, Carl S, John WH, Julia AW, and Jim AT (2014). Dinuclear ruthenium(II) complexes as two-photon, time-resolved emission microscopy probes for cellular DNA, *Angew Chem Int Ed*, **126**: 3435–3439.

Beer PD, Hayes EJ (2003). Transition metal and organometallic anion complexation agents, *Coord Chem Rev*, **240**: 167–189.

Bhaumik C, Saha D, Das S, and Baitalik S (2011). Synthesis, structural characterization, photophysical, electrochemical, and anion-sensing studies of luminescent homo- and heteroleptic ruthenium(II) and osmium(II) complexes based on terpyridyl-imidazole ligand, *Inorg Chem*, **50**: 12586–12600.

Boal AK, and Rosenzweig AC (2009). Structural biology of copper trafficking, *Chem Rev*, **109**: 4760–4779.

Chouai A, Wicke SE, Turro C, Bacsa J, Dunbar KR, Wang D, and Thummel RP (2005). 2D NMR spectroscopic evidence for unprecedented interactions of *cis*-$[Rh_2(dap)\text{-}(m\text{-}O_2CCH_3)_2(h_1\text{-}O_2CCH_3)(CH_3OH)](O_2CCH_3)$ with a DNA oligonucleotide: combination of intercalative and coordinative binding, *Inorg Chem*, **44**: 5996–6003.

Chow CF, Lam MH, and Wong WY (2004). A chemodosimetric probe based on a conjugated oxidized bis-indolyl system for selective naked-eye sensing of cyanide ions in water, *Inorg Chem*, **43**: 8387–8393.

Chun-Lian H, Feng-Lian R, Xiao-Bing Z, Yan-Yan D, and Yan Z (2006). A fluorescent chemosensor for copper(II) based on a carboxylic acid-functionalized *tris*(2,2′-bipyridine)-ruthenium(II) complex, *Anal Sci*, **22**: 1547–1551.

Cook NP, Torres V, Jain D, and Marti AA (2011). Sensing amyloid-β aggregation using luminescent dipyridophenazine ruthenium(II) complexes, *J Am Chem Soc*, **133**: 11121–111213.

Coronado E, Galan-Mascaros JR, Marti-Gastaldo C, Palomares E, Durrant JR, Vilar R, Gratzel M, and Nazeeruddin MK (2005). Reversible colorimetric probes for mercury sensing, *J Am Chem Soc*, **127**: 12351–12356.

Coury JE, Anderson JR, McFail-Isom L, Williams LD, and Bottomley LA (1997). Scanning force microscopy of small ligand–nucleic acid complexes: Tris(ophenanthroline)ruthenium(II) as a test for a new assay, *J Am Chem Soc*, **119**: 3792–3796.

Coutant MA, Payra P, and Dutta PK (2003). Oxygen transport in zeolite Y measured by quenching of encapsulated *tris*(bipyridyl)ruthenium, *Microporous Mesoporous Mater*, **60**: 79–90.

Das S, Karmakar S, Mardanya S, and Baitalik S (2014). Synthesis, structural characterization, and multichannel anion and cation sensing studies of a bifunctional Ru(II) polypyridyl–imidazole based receptor, *Dalton Trans*, **43**: 3767–3782.

Demas JN, and DeGraff BA (1992). Applications of luminescent transition platinum group metal complexes to sensor technology and molecular probes, *Coord Chem Rev*, **211**: 317–351.

Demas JN, Harris EW, and McBride RP (1977). Energy transfer from luminescent transition metal complexes to oxygen, *J Am Chem Soc*, **99**: 3547–3551.

Elmes RBP, and Gunnlaugsson T (2010). Luminescence anion sensing via modulation of MLCT emission from a naphthalimide–Ru(II)–polypyridyl complex, *Tetrahedron Lett*, **51**: 4082–4087.

Elsworth JD, and Roth RH (1997). Dopamine synthesis, uptake, metabolism, and receptors: relevance to gene therapy of Parkinson's disease, *Exp Neurol*, **144**: 4–9.

Eththilu B, Paulpandian MM, Veerasamy S, Subramanian S, and Seenivasan R (2015). Sensing and inhibition of amyloid-β based on the simple luminescent aptamer–ruthenium complex system, *Talanta*, **134**: 348–353.

Ghosh A, Das P, Saha S, Banerjee T, Bhatt HB, and Das A, (2011). Diamine derivative of Ru(II)-polypyridyl complex for chemodosimetric detection of nitrate ion in aqueous solution, *Inorg Chim Acta,* **372**: 115–119.

Gill MR, and Thomas JA (2012). Ruthenium(II) polypyridyl complexes and DNA–from structural probes to cellular imaging and therapeutics, *Chem Soc Rev,* **41**: 3179–3192.

Greiner G, and Maier I (2002). Products of low potential energy in mass spectra as a consequence of ion–dipole attractions: some neopentyl compounds, *J Chem Soc Perkin Trans,* **2**: 1005–1007.

Grigg R, and Norbert, WDJA (1992). Luminescent pH sensors based on di(2,2′-bipyridyl)(5,5′-diaminomethyl-2,2′-bipyridyl)-ruthenium(II) complexes, *J Chem Soc Chem Commun,* 1300–1302.

Guo NC, Xiao PW, Jian PD, and Hong QC (1999). A study on electrochemistry of histidine and its metabolites based on the diazo coupling reaction, *Talanta,* **49**: 319–330.

Hamid AA, Amer AG, Mohammad AK, Ziyad A, Mahmoud Q, Safwan O, Rawashdeh M, and Monem A (2011). A selective chemosensor for mercuric ions based on 4-aminothiophenol-ruthenium(II) bis(bipyridine) complex, *Int J Inorg Chem*: Article ID 843051, doi:10.1155/2011/843051.

Happ B, Winter A, Hager MD, and Schubert US (2012). Photogenerated avenues in macromolecules containing Re(I), Ru(II), Os(II), and Ir(III) metal complexes of pyridine-based ligands, *Chem Soc Rev,* **41**: 2222–2255.

Harrigan RW, Hager GD, and Crosby GA (1973). Evidence for multiple-state emission from ruthenium (II) complexes, *Chem Phys Lett,* **21**: 487–490.

Hartshorn RM, and Barton JK (1992). Novel dipyridophenazine complexes of ruthenium(II): Exploring luminescent reporters of DNA, *J Am Chem Soc,* **114**: 5919–5925.

Higgins B, DeGraff BA, and Demas JN, (2005). Luminescent transition metal complexes as sensors: Structural effects on pH response, *Inorg Chem,* **44**: 6662–6669.

Höckel M, and Vaupe P (2001). Tumor hypoxia: Definitions and current clinical, biologic, and molecular aspects, *J Natl Cancer Inst.,* **93**: 266–276.

Huber C, Werner T, Krause C, Klimant I, and Wolfbeis OS (1998). Analysis of questioned documents: A review, *Anal Chim Acta,* **364**: 143–166.

Ji S, Wu W, Wu W, Song P, Han K, Wang Z, Liu S, Guo H, and Zhao J (2010). Tuning the luminescence lifetimes of ruthenium(II) polypyridine complexes and its application in luminescent oxygen sensing, *J Mater Chem*, **20**: 1953–1963.

Kalinowski J, Fattori V, Cocchi M, and Williams JAG (2011). Light-emitting devices based on organometallic platinum complexes as emitters, *Coord Chem Rev*, **255**: 2401–2425.

Keefe MH, Benkstein KD, and Hupp JT (2000). Luminescent sensor molecules based on coordinated metals: A review of recent developments, *Coord Chem Rev,* **205**: 201–228.

Kitchen JA, Boyle EM, and Gunnlaugsson T (2012). Synthesis, structural characterisation and luminescent anion sensing studies of a Ru(II) polypyridyl complex featuring an aryl urea derivatised 2,2′-bpy auxiliary ligand, *Inorg Chim Acta,* **381**: 236–242.

Kosch U, Klimant I, Werner T, and Wolfbeis OS (1998). Strategies to design pH optodes with luminescence decay times in the microsecond time regime, *Anal Chem*, **70**: 3892–3897.

Kung HF, Choi SR, Qu W, Zhang W, and Skovronsky D (2009). 18F stilbenes and styrylpyridines for PET imaging of Aβ plaques in Alzheimer's disease: A mini perspective, *J Med Chem*, **53**: 933–941.

Li Y, and Zamble DB (2009). Nickel homeostasis and nickel regulation: An overview, *Chem Rev*, **109**: 4617–4643.

McEvoy AK, McDonagh CM and MacCraith, BD (1996). Dissolved oxygen sensor based on fluorescence quenching of oxygen-sensitive ruthenium complexes immobilized in sol–gel-derived porous silica coatings, *Analyst*, **121**: 785–788.

McKinley AW, Lincoln P, and Tuite EM (2011). Environmental effects on the photophysics of transition metal complexes with dipyrido[2,3-a:3′, 2′-c]phenazine (dppz) and related ligands, *Coord Chem Rev*, **255**: 2676–2692.

Meggers E (2009). Targeting proteins with metal complexes, *Chem Commun*, 1001–1010.

Mei HY, and Barton JK (1986). Chiral probe for A-form helixes of DNA and RNA: Tris(tetramethylphenanthroline)ruthenium(II), *J Am Chem Soc,* **108**: 7414–7416.

Meier B, Werner T, Klimant I, and Wolfbeis OS (1995). Novel oxygen sensor material based on a ruthenium bipyridyl complex encapsulated in zeolite Y: Dramatic differences in the efficiency of luminescence quenching by oxygen on going from surface-adsorbed to zeolite-encapsulated fluorophores, *Sens Actuators B: Chem*, **29**: 240–245.

Mo HJ, Niu YL, Zhang M, Qiao ZP, and Ye BH (2011). Photophysical, electrochemical and anion sensing properties of Ru(II) bipyridine complexes with 2,2′-biimidazole-like ligand, *Dalton Trans.*, **40**: 8218–8225.

Moragues ME, Martinez-Manez R, and Sancenon F (2011). Chromogenic and fluorogenic chemosensors and reagents for anions. A comprehensive review, *Chem Soc Rev* , **40**: 2593–2643.

Niu Y, Han F, Zhang Q, Xie T, Lu L, Li S, and Xia H (2013). Off/On fluorescent chemosensors for organotin halides based on binuclear ruthenium complexes, *Angew Chem Int Ed*, **52**: 5599–5603.

Palomares E, Vilar R, and Durrant JR (2004). Heterogenious colorimetric sensor for mercuric salts, *Chem Commun*, 362–363.

Park Y, Kim JH, Lee KT, Jeon KS, Na HB, Yu JH, Kim HM, Lee N, Choi SH, Aik SB (2009). Nonblinking and nonbleaching upconverting nanoparticles as an optical imaging nanoprobe and T1 magnetic resonance imaging contrast agent, *Adv Mater*, **21**: 4467–4471.

Peterson S, and Demas JN (1976). Excited state acid-base reactions of transition metal complexes: Dicyanobis(2,2′-bipyridine) ruthenium(III) in aqueous acid, *J Am Chem Soc*, **98**: 7880–7881.

Pyle AM, Rehmann JP, Meshoyrer R, Kumar CV, Turro NJ, and Barton JK (1989). Mixed-ligand complexes of ruthenium: Factors governing binding to DNA, *J Am Chem Soc*, **111**: 3051–3058.

Qian L, Juanjuan P, Lining S, and Li F (2011). High-efficiency upconversion luminescent sensing and bioimaging of Hg(II) by chromophoric ruthenium complex-assembled nanophosphors, *ACSNANO*, **5**: 8040–8048.

Rogers CW, and Wolf MO (2002). Luminescent molecular sensors based on analyte coordination to transition-metal complexes, *Coord Chem Rev*, **341**: 233–234.

Saha D, Das S, Bhaumik C, Dutta S, and Baitalik S (2010). Metal-controlled assembly and selectivity of a urea-based anion receptor, *Inorg Chem*, **49**: 2334–2348.

Shang XF, Li J, Lin H, Jiang P, Pin, Cai ZS, and Lin KH (2009). Anion recognition and sensing of ruthenium(II) and cobalt(II) sulfonamido complexes, *Dalton Trans*, 2096–2102.

Stefani M, and Dobson C (2003). Protein aggregation and aggregate toxicity: New insights into protein folding, misfolding diseases and biological evolution, *J Mol Med*, **81**: 678–699.

Sun Y, Hudson ZM, Rao Y, and Wang S (2011). Tuning and switching MLCT phosphorescence of $[Ru(bpy)_3]^{2+}$ complexes with triarylboranes and anions, *Inorg Chem*, **50**: 3373–3378.

Vaidyalingam AS, Coutant MA, and Dutta PK (2001). In: Balzani V (ed.), *Electron Transfer in Chemistry,* Wiley-VCH, pp. 412–486.

Wang C, Chen L, and Liu Z (2010). Drug delivery with upconversion nanoparticles for multi-functional targeted cancer cell imaging and therapy, *Biomaterials*, **32**: 1110–1120.

Watanabe S, Onogawa O, Komatsu Y, and Yoshida K (1998). Luminescent metaloreceptor with a neutral bis(acylaminoimidazoline) binding site: Optical sensing of anionic and neural phosphodiesters, *J Am Chem Soc*, **120**: 229–230.

Wenzel M, Hiscock JR, and Gale PA (2012). Anion receptor chemistry: Highlights from 2010, *Chem Soc Rev*, **41**: 480–520.

Wu W, Ji S, Wu W, Guo H, Wang X, Zhao J, and Wang Z (2010). Enhanced luminescence oxygen sensing property of Ru(II) bispyridine complexes by ligand modification, *Sens Actuators B Chem*, **149**: 395–406.

Yam VWW, and Kai ASF (2000). Synthesis, luminescence and ion-binding studies of a trinuclear copper(I) acetylide complex containing benzo-15-crown-5. X-ray crystal structure of $[Cu_3(\mu\text{-dppm})_3(\mu_3\text{-}\eta_1\text{-}C{=}C\text{-benzo-15-crown-5})_2]PF_6$, *Inorg Chim Acta*, **300–302**: 82–90.

Yang Y, Zhao Q, Feng W, Li F (2012). Luminescent chemodosimeters for bioimaging, *Chem Rev*, **113**: 192–270.

Zhang Y, Liu Z, Yang K, Zhang Y, Xu Y, Li H, Wang C, Lu A, and Sun S (2015). Aruthenium(II) complex as turn-on Cu(II) luminescent sensor based on oxidation cyclization mechanism and its application in vivo, *Sci Rep*, DOI: 10.1038/srep08173.

Zhang R, Yu X, Ye Z, Wang G, Zhang W, and Yuan J (2010). Turn-on luminescent probe for cysteine/homocysteine based on a ruthenium(II) complex, *Inorg Chem*, **49**: 7898–7903.

Zhao J, Ji S, Wu W, Wu W, Guo H, Sun J, Sun H, LiuY, Li Q, and Huang L (2012). Transition metal complexes with strong absorption of visible light and long-lived triplet excited states: From molecular design to applications, *RSC Adv*, **2**: 1712–1728.

Zheng ZB, Duan ZM, Ma YY, and Wang KZ (2013). Highly sensitive and selective difunctional ruthenium(II) complex-based chemosensor for dihydrogen phosphate anion and ferrous cation, *Inorg Chem*, **52**: 2306–2316.

Zheng ZB, Wu YQ, Wang KZ, and Li F (2014). pH luminescence switching, dihydrogen phosphate sensing, and cellular uptake of a heterobimetallic ruthenium(II)–rhenium(I) complex, *Dalton Trans*, **43**: 3273–3284.

Chapter 4

Ruthenium Compounds: A New Approach in Nanochemistry

Pradeep Pratap Singh[a] **and Ambika**[b]

[a]*Department of Chemistry, Swami Shraddhanand College, University of Delhi, Delhi 110036, India*
[b]*Department of Chemistry, Hans Raj College, University of Delhi, Delhi 110007, India*

ambikasinghhrc@gmail.com

Transition-metal nanoparticles have attracted attention in the last few decades due to their potential applications in biomedical, optical, electronic areas and catalysis. Ruthenium (Ru) compounds such as Ru oxide (RuO_2) and Ru oxide hydrate ($RuO_2 \cdot nH_2O$) have been investigated for applications such as electrocatalysts, materials for electrochemical supercapacitors, catalysts for hydrogen production, or CO oxidizing catalysts. Supported Ru nanoparticles (NPs) such as Ru oxide nanotubes, ruthenic acid nanosheets, and Ru oxide-based nanocomposites are highly efficient in various reactions, such as ammonia synthesis, Fischer–Tropsch synthesis, selective hydrogenation, and cellulose hydrolysis. Ru nanoparticle have also found medical applications and as a probe for upconversion luminescence sensing and bioimaging of intracellular metal ions. This chapter will focus on

Ruthenium Chemistry
Edited by Ajay Kumar Mishra and Lallan Mishra

ISBN 978-981-4774-39-0 (Hardcover), 978-1-315-11058-5 (eBook)
www.panstanford.com

the emerging concept of Ru NPs and their diverse applications in different areas.

4.1 Introduction

Nanotechnology has become very important in daily life and is constantly broadening its borders and complexity. Nanotechnology involves designing and using systems and devices in the length scale range of approximately 1–100 nm (Frohlich, 1937; Giaever and Zeller, 1968; Kubo, 1962). Nanomaterials can be generated from any material, such as metals, semiconductors and polymers with specific sizes, shapes, surface properties, defects, self-assembly properties by utilizing different methodologies from synthetic and materials chemistry (Matsoukas et al., 2015). They possess unique properties, such as high surface to volume ratio and high catalytic activities, due to which they have been employed in pharmaceutical, cosmetic, electronic and environmental industry (Liu, 2006).

Ruthenium (Ru) is a transition metal in VIII group and is inert to most chemicals. The chemistry of Ru is currently receiving a lot of attention, primarily because its complexes have good electron transfer and energy transfer properties (Collin et al., 1994). Ru is a versatile catalyst, and have been investigated for applications such as electrocatalysts, materials for electrochemical supercapacitors, catalysts for H_2 production, or CO oxidizing catalysts, in ammonia synthesis and cellulose hydrolysis (Abu-Dief and Mohamed, 2015). Ru can exist in a range of oxidation states (II, III, and IV) under physiologically relevant conditions due to which it can be utilized in pharmacological applications, for example Ru coordination compounds have shown promising application as anticancer agents, in the treatment of eye melanomas. New Ru-based compounds with fewer and less severe side effects, could replace longstanding platinum (Pt)-based anticancer drugs (Allardyce and Dyson, 2001). Ru compounds are being researched for use in a number of developing solar energy technologies (Nosheen et al., 2016). They have also been utilized as a probe for upconversion luminescence sensing and bioimaging of intracellular metal ions. This article will focus on the emerging concept of Ru nanoparticles and their diverse applications in different areas.

4.2 Synthesis of Metallic Nanoparticles

There are usually two general approaches for the production of NPs: (a) chemical method, also known as bottom-up (Gao et al., 2000; Yang et al., 2004) and (b) physical methods known as top-down method (Rawat, 2013; Yu et al., 2013). Both approaches can be carried out in gas, liquid, supercritical fluids, solid states, or vacuum.

4.2.1 Physical Method (Top Down)

In this process, the bulk materials are used as the starting materials and are broken down into smaller pieces using chemicals and/or by using physical means such as mechanical grinding (Gubin et al., 2005), mechanical alloying (Gaffet et al., 1996), and sputtering techniques (Chung et al., 2004; Suzuki et al., 2012; Bouchat et al., 2011) followed by subsequent stabilization of the resulting nanosized metal particles by the addition of colloidal protecting agents to synthesize nanomaterials. The metal NPs synthesized by this method are generally larger and cannot be reproduced, resulting in irreproducible catalytic activity.

4.2.2 Chemical Method (Bottom Up)

In this process, the single atoms or ions are allowed to grow into clusters or NPs using wet chemical synthetic methods such as chemical reduction of metal salts (Guzman et al., 2009; Khanna, 2007) and the decomposition of precursors using thermal (Logvinenko et al., 2007; Li et al., 2001), photolytic or sonochemical treatment (Bunker et al., 2007). The particles synthesized by this method have narrow size distribution and controlled shape.

4.2.3 Wet Chemical Synthesis

The bottom up methods of wet chemical NPs preparation involve the chemical reduction of metal salts, electrochemical pathways, or the controlled decomposition of metastable organometallic compounds (Scheme 4.1).

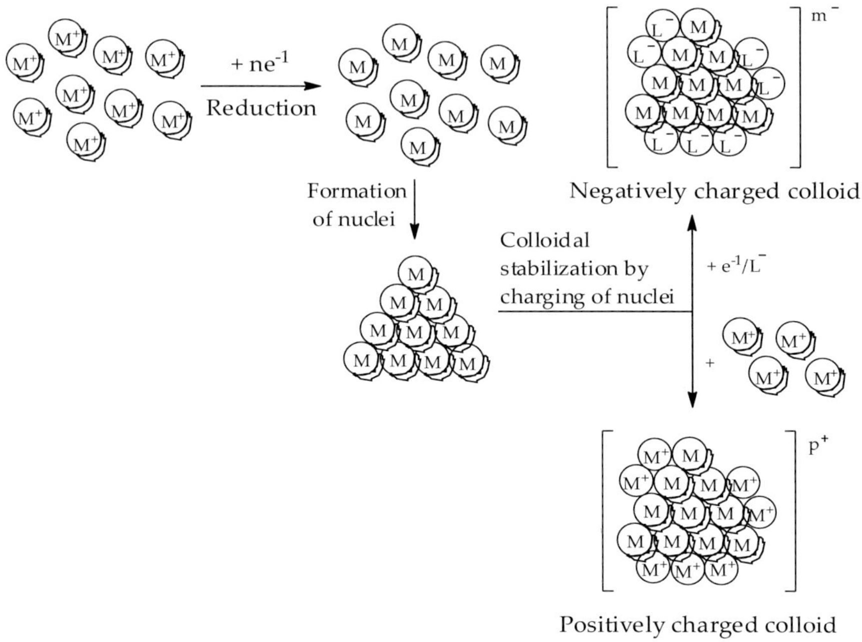

Scheme 4.1 Synthesis of metal nanoparticle by wet chemical synthesis.

A large variety of stabilizers, e.g., donor ligands, polymers, and surfactants, are used to control the growth of the primarily formed nanoclusters and to prevent their agglomeration (Faraday, 1857). In the initial stage of nucleation, metal salt is reduced to give zero valent metal atoms (Leisner et al., 1996). These can collide in solution with further metal ions, metal atoms, or clusters to form an irreversible seed of stable metal nuclei. The diameter of the seed nuclei depends on the metal–metal bond strength and difference of the redox potentials between the metal salt and the reducing agent applied. Protective agents are required for the stabilization and prevention of agglomeration of nanostructured colloidal metals. Nanometallic cores are separated from each other by electrostatic and steric stabilization and prevent agglomeration (Bradley, 1994). The main classes of protective groups are polymers and block copolymers (Hirai et al., 1976; Hirai et al., 1978). Lipophilic protective agents give metal colloids that are soluble in organic media (organosols), while hydrophilic agents yield water-soluble colloids (hydrosols).

4.3 Applications of Ru Nanoparticles

Ru NPs have been utilized in catalytic, analytical, and biomedical applications.

4.3.1 Catalytic Application

The catalytic activity of NPs represents an important resource for chemical processes employed in both industry and research. The functionalized NPs, including graphene, silica, carbon nanotubes-based, magnetite-supported nanocatalysts, and various other nanostructures, have been synthesized because of their catalytic properties.

4.3.1.1 Oxidation

Pt-Ru/1-aminopyrene functionalized carbon nanofibers have been used in electrocatalytic oxidation of methanol (Lin et al., 2010). Selenium functionalized carbon as support, Pt-Ru NPs were highly dispersed on the carbon surface, and showed improved electrochemical properties for methanol electrooxidation (Wang et al., 2013). Aerial oxidation of various aliphatic, aromatic, alicyclic, benzylic, allylic, amino, and heterocyclic alcohols has been carried out using Ru NP decorated on graphene nanosheets (GNS-RuNPs) with high chemoselectivity and reusability (Scheme 4.2) (Gopiraman et al., 2013).

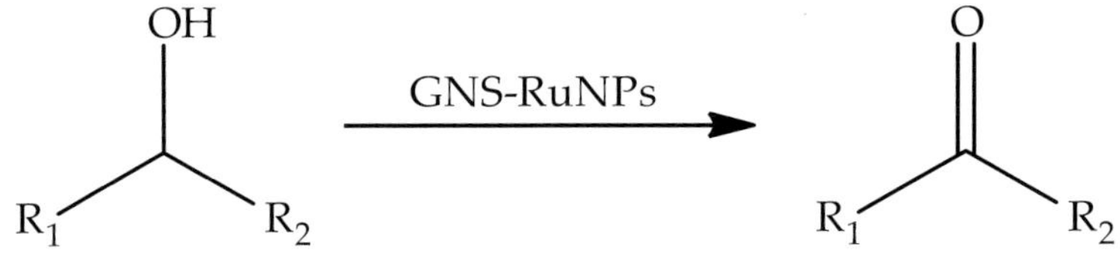

Scheme 4.2 Ru NPs catalyzed oxidation of alcohols.

A novel composite material of hierarchically structured Pt-Ru NPs grown on large surface area three dimensional graphene foam (3D GF) was utilized for methanol and ethanol oxidation with higher tolerance to poisoning by CO and exhibited improved catalytic activity. Ru/3D GF exhibits excellent catalytic activity towards oxidation reaction as compared with Pt-Ru/

Graphene, Pt-Ru/C (Vulcan XC-72R carbon) and PtRu alone. Pt-Ru catalysts deposited on carbon nanotubes (CNTs) surfaces have been utilized for methanol oxidation.

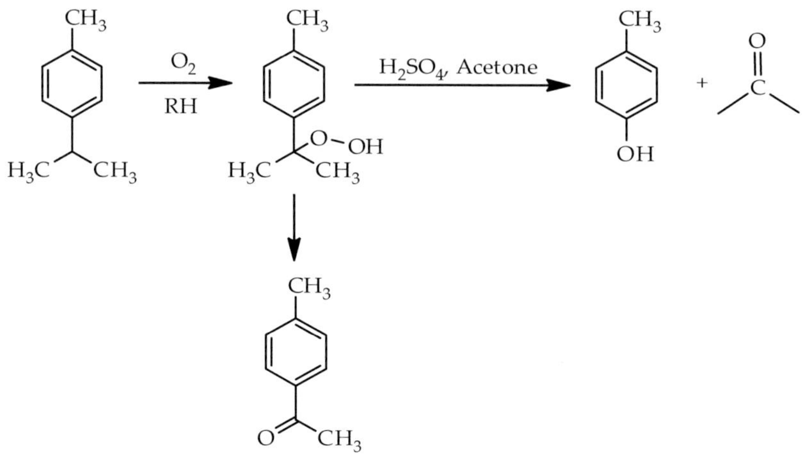

Scheme 4.3 Ru-cerium oxide nanocomposite catalyzed oxidation of *p*-cymene.

The catalyst efficiency and suppression of CO adsorption get influenced by the concentrations of sulfuric acid, EG, and Ru precursor salt during the electrodeposition processes (Tsai et al., 2008). Ru oxide NP deposited on the surface of a glassy carbon electrode (RuON-GCE) plays the role of an excellent bifunctional electrocatalyst for the oxidation of adrenaline and uric acid. Ru-cerium oxide (Ru/CeO_2) composite was employed to catalyze liquid-phase selective oxidation of *p*-cymene to tertiary cymene hydroperoxide (TCHP) and its derivative *p*-methylacetophenone (PMAP) using a green molecular oxygen oxidant (Scheme 4.3) (Makgwanea and Ray, 2016).

The catalytic activity toward the oxidation of ethanol using bimetallic Pt-Ru NP gets enhanced on dispersing over mixed-valent tungsten(VI,V) oxide and zirconium(IV) oxide under different reaction conditions (Rutkowska et al., 2016). Mesoporous carbon loaded with Pt and Pt-Ru NPs electrocatalyst was not as effective for methanol oxidation (Ding et al., 2005). Au-Ru bimetallic NPs have been utilized in methanol oxidation fuel cells (Sathish et al., 2010). Palladium (Pd)-Ru NPs supported on carbon Pd_xRu_x/C

(x = 1, 0.99, 0.95, 0.90, 0.80, 0.50) have been employed for ethanol electro-oxidation in alkaline media (Monyoncho et al., 2015). A heterogeneous catalytic permanganate oxidation system with three molecular sieves, i.e., nanosized ZSM-5 (ZSM-5A), microsized ZSM-5 (ZSM-5B), and MCM-41, supported Ru NP as catalyst has been used for the oxidation of sulfamethoxazole (SMX). The catalytic performance was strongly dependent on Ru loading on the catalysts (Zhang et al., 2015). The effect of particle size on CO oxidation over two-dimensional Ru NPs catalysts has been reported. The catalytic activity of Ru NPs depends on its size. As the size of NPs increases, the activity of the catalyst also increases (Joo et al., 2010).

4.3.1.2 Reduction

u-GNS-RuNPs catalyst were modified to RuO_2 nanorod hybrid GNSs (*u*-GNS-RuO_2NRs) and were used for transfer hydrogenation of various aromatic, alicyclic, and heterocyclic ketones with excellent catalytic activity (Scheme 4.4) (Gopiraman et al., 2013). Ru NPs have been prepared by adsorption-bioreduction method using the *Cacumen platycladi* extract and supported on the activated carbon, which was modified with nitric acid. The Ru/AC-HNO_3 catalyst was used in the liquid phase hydrogenation of 2,2,4,4-tetramethylcyclobutane-1,3-dione (Huang et al., 2015).

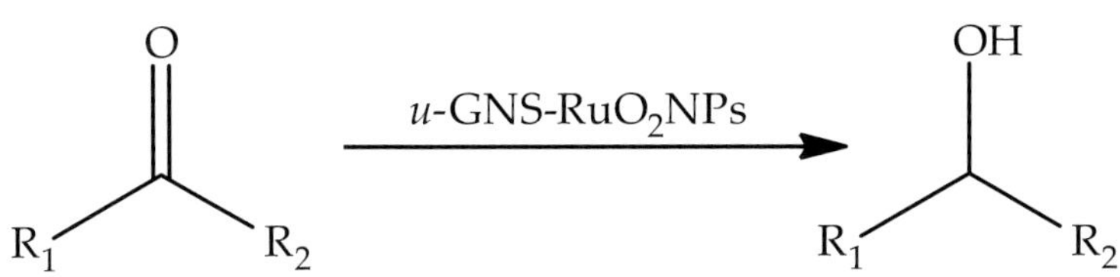

Scheme 4.4 *u*-GNS-RuO_2NRs catalyzed hydrogenation of different ketones.

Efficient catalytic partial hydrogenation of 1,1′-bi-2-naphthol and β-naphthylamine derivatives in optically pure 5,5′,6,6′,7,7′, 8,8′-octahydro-1,1′-binaphthyls in good to high yields using nano Ru particles dispersed on a nanocarbon fiber support (Ru/CNF-P) without racemization of the axial chirality has been reported (Scheme 4.5) (Takasaki et al., 2007).

RuO_2 nanorods (RuNRs) supported on G platelets (GPs) were employed for the reduction of aromatic aldehydes and ketones using isopropanol as the hydrogen donor with high selectivity and reusablity (Scheme 4.6) (Gopiraman et al., 2013).

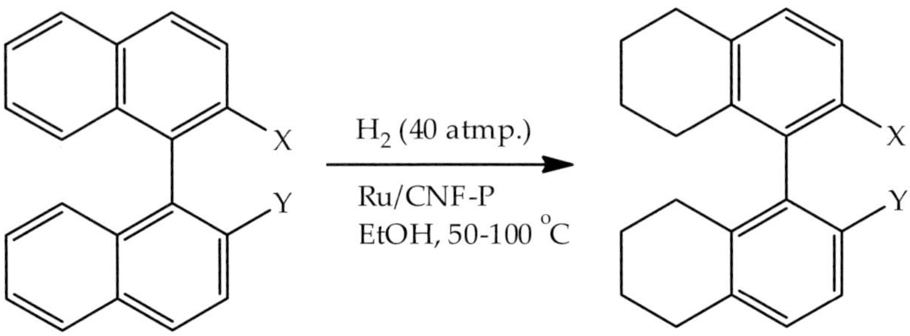

X, Y = OH, OR, NH_2

Scheme 4.5 Ru/CNF-P-catalyzed hydrogenation of 1,1′-binaphthyls.

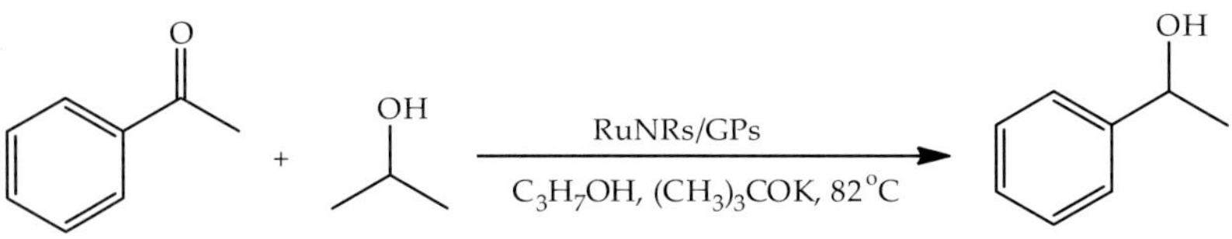

Scheme 4.6 RuO_2 nanorod-catalyzed reduction of aromatic aldehydes and ketones.

Ru/CNT catalyst has been utilized for the hydrogenation of benzene to cyclohexane in high yields under solvent-free mild conditions and can be reused for several cycles without any loss in activity (Ma et al., 2014). Three different Ru catalysts were prepared by treating Ru NPs with CNTs. Out of three catalysts, one has most of Ru NPs confined inside the channels, and the other two have most of Ru NPs outside through different preparation methods. The three Ru catalysts were used for the hydrogenation of benzene, *p*-chloronitrobenzene, and cinnamaldehyde. The electronic effect plays an important role in catalytic performance. The electron-rich Ru favors hydrogenation of *p*-chloronitrobenzene but is unfavorable for benzene hydrogenation. For cinnamaldehyde hydrogenation, the electron-rich Ru would be favorable for the adsorption of C@O bond rather than C@C bond and thus promoted the selectivity to cinnamyl alcohol. A confinement effect induced by the electronic effect has different effects on these substrates, and it can be enhanced by heat treating for all the reactions (Wang et al., 2015).

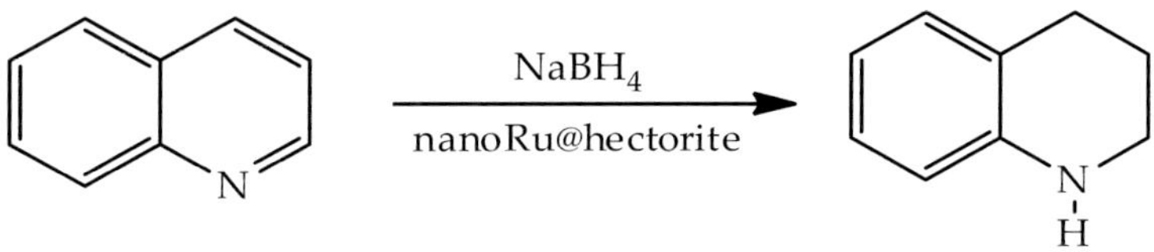

Scheme 4.7 Nano-Ru@hectorite catalyzed reduction of quinoline.

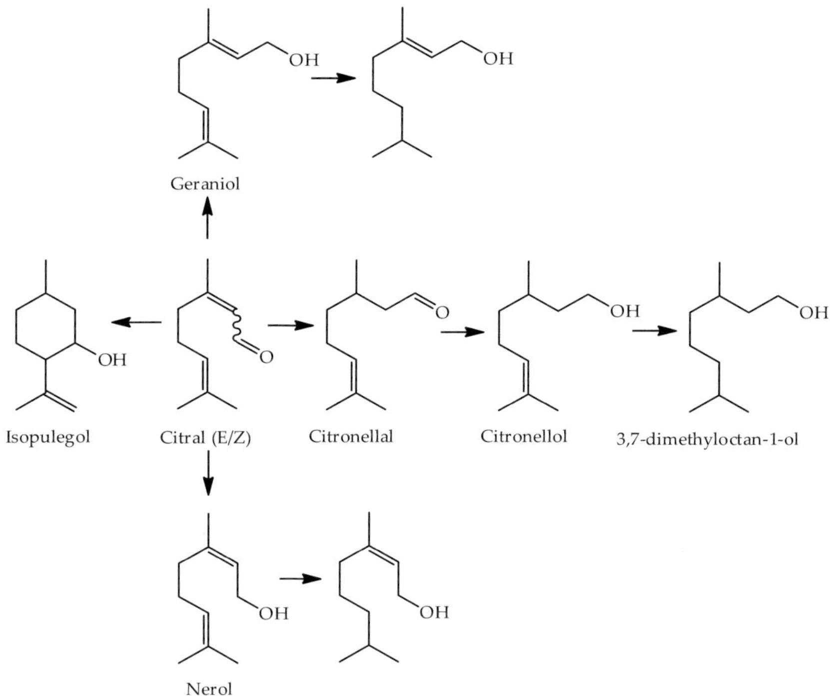

Scheme 4.8 Ru NP-catalyzed hydrogenation of citral in ionic liquids.

Nano Ru@hectorite was found to catalyze the hydrogenation of aromatic amino acids in aqueous solution with retention of chirality under mild conditions, with high conversion and selectivity. The catalyst can be recycled and reused at least three times without loss in activity (Sun and Suss-Fink, 2016). Also, nano-Ru@hectorite can catalyze the reduction of quinoline and its derivatives by $NaBH_4$ in aqueous solution to give selectively the corresponding1,2,3,4-tetrahydroquinolines(*N*-cyclehydrogenation) under mild conditions (Scheme 4.7) (Sun et al., 2016).

Zeolite Y (HYZ)-supported Ru NP catalyst was utilized in hydrogenation of xylose to xylitol under green aqueous phase system with high yield and reusability (Mishra et al., 2013). Silica supported dendrimer-encapsulated Ru NPs catalyst has been utilized in the hydrogenation of citral in ionic liquids (Scheme 4.8). [BMIM][NTf_2] resulted in enhanced catalytic activity with selectivity toward citronellal (Antonels and Meijboom, 2014).

RuTsDPEN complex (TsDPEN=*N*-(ptoluenesulfonyl)1,2-diphenyl ethylenediamine) that was immobilized onto a magnetic material as catalyst was employed for the asymmetric hydrogenation supported on siliceous meso-cellular foam filled magnetite. This heterogeneous catalyst afforded 97–99% conversion and an ee value of 94% in the asymmetric transfer hydrogenation of imines in a HCOOH-Et_3N system (Scheme 4.9).

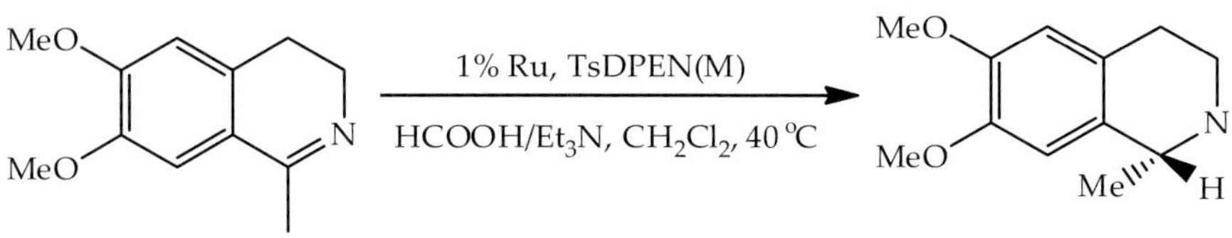

Scheme 4.9 RuTsDPEN complex-catalyzed asymmetric hydrogenation of imines.

Also, this catalyst can be used for the asymmetric transfer hydrogenation (ATH) of aromatic ketones, with HCOOH-Et_3N azeotrope as a hydrogen donor in CH_2Cl_2 as a solvent with good ee values (Li et al., 2009; Polshettiwar et al., 2011).

Ruthenium NPs have been prepared using [$Ru_3(CO)_{12}$] as precursor (Scheme 4.10). These NPs were used to catalyze the hydrogenation and cyclization of levulinic acid to selectively produce γ-valerolactone in high yields, either using molecular hydrogen (H_2) or formic acid in water as reducing agents (Ortiz-Cervantes and García, 2013).

A magnetically recoverable Ru/$NiFe_2O_4$ catalyst for chemoselective hydrogenation of alkynes to their respective alkanes at room temperature has been reported. Ru/$NiFe_2O_4$ catalyst can also be utilized in microwave (MW)-assisted hydrogenation of a series of carbonyl compounds with excellent conversions and reusability (Baruwati et al., 2009).

Magnetic chiral Ru complex-based catalyst can be used for the hydrogenation of a wide range of aromatic ketones to their corresponding secondary alcohols and exhibited high reactivity and enantioselectivity (Scheme 4.11). This catalyst could be recycled 14 times with no decrease in conversion and/or enantiomeric excess (Hu et al., 2005).

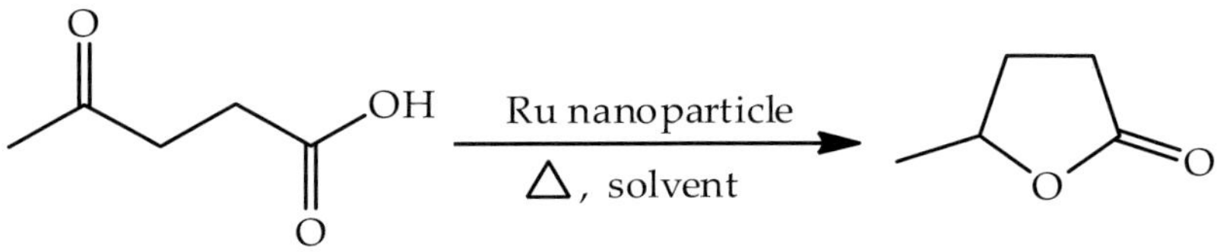

Scheme 4.10 Ru NP-catalyzed hydrogenation and cyclization of levulinic acid.

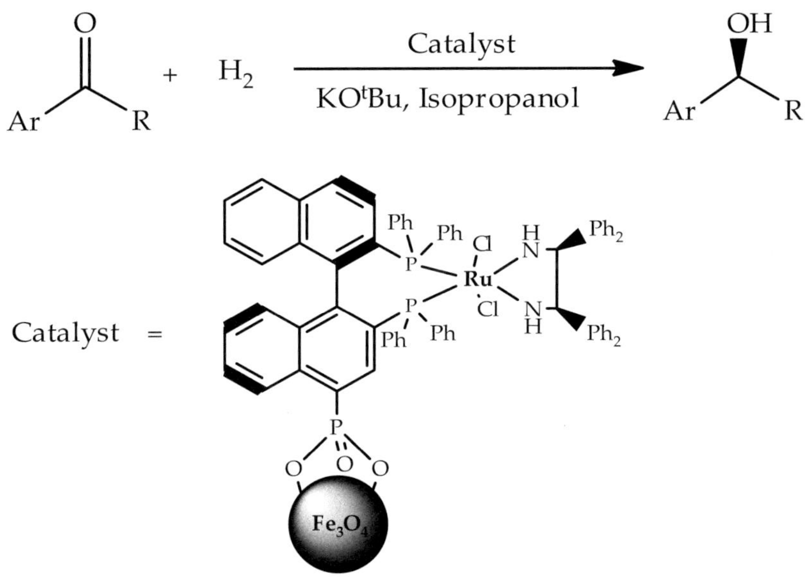

Scheme 4.11 Asymmetric hydrogenation of aromatic ketones using Ru complex-based catalyst.

The hydrogenation of benzene and its derivatives was reported using Ru(0) NPs in imidazolium ionic liquids with high total turnovers under solventless conditions (Rossi and

Machadob, 2009). Ru NPs supported on rGO (Ru/rGO) exhibit higher catalytic activity than analogous Ru/AC during the reduction of various nitroaromatics containing electron-donating or electron-withdrawing groups (Scheme 4.12). The catalyst could be reused without any significant loss of activity after five runs (Fan and Huang, 2014).

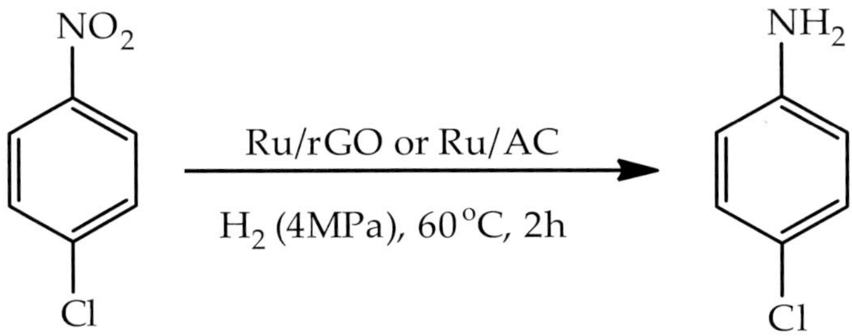

Scheme 4.12 Ru/rGO-catalyzed reduction of nitroaromatic compounds.

4.3.1.3 Hydrolysis

Ru/graphene catalyst was used for hydrogen generation from the hydrolysis of ammonia borane (AB), with the turnover frequency (TOF) value of 600 mol H_2 min^{-1} (mol $Ru)^{-1}$ (Du et al., 2015). Ru/g-C_3N_4 was utilized for the hydrolysis of AB in an aqueous solution under mild conditions with a high TOF at room temperature (Fan et al., 2014). Ru NPs immobilized in montmorillonite (MMT) catalyst have been utilized for the effective methanolysis reaction of AB (Dai et al., 2010). Ru(0) NPs supported on the surface of titania nanospheres (Ru(0)/TiO_2) showed high catalytic activity in hydrogen generation from the hydrolysis of ammonia borane with high turnover frequency (Akbayrak et al., 2014). Ru(0) NPs supported on magnetic silica-coated cobalt ferrite (Ru(0)/SiO_2-$CoFe_2O_4$) can hydrolyze AB and could be reused even after the 10th run as compared to the other Ru catalysts (Akbayraka et al., 2014). Ru NPs (Ru/$Ti_3C_2X_2$; X = OH, F) exhibited excellent catalytic activity toward the hydrolysis of sodium borohydride with a hydrogen generation rate of 59.04 L H_2/gRu/min and an activation energy of 22.1 kJ/mol (Li et al., 2014). Ru(0) NPs supported on xonotlite nanowire (Ru(0)/X-NW) showed high catalytic activity and long life time in hydrogen generation from the hydrolysis of sodium borohydride with a high TOF. Moreover, the catalytic activity of

Ru(0)/X-NW shows an inverse dependence on the catalyst concentration (Akbayrak and Ozkar, 2016).

4.3.1.4 Hydrogenolysis

The fibrous nanosilica/Ru NPs (KCC-1/Ru) catalyst has been utilized for the hydrogenolysis of propane and ethane at atmospheric pressure and at low temperature. The catalyst is stable with no sign of deactivation, even after eight days; this could be attributed to the fibrous nature of KCC-1, which restricts Ostwald ripening of Ru nanoparticles (Fihri et al., 2012). Ru NPs on magnetic silica have been used for the transfer hydrogenation of carbonyl compounds in high yields along with excellent selectivity under microwave irradiation conditions (Scheme 4.13) (Nasir Baig and Varma, 2013).

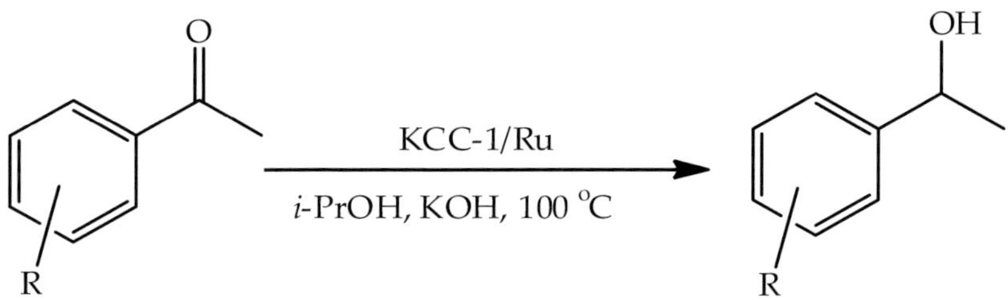

Scheme 4.13 Magnetic silica supported Ru NPs catalyzed transfer hydrogenation of carbonyl compounds.

4.3.1.5 Olefin metathesis

The reaction of a zero valent Ru precursor, Ru(σ^4-cyclooctadiene) (σ^6-cyclooctatriene), with different coordinating ligands such as alcohol has been employed to initiate olefin metathesis when they react with diazoalkanes and then strained olefins such as dicyclopentadiene. The rates of reaction depend on the nature of the ligands (Ren et al., 2007).

4.3.1.6 Isotope exchange

Ru(0) NPs, stabilized in the ionic liquid agent, trihexyltetradecylphosphonium dodecylbenzenesulfonate, were used to catalyze the isotope exchange reaction between 10B enriched diborane and natural abundant $B_{10}H_{14}$ to produce highly 10B enriched decaborane products (Yinghuai et al., 2007).

4.3.1.7 Miscellaneous catalytic reactions

NP of PdRu, Pd_3Ru, and Pd_9Ru were used for the electrocatalytic activity for oxygen reduction reaction (ORR). The Pd_9Ru@Pt reveals better ORR performance than that of Pt as compared to that of commercially available Pt nanoparticles (Sun et al., 2015). Ru NPs stabilized by the ZK-4 zeolite framework, for the chemical hydrogen storage and production, have been reported. The catalyst is highly active, long-lasting, and reusable for H_2 production at room temperature under air. Ru and Pt-based catalysts using hydroxyapatite (HAP) as support have been utilized for the catalytic transformation of methane and CO_2 into syngas (mixture of CO and H_2) (De et al., 2016).

4.3.2 As Analytical Tool

Chemo/biosensors have important applications in environmental monitoring and chemical process controlling, as well as in agricultural and biomedical fields.

Modified glassy carbon electrode prepared by potentiostatic electrodeposition of Pt-Ru NPs onto a multi-walled carbon nanotube (MWCNT) layer was used for the accurate determination of trace amounts of *m*-dopa in pharmaceutical and clinical preparations (Shahrokhiana and Rastgara, 2011). A new approach to enhance the electrochemiluminescence (ECL) emission of the (+) $Rubpy_3^{2-}$ tripropyl amine (TPrA) system for ultrasensitive determination of ochratoxin A (OTA) has been developed. (+) $Rubpy_3^{2-}$ doped silica NPs (RuSi NPs) act as ECL materials, which are immobilized on the surface of electrode by chitosan to fabricate a solid-state ECL sensor. CdTe quantum dots (QDs) can improve the sensitivity of the sensor that could be attributed to enhancing the ECL emission of the (+) $Rubpy_3^{2-}$ TPrA ECL system. To improve the selectivity of this sensor, this ECL approach was combined with the molecular imprinting technique. The cavities could also work as the tunnel for the transfer of co-reactant TPrA to produce responsive signal (Wang et al., 2016). A rapid, facile, sensitive ECL sensor has been developed based on a 3D graphene/CdSeTe/$Ru(bpy)_3^{2+}$-doped silica nanocomposite-modified electrode for the detection of folic acid (FA) in drugs (Li et al., 2016). Bifunctional Ru(II) polypyridine complex-based core–shell magnetic silica nanocomposites were reported for the detection of TPrA and other polyamines (Zhang et al., 2007). An encoding ECL-based method, generated by a dual-dye nanosystem, including Ru-derivatized NPs, as an ECL emitter and fluorescein isothiocyanate (FITC) as a coding dye, both entrapped in the silica matrix, has been reported (Luan et al., 2011). The resulting emitted color could be changed by tuning two orthogonal stimuli, namely the electrode potential and the 488 nm illumination intensity. Ru-loaded NiO NPs were synthesized for gas-sensing applications. Increasing Ru loading content considerably improved C_2H_5OH response and selectivity against H_2S, NO_2, SO_2, H_2, and NH_3 at an operating temperature of 350°C (Kruefu et al., 2016). Phosphonate-functionalized Au NPs (P-Au-NPs) which can effectively bind Ru(III) ions via coordination interaction have been used as an iodate amperometric sensor with the wide linear range and low detection limit (Li et al., 2015). Highly luminescent gels based on carboxylated nanocellulose as gelator and Ru(bpy)$_2$(α-bpy)](PF$_6$)$_2$ as the luminophore and sensitizer exhibit a significant enhancement of photoluminescence (PL) as well as a strong sensing response toward Ag NPs (Ruiz-Palomero et al., 2016). Prussian blue NP (PBNP) sol has been utilized in functional activity for making nanocomposite sol with *tris*(2,2-β-bipyridyl)Ru nanocomposite that shows high sensitivity for H_2O_2 electrochemical sensing and storage stability of the materials for more than 3 months. Apart from electrocatalytic application, shows photoluminescent ability for many opto-electroanalytical applications (Pandey and Pandey, 2013).

A new stable solid-state ECL sensor was fabricated by electropolymerizing *tris*(4-(2-pyrrol-1-ylethyl)-40-methyl-2,20-bipyridine) Ru(II)complex (Ru(L)$_3$) upon glassy carbon electrode (GCE). ECL quenching was observed by introducing gold NPs onto the polymer film. Based on the efficient quenching, a new glucose ECL biosensor as a model was accomplished by in situ growing-up gold seeds in $AuCl_4$ solution induced by biologically generated H_2O_2 (Xia et al., 2013). Pyrene-functionalized Ru NPs were synthesized and utilized as sensitive chemosensors for the detection of nitroaromatic compounds, such as 2,4,6-trinitrotoluene (TNT), 2,4-dinitrotoluene (2,4-DNT), 2,6-dinitrotoluene (2,6-DNT), 1-chloro-nitrobenzene (CNB), and nitrobenzene (NB), by their effective quenching of the NP fluorescence. As the nitration of the molecules increases, the sensitivity of chemosensor also increases (Chen et al., 2010). A new amphiphilic poly(propylene imine) dendrimer (APPI(G2))-encapsulated Ru NP (APPI(G2)RuNPs) catalyst was synthesized for applications such as sensing of H_2O_2 and reduction of 4-nitrophenol through newly fabricated GC-APPI(G2)-RuNP electrode (Murugan and Pakrudheen, 2015) (Scheme 4.14).

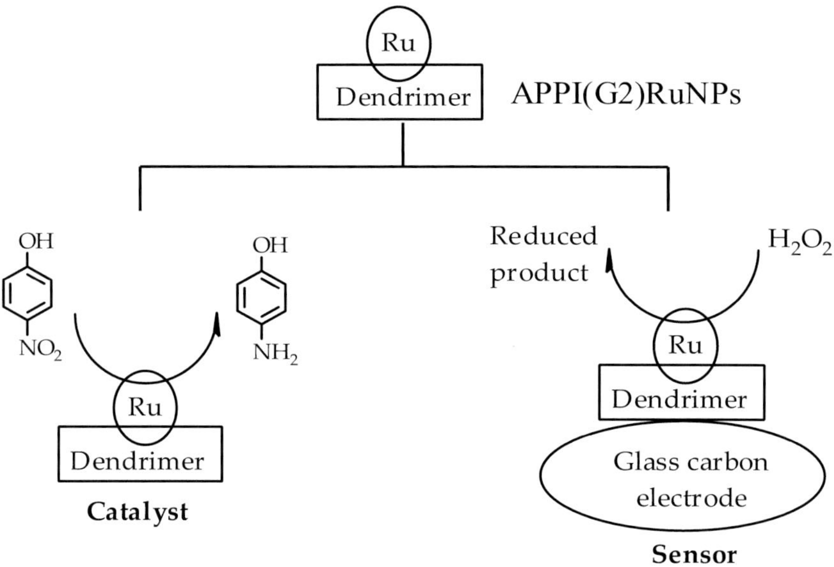

Scheme 4.14 Dendrimer-based GC-APPI(G2)-RuNPs for (a) reduction of 4-nitrophenol and (b) sensing of H_2O_2.

Nonenzymatic glucose sensors employing MWNTs with highly dispersed Pt-M (M = Ru and Sn) NPs (Pt-M@PVP-MWNTs) have been reported which can be utilized for the detection of glucose in commercial red wine samples (Kwon et al., 2012) (Scheme 4.15).

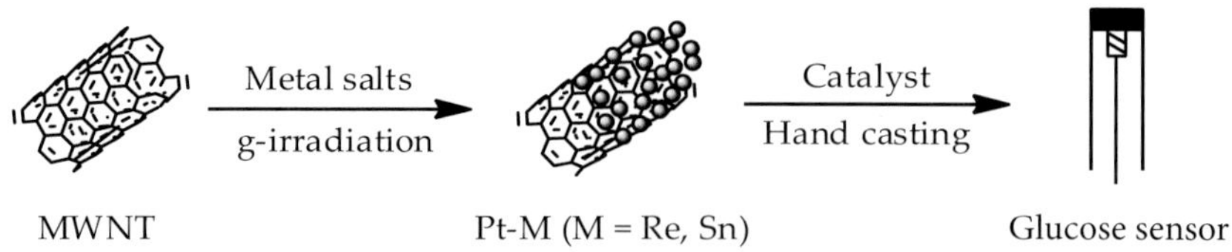

Scheme 4.15 CNT-supported Ru NPs as glucose sensors.

SnO_2 NPs with high-specific-surface-area (SSA_{BET}: 141.6 m^2/g) doped with 0.2–3 wt% Ru with high sensitivity and fast response time compared to a pure SnO_2 sensing film was used for hydrogen gas (Liewhiran et al., 2009). A bimetallic Pt-Ru NPs catalyst was utilized for the enhancement of H_2O_2 detection in biosensing applications (Janyasupab et al., 2011). Au NPs modified by the new Ru(II) complexes have been utilized for the detection of cysteine. Au NPs quench the intense emission of the modified Ru(II) complexes efficiently. Upon addition of cysteine, the emission of the Ru(II) complexes enhances significantly due to the release of the Ru(II) complexes from the surface of the gold NPs, which leads to the detection of cysteine-based probes (Li et al., 2011). The electronic conductivity of solid films of nitrene-functionalized Ru NPs was measured and compared in the presence of organic vapors of various relative polarity that included ethanol, acetone, n-hexane, 1-hexene, toluene, and styrene. The NP conductivity was found to decrease upon the exposure to the vapors tested; conductivity was found to be minimum in styrene and toluene. Functional NPs may be utilized for chemical vapor detection with enhanced sensitivity and selectivity (Chritopher et al., 2014). Metal-oxide NPs were synthesized and utilized for field-effect sensor devices. RuO_2 has been identified as prudent gate material for the field-effect sensor devices that can change its resistivity in different gaseous ambients (Salomonssona et al., 2005). Pt-Ru/3D GF nanocatalyst has been utilized for the development of electrochemical sensors in biosensing and H_2O_2 detection. Ascorbic acid and uric acid have negligible interference for the detection of H_2O_2 using the above catalyst (Kung et al., 2014). Porous $Ru(bpy)_3^{2+}$/silica NPs have been developed as a novel electrogenerated chemiluminescence sensor for tripropylamine (Li et al., 2015).

4.3.3 Biomedical Applications

The plant extract (*Acalypha indica*) has been utilized for the synthesis of RuO_2 NPs. These NPs possess high antibacterial activity against four different strains of the bacteria, *Escherichia coli*, *Pseudomonas aeruginosa*, *Serratia marcescens*, and *Staphylococcus aureus* (Kannan and Sundrarajan, 2015). The synthesis and the photophysical properties of Ru(II) and Ir(III) complex-doped gold core-polysiloxane shell particles have been reported for theranostic purposes (Ricciardi et al., 2014). Luminescent Ru(II) complex covalently bound silica NPs with a longer luminescence lifetime and decreased leakage of dye have been prepared which can be used as a probe for time-gated luminescence bioimaging. The nanoparticle-labeled streptavidin was prepared and used for time-gated luminescence imaging detection of an environmental pathogen, *Giardia lamblia*, with high specificity and sensitivity (Song et al., 2009). The dual-target inhibitor with high efficacy of luminescent Ru(II)-thiols protected selenium NPs (Ru-MUA@Se) that not only suppress the tumor growth but also block blood vessel growth with low side effects using metronomic chemotherapy has been developed (Sun et al., 2014). The synthesis of stable Se/Ru alloy NPs (GA-Se/RuNPs) using gallic acid as both a reducing and a capping agent with enhanced anticancer effects was utilized to suppress HeLa cell proliferation through the induction of apoptosis at concentrations that were nontoxic in normal cells. Furthermore, GA-Se/RuNPs effectively inhibits migration and invasion in HeLa cells via the inhibition of MMP-2 and MMP-9 proteins (Zhou et al., 2016). The anti-inflammatory effects of CO via sustained release of CO have been enhanced from CO-releasing molecule-2-loaded lipid Ru-based NPs (CORM-2-NPs). CORM-2-NPs showed enhanced in vitro anti-inflammatory effects by inhibition of NO production (Scheme 4.16). Edema volume in rat paw was significantly reduced after treatment with CORM-2-NPs (Salman et al., 2016).

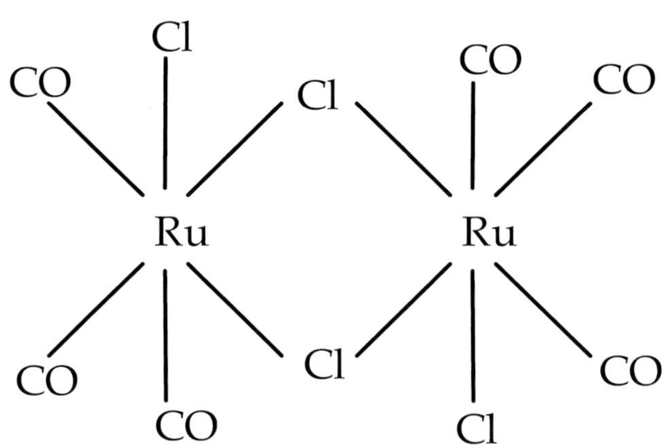

Scheme 4.16 Structure of CORM-2-NPs.

Chiral selenium NPs (L-SeNPs/D-SeNPs) modified with a dinuclear Ru(II) complex, were used as vectors for the delivery of siRNA. The dinuclear Ru(II) complex can act as a gene carrier and enhanced antitumor drug with luminescent imaging, to follow the intracellular trafficking with decreased systemic toxicity (Chen et al., 2015). A chromophoric Ru complex-assembled nanophosphor (N719-UCNPs) was used as a highly selective water-soluble probe for upconversion luminescence sensing and bioimaging of intracellular Hg^{2+} ions (Liu et al., 2011). The binding properties of AziRu, a Ru(III) complex with high antiproliferative activity, toward a hen egg white lysozyme have been investigated (Vergara et al., 2013). In order to utilize macromolecules for drug targeting and delivery, a strategy to tether organometallic Ru-arene drugs to carrier protein molecules was developed (Ang et al., 2007). DNA-sensing architectures based on Au NPs in conjunction with an in situ prepared Ru complex (Ru[3-(2-phenanthren-9-yl-vinyl)-pyridine]; ($Ru(NH_3)_5L$)) as a new route to improve the analytical properties of genosensors have been developed. This complex, due to its intercalative character, is able to bind to double-stranded DNA more efficiently than to single-stranded DNA. In addition, the metal provides a redox center that can be used as an electrochemical indicator (Garcia et al., 2008).

4.4 Future Prospects

Ru NPs have diverse applications; however, only selected Ru complexes and their NPs have been utilized as catalysts for performing different reactions, and not all catalysts have provided satisfactory results. The development of more efficient protocols in terms of sustainability and environmental friendliness, allowing the development of active and selective materials for a wide variety of applications, could be an important area of research. Nanotechnology can provide opportunities to obtain methods and materials for the development of synthetic strategies for Ru NPs and their application in nanosized devices that can result in the reduction of costs and environmental impact. It can also be used to develop and improve the sensitivity of sensors. New Ru-based NPs with multifunctionalities, sensitivity, and stability could be designed and developed that offer the possibility to utilize the materials in various fields at the industrial scale.

4.5 Conclusions

Ruthenium is an inert transition metal that possesses a number of properties due to which it has drawn the attention of scientists and researchers. The application of different support-based Ru NPs has been utilized for catalytic reactions, analytical techniques, and biomedical applications. Ru NPs exhibit a great potential as enantioselective catalysts, which are used extensively for the synthesis of medicines, drugs, and other bioactive molecules. By functionalizing these materials using chiral ligands, a series of chiral nanocatalysts can be designed, which offer great potential to reuse these expensive catalyst systems. Ru NPs have also found applications in ultrasensitive detection methods in biological assays, due to the small sample volumes required and the wide dynamic range. Ru NPs find increasing applications in medicinal chemistry as drug delivery agents, medicinal imaging tools or as diagnostic agents. In addition, these NPs can also benefit from the enhanced permeability and retention effect and can be tunable to present specific properties. Ru-based drugs are particularly important in the clinic due to their low toxicity. Ru NPs have been applied to treat different diseases such as cancer.

References

Abu-Dief AM, Mohamed IMA. (2015). A review on versatile applications of transition metal-complexes incorporating Schiff bases. *Beni-Suef Univ J Bas Appl Sci*, **4**: 119–133.

Akbayraka S, Kayab M, Volkana M, Ozkar S. (2014). Ruthenium(0) nanoparticles supported on magnetic silica coated cobalt ferrite: Reusable catalyst in hydrogen generation from the hydrolysis of ammonia-borane. *J Mol Catal A: Chem*, **394**: 253–261.

Akbayrak S, Ozkar S. (2016). Inverse relation between the catalytic activity and catalystconcentration for the ruthenium(0) NP supported on xonotlite nanowire in hydrogen generation from the hydrolysis of sodium borohydride. *J Mol Cat A: Chem*, **424**: 254–260.

Akbayrak S, Tanyýldýzi S, Morkan I, Ozkar S. (2014). Ruthenium(0) nanoparticles supported on nanotitania as highly active and reusable catalyst in hydrogen generation from the hydrolysis of ammonia borane. *Int J Hydrogen Energy*, **39**: 9628–9637.

Allardyce CS, Dyson PJ. (2001). Ruthenium in medicine: Current clinical uses and future prospects, *Platium Metals Rev*, **45**: 62–69.

Ang WH, Daldini E, Juillerat-Jeanneret L, Dyson PJ. (2007). Strategy to tether organometallic ruthenium arene anticancer compounds to recombinant human serum albumin. *Inorg Chem*, **46**: 9048–9050.

Antonels NC, Meijboom R. (2014). Preparation of well-defined dendrimer encapsulated ruthenium nanoparticles and their application as catalyst and enhancement of activity when utilised as SCILL catalysts in the hydrogenation of citral. *Catal Commun*, **57**: 148–152.

Baruwati B, Polshettiwar V, Varma RS. (2009). Magnetically recoverable supported ruthenium catalyst for hydrogenation of alkynes and transfer hydrogenation of carbonyl compounds. *Tet Lett*, **50**: 1215–1218.

Bouchat V, Feron O, Gallez B, Masereel B, Michiels C, Borght TV, Lucas S. (2011). Carbon nanoparticles synthesized by sputtering and gas condensation inside a nanocluster source of fixed dimension. *Surf Coat Tech*, **205**: S577–S581.

Bradley JS. (1994). In *Clusters and Colloids*, ed., Schmid G, VCH, Weinheim, pp. 469–473.

Bunker CE, Novak KC, Guliants EA, Harruff BA, Meziani MJ, Lin Y, Sun YP. (2007). Formation of protein-metal oxide nanostructures by the sonochemical method: Observation of nanofibers and nanoneedles. *Langmuir*, **23**: 10342–10347.

Chen Q, Yu Q, Liu Y, Bhavsar D, Yang L, Ren X, Sun D, Zheng W, Liu J, Chen LM. (2015). Multifunctional selenium nanoparticles: Chiral selectivity of delivering MDR-siRNA for reversal of multidrug resistance and real-time biofluorescence imaging. *Nanomed Nanotech Biol Med*, **11**: 1773–1784.

Chen W, Zuckerman NB, Konopelski JP, Chen S. (2010). Pyrene-functionalized ruthenium nanoparticles as effective chemosensors for nitroaromatic derivatives. *Anal Chem*, **82**: 461–465.

Chung BX, Liu CP. (2004). Synthesis of cobalt nanoparticles by DC magnetron sputtering and the effects of electron bombardment. *Mater Lett*, **58**: 1437–1440.

Collin JP, Harriman A, Heitz V, Odobel F, Sauvage JP. (1994). Photoinduced electron and energy-transfer processes occurring within porphyrin-metal-bisterpyridyl conjugates, *J Am Chem Soc*, **116**: 5679–5690.

Dai HB, Kang XD, Wang P. (2010). Ruthenium NP immobilized in montmorillonite used as catalyst for methanolysis of ammonia borane. *Int J Hydrogen Energy*, **35**: 10317–10323.

De Vasconcelos, BR, Zhao, L, Sharrock, P, Nzihou, A, Minh DP. (2016). Catalytic transformation of carbon dioxide and methane into syngasover ruthenium and platinum supported hydroxyapatites. *Appl Surf Sci*, **390**: 141–156.

Deming CP, Kang X, Liu K, Chen S. (2014). Nitrene-functionalized ruthenium nanoparticles: Selective manipulation of nanoparticles electronic conductivity by vinyl derivatives. *Sens Actuators B Chem*, **194**: 319–324.

Ding J, Chan KY, Ren J, Xiao FS. (2005). Platinum and platinum-ruthenium nanoparticles supported on ordered mesoporous carbon and their electrocatalytic performance for fuel cell reactions. *Electrochim Acta*, **50**: 3131–3141.

Du C, Ao Q, Cao N, Yang L, Luo W, Cheng G. (2015). Facile synthesis of monodisperse ruthenium nanoparticles supported on graphene for hydrogen generation from hydrolysis of ammonia borane. *Int J Hydrogen Energy*, **40**: 6180–6187.

Fan GY, Huang WJ. (2014). Synthesis of ruthenium/reduced graphene oxide composites and application for the selective hydrogenation of halonitroaromatics. *Chin Chem Lett*, **25**: 359–363.

Fan Y, Li X, He X, Zeng C, Fan G, Liu Q, Tang D. (2014). Effective hydrolysis of ammonia borane catalyzed by ruthenium nanoparticles immobilized on graphic carbon nitride. *Int J Hydrogen Energy*, **39**: 19982–19989.

Faraday M. (1857). The Bakerian lecture: Experimental relations of gold (and other metals) to light. *Philos Trans R Soc London*, **147**: 145–153.

Fihri A, Bouhrara M, Patil U, Cha D, Saih Y, Polshettiwar V. (2012). Fibrous nano-silica supported ruthenium (KCC-1/Ru): A sustainable catalyst for the hydrogenolysis of alkanes with good catalytic activity and lifetime. *ACS Catal*, **2**: 1425–1431.

Fröhlich H. (1937). Die spezifische warme der elektronen kleiner metallteilchen bei tiefen temperaturen. *Physica*, **4**: 406–412.

Gaffet E, Tachikart, M, Kedim OE, Rahouadj, R. (1996). Nanostructural materials formation by mechanical alloying-morphologic analysis basis on transmission and scanning electron microscopic observations. *Mater Charact*, **36**: 185–190.

Gao S, Zhang J, Zhub YF, Che CM. (2000). A convenient solvothermal route to ruthenium nanoparticles. *New J Chem*, **24**: 739–740.

Garcia T, Casero E, Revenga-Parra M, Martín-Benito J, Pariente F, Vazquez L, Lorenzo E. (2008). Architectures based on the use of gold nanoparticles and ruthenium complexes as a new route to improve genosensor sensitivity. *Biosens Bioelect*, **24**: 184–190.

Giaever I, Zeller HR. (1968). Superconductivity of small tin particles measured by tunneling. *Phys Rev Lett*, **20**: 1504–1507.

Gopiraman M, Babu SG, Khatri Z, Kai W, Kim YA, Endo M, Karvembu R, Kim IS. (2013). dry synthesis of easily tunable nano ruthenium supported on graphene: Novel nanocatalysts for aerial oxidation of alcohols and transfer hydrogenation of ketones. *J Phys Chem C*, **117**: 23582–23596.

Gopiraman M, Babu SG, Khatri Z, Wei K, Endo M, Karvembu R, Kim IS. (2013). Facile and homogeneous decoration of RuO_2 nanorods on graphene nanoplatelets for transfer hydrogenation of carbonyl compounds. *Catal Sci Technol*, **3**: 1485–1489.

Gubin, SP, Koksharov, YA, Khomutov GB, Yurkov GY. (2005). Magnetic nanoparticles: Preparation, structure and properties. *Russ Chem Rev*, **74**: 489–520.

Guzman MG, Dille J, Godet S. (2009). Synthesis of silver nanoparticles by chemical reduction method and their antibacterial activity. *Internat J Chem Bio Eng*, **2**: 104–111.

Hirai H, Nakao Y, Toshima N. (1978). Colloidal rhodium in poly(vinylpyrrolidone) as hydrogenation catalysts for internal olefins, *Chem Lett*, **5**: 545.

Hirai H, Nakao Y, Toshima N, Adachi K. (1976). Colloidal rhodium in poly-vinylalcohol as hydrogenation catalyst of olefins. *Chem Lett*, **9**: 905.

Hu A, Yee GT, Lin W. (2005). Magnetically recoverable chiral catalysts immobilized on magnetite nanoparticles for asymmetric hydrogenation of aromatic ketones. *J Am Chem Soc*, **127**: 12486–12487.

Huang Y, Ma Y, Cheng Y, Wang L, Li X. (2015). Biosynthesis of ruthenium nanoparticles supported on nitric acid modified activated carbon for liquid-phase hydrogenation of 2,2,4,4-tetramethylcyclobutane-1,3-dione. *Catal Commun*, **72**: 20–23.

Janyasupab M, Zhang Y, Lin PY, Bartling B, Xu J, Liu CC. (2011). Bimetallic Pt-Ru nanoparticles catalyst for hydrogen peroxide detection. *J Nanotech*, Article ID 506862, 6 pages doi:10.1155/2011/506862.

Joo SH, Park JY, Renzas JR, Butcher DR, Huang W, Somorjai GA. (2010). Size effect of ruthenium nanoparticles in catalytic carbon monoxide oxidation. *Nano Lett,* **10**: 2709–2713.

Kannan SK, Sundrarajan M. (2015). Green synthesis of ruthenium oxide nanoparticles: Characterization and its antibacterial activity. *Adv Power Tech*, **26**: 1505–1511.

Khanna PK. (2007). Reduction of transition metal salts by SFS: Synthesis of copper and silver sulphides. *J Synth React Inorg Metal Org Nano Metal Chem*, **37**: 805–808.

Kruefu V, Wisitsoraat A, Phokharatkul D, Tuantranont A, Phanichphant S. (2016). Enhancement of p-type gas-sensing performances of NiO nanoparticles prepared by precipitation with RuO_2 impregnation. *Sens Actuators B*, **236**: 466–473.

Kubo R. (1962). Electronic properties of small metallic particles. *Phys Lett*, **1**: 49–50.

Kung CC, Lin PY, Buse FJ, Xue Y, Yu X, Dai L, Liu CC. (2014). Preparation and characterization of three dimensional graphene foam supported platinum-ruthenium bimetallic nanocatalysts for hydrogen peroxide based electrochemical biosensors. *Biosens Bioelect*, **52**: 1–7.

Kung CC, Lin PY, Xue Y, Akolkar R, Dai L, Yu X, Liu CC. (2014). Three dimensional graphene foam supported platinum-ruthenium bimetallic nanocatalysts for direct methanol and direct ethanol fuel cell applications. *J Power Source*, **256**: 329–335.

Kwon SY, Kwen HD, Choi SH. (2012). Fabrication of nonenzymatic glucose sensors based on multiwalled carbon nanotubes with bimetallic Pt-M (M = Ru and Sn) catalysts by radiolytic deposition. *J Sens*, 2012: 1–8.

Leisner T, Rosche C, Granzer F, Woste L. (1996). The catalytic role of small-coinage clusters in photography, *Surf Rev Lett*, **3**: 1105–1108.

Li X, Fan G, Zeng C. (2014). Synthesis of ruthenium nanoparticles deposited on graphene-like transition metal carbide as an effective catalyst for the hydrolysis of sodium borohydride. *Int J Hydrogen Energy*, **39**: 14927–14934.

Li Y, Liu J, Wang, Y, Wang ZL. (2001). Preparation of monodispersed Fe-Mo nanoparticles as the catalyst for cvd synthesis of carbon nanotubes. *Chem Mater*, **13**: 1008–1014.

Li Y, Liu W, Zheng X. (2015). Enhanced sensitivity of the tris(bipyridine) ruthenium(II) chloride electrogenerated chemiluminescence sensor with porous silica nanoparticles. *Anal Lett*, **48**: 1907–1917.

Li P, Lu Z, Sun D, Cai C, Wei S, Chen Y, Lee JM. (2015). Immobilization and electrocatalysis of Ru(III) ions on phosphonate functionalized gold nanoparticles. *Electrochim Acta*, **157**: 199–204.

Li X, Tan X, Yan J, Hu Q, Wu J, Zhang H, Chen X. (2016). A sensitive electrochemiluminescence folic acid sensor based on a 3D graphene/CdSeTe/$Ru(bpy)_3^{2+}$-doped silica nanocomposite modified electrode. *Electrochim Acta*, **187**: 433–441.

Li M, Zhan CQ, Zheng YM, Chen GN, Chen X. (2011). Simple and selective sensing of cysteine using gold nanoparticles modified by ruthenium(II) complexes. *J Nanosci Nanotech*, **11**: 3578–3585.

Li J, Zhang Y, Han D, Gao Q, Li C. (2009). Asymmetric transfer hydrogenation using recoverable ruthenium catalyst immobilized into magnetic mesoporous silica. *J Mol Catal A: Chem*, **298**: 31–35.

Liewhiran C, Tamaekong N, Wisitsoraat A, Phanichphant S. (2009). H_2 Sensing response of flame-spray-made Ru/SnO_2 thick films fabricated from spin-coated nanoparticles. *Sensors (Basel)*, **9**: 8996–9010.

Lin Z, Ji L, Krause WE, Zhang XW. (2010). Synthesis and electrocatalysis of 1-aminopyrene-functionalized carbon nanofiber-supported platinum-ruthenium nanoparticles. *J Power Source*, **195**: 5520–5526.

Liu WT. (2006). Nanoparticles and their biological and environmental applications. *J Biosci Bioeng*, **102**: 1–7.

Liu Q, Peng J, Sun L, Li F. (2011). High-efficiency upconversion luminescent sensing and bioimaging of Hg(II) by chromophoric ruthenium complex-assembled nanophosphors. *ACS Nano*, **5**: 8040–8048.

Logvinenko, V, Polunina O, Mikhailov Y, Mikhailov K, Bokhonov B. (2007). Study of thermal decomposition of silver acetate. *J Therm Anal Calal*, **90**: 813–816.

Luan L, Lin ZJ, Wu G, Huang XL, Cai ZM, Chen X. (2011). Encoding electrochemiluminescence using $Ru(bpy)_3^{2+}$ and fluorescein isothiocyanate co-doped silica nanoparticles. *Chem Commun*, **47**: 3963–3965.

Ma Y, Huang Y, Cheng Y, Wang L, Li X. (2014). Biosynthesized ruthenium Nanoparticles supported on carbonnanotubes as efficient catalysts for hydrogenation of benzene tocyclohexane: An eco-friendly and economical bioreduction method. *Appl Catal A: General*, **48**: 4154–160.

Makgwanea PR, Ray SS. (2016). Interface structural effect of ruthenium-cerium oxide nanocompositeon its catalytic activity for selective oxidation of bioterpenes-derivedp-cymene. *J Mol Catal A: Chem*, **418–419**: 19–29.

Matsoukas T, Desai T, Lee K. (2015). Engineered nanoparticles and their applications. *J Nanomater*, 2015: 1–2.

Monyoncho EA, Ntais S, Soares F, Woo TK, Baranova EA. (2015). Synergetic effect of palladiumeruthenium nanostructures for ethanol electrooxidation in alkaline media. *J Power Source*, **287**: 139–149.

Mishra DK, Dabbawala AA, Hwang JS. (2013). Ruthenium nanoparticles supported on zeolite Y as an efficient catalyst for selective hydrogenation of xylose to xylitol. *J Mol Catal A: Chem*, **376**: 63–70.

Murugan E, Pakrudheen I. (2015). Efficient amphiphilic poly(propylene imine) dendrimer encapsulated ruthenium NP for sensing and catalysis applications. *Sci Adv Mater*, **7**: 891–901.

Nasir Baig RB, Varma RS. (2013). Magnetic silica-supported ruthenium nanoparticles: An Efficient catalyst for transfer hydrogenation of carbonyl compounds. *ACS Sustainable Chem Eng*, **1**: 805–809.

Nosheen E, Shah SM, Hussain H, Murtaza G. (2016). Photo-sensitization of ZnS nanoparticles with renowned ruthenium dyes N3, N719 and Z907 for application in solid state dye sensitized solar cells: A comparative study. *J Photochem Photobiol B Biol*, **162**: 583–591.

Ortiz-Cervantes C, García JJ. (2013). Hydrogenation of levulinic acid to c-valerolactone using ruthenium nanoparticles. *Inorg Chim Acta*, **397**: 124–128.

Pandey PC, Pandey AK. (2013). Novel synthesis of Prussian blue NP and nanocomposite sol: Electro-analytical application in hydrogen peroxide sensing. *Electrochim Acta*, **87**: 1–8.

Polshettiwar V, Luque R, Fihri A, Zhu H, Bouhrara M, Basset JM. (2011). Magnetically recoverable nanocatalysts. *Chem Rev*, **111**: 3036–3075.

Rawat RS. (2013). High-energy-density pinch plasma: A unique nonconventional tool for plasma nanotechnology. *IEEE Trans Plasma Sci*, **41**: 701–715.

Ren F, Feldman AK, Carnes M, Steigerwald M, Nuckolls C. (2007). Polymer growth by functionalized ruthenium nanoparticles. *Macromolecules*, **40**: 8151–8155.

Ricciardi L, Martini M, Tillement O, Sancey L, Perriat P, Ghedini M, Szerb EI, Yadav YJ, Deda ML. (2014). Multifunctional material based on ionic transition metal complexes and gold-silica nanoparticles: Synthesis and photophysical characterization for application in imaging and therapy. *Photochem Photobiol B Biol*, **140**: 396–404.

Rossi LM, Machadob G. (2009). Ruthenium nanoparticles prepared from ruthenium dioxide precursor: Highly active catalyst for hydrogenation of arenes under mild conditions. *J Mol Catal A: Chem*, **298**: 69–73.

Ruiz-Palomero C, Soriano ML, Valcarcel M. (2016). Gels based on nanocellulose with photosensitive rutheniumbipyridine moieties as sensors for silver NP in real samples. *Sens Actuators B*, **229**: 31–37.

Rutkowska IA, Wadas A, Kulesza PJ. (2016). Mixed layered WO_3/ZrO_2 films (with and without rhodium) as active supports for PtRu nanoparticles: Enhancement of oxidation of ethanol. *Electrochim Acta*, **210**: 575–587.

Salman Qureshi O, Zeb A, Akram M, Kim MS, Kang JH, Kim HS, Majid A, Han I, Chang SY, Bae ON, Kim JK. (2016). Enhanced acute anti-inflammatory effects of CORM-2-loaded NP via sustained carbon monoxide delivery. *Eur J Pharm Biopharm*, **108**: 187–195.

Salomonssona A, Roya S, Aulinb C, Cerda J, Kall P, Ojamae L, Strand M, Sanatic M, Spetz AL. (2005). Nanoparticles for long-term stable, more selective MISiCFET gas sensors. *Sens Actuators B*, **107**: 831–838.

Sathish Kumar PS, Manivela A, Anandana S, Zhou M, Grieser F, Ashok Kumar M. (2010). Sonochemical synthesis and characterization of gold-ruthenium bimetallic nanoparticles. *Coll Surf A Physicochem Eng Aspects*, **356**: 140–144.

Shahrokhiana S, Rastgara S. (2011). Electrodeposition of Pt-Ru nanoparticles on multi-walled carbon nanotubes: Application in sensitive voltammetric determination of methyldopa. *Electrochim Acta*, **58**: 125–133.

Song C, Ye Z, Wang G, Jin D, Yuan J, Guan Y, Piper J. (2009). Preparation and time-gated luminescence bioimaging application of ruthenium complex covalently bound silica nanoparticles. *Talanta*, **79**: 103–108.

Sun B, Carnevale D, Suss-Fink G. (2016). Selective N-cycle hydrogenation of quinolines with sodium borohydride in aqueous media catalyzed by hectorite-supported ruthenium nanoparticles. *J Organometallic Chem*, **821**: 197–205.

Sun Y, Hsieh YC, Chang LC, Wu PW, Lee JF. (2015). Synthesis of $Pd_9Ru@Pt$ nanoparticles for oxygen reduction reaction in acidic electrolytes. *J Power Source*, **277**: 116–123.

Sun D, Liu Y, Yu Q, Qin X, Yang L, Zhou Y, Chen L, Liu J. (2014). Inhibition of tumor growth and vasculature and fluorescence imaging using functionalized ruthenium-thiol protected selenium nanoparticles. *Biomaterials*, **35**: 1572–1583.

Sun B, Suss-Fink G. (2016). Ruthenium-catalyzed hydrogenation of aromatic amino acids in aqueous solution. *J Organometallic Chem*, **812**: 81–86.

Suzuki S, Suzuki T, Tomita Y, Hirano M, Okazaki K, Kuwabata S, Torimoto T. (2012). Compositional control of AuPt nanoparticles synthesized in ionic liquids by the sputter deposition technique. *Cryst Eng Commun*, **14**: 4922–4926.

Takasaki M, Motoyama Y, Yoon SH, Mochida I, Nagashima H. (2007). Highly efficient synthesis of optically pure 5,5′,6,6′,7,7′,8,8′-octahydro-1,1′-bi-2-naphthol and -naphthylamine derivatives by partial hydrogenation of 1,1′-Binaphthyls with carbon nanofiber supported ruthenium nanoparticles. *J Org Chem,* **72**: 10291–10293.

Teranishi, T, Miyake, M. (1998). Size control of palladium nanoparticles and their crystal structures. *Chem Mater*, **10**: 594–600.

Tsai MC, Yeh TK, Tsai CH. (2008). Electrodeposition of platinum-ruthenium nanoparticles on carbon nanotubes directly grown on carbon cloths for methanol oxidation. *Mater Chem Phys*, **109**: 422–428.

Vergara A, D'Errico G, Montesarchio D, Mangiapia G, Paduano L, Merlino A. (2013). Interaction of anticancer ruthenium compounds with proteins: High-resolution X-ray structures and raman microscopy studies of the adduct between hen egg white lysozyme and AziRu. *Inorg Chem,* **52**: 4157–4159.

Wang Q, Chen, M, Zhang H, Wen W, Zhang X, Wang S. (2016). Enhanced electrochemiluminescence of RuSi NP for ultrasensitive detection of ochratoxin A by energy transfer with CdTe quantum dots. *Biosens Bioelectronics*, **79**: 561–567.

Wang R, Da H, Wang H, Ji S, Tian Z. (2013). Selenium functionalized carbon for high dispersion of platinum-ruthenium nanoparticles and its effect on the electrocatalytic oxidation of methanol. *J Power Source*, **233**: 326–330.

Wang Y, Rong Z, Wang Y, Zhang P, Wang Y, Qu J. (2015). Ruthenium nanoparticles loaded on multiwalled carbon nanotubes for liquid-phase hydrogenation of fine chemicals: An exploration of confinement effect. *J Catal*, **329**: 95–106.

Xia J, Ding SN, Gao BH, Sun YM, Wang YH, Cosnier S, Guo X. (2013). A biosensing application based on quenching the enhanced electrochemiluminescence of poly[tris(N-bipyridylethyl)pyrrole] ruthenium(II) film by Au nanoparticles. *J Electroanal Chem*, **692**: 60–65.

Yang J, Lee JY, Deivaraj TC, Too HP. (2004). A highly efficient phase transfer method for preparing alkylamine-stabilized Ru, Pt, and Au nanoparticles. *J Colloid Interface Sci*, **277**: 95–99.

Yinghuai Z, Widjaja E, Sia SL, Zhan W, Carpenter K, Maguire JA, Hosmane NS, Hawthorne MF. (2007). Ruthenium(0) nanoparticle-catalyzed isotope exchange between 10B and 11B nuclei in decaborane(14). *J Am Chem Soc*, **129**: 6507–6512.

Yu HD, Regulacio MD, Ye E, Han MY. (2013). Chemical routes to top-down nanofabrication. *Chem Soc Rev*, **42**: 6006–6018.

Zahmakiran M. (2012). Preparation and characterization of LTA-type zeolite framework dispersed ruthenium NP and their catalytic application in the hydrolytic dehydrogenation of ammonia-borane for efficient hydrogen generation. *Mater Sci Eng B*, **177**: 606–613.

Zare HR, Ghanbari Z, Nasirizadeh N, Benvidi A. (2013). Simultaneous determination of adrenaline, uric acid, and cysteine using bifunctional electrocatalyst of ruthenium oxide nanoparticles. *C R Chim*, **16**: 287–295.

Zhang LH, Liu BF, Dong SJ. (2007). Bifunctional nanostructure of magnetic core luminescent shell and its application as solid-state electrochemiluminescence sensor material. *J Phys Chem B*, **111**: 10448–10452.

Zhang J, Sun B, Huang Y, Guan X. (2015). Catalyzing the oxidation of sulfamethoxazole by permanganate using molecular sieves supported ruthenium nanoparticles. *Chemosphere*, **141**: 154–161.

Zhou Y, Xu M, Liu Y, Bai Y, Deng Y, Liu J, Chen L. (2016). Green synthesis of Se/Ru alloy nanoparticles using gallic acid and evaluation of their anti-invasive effects in HeLa cells. *Colloids Surf B Biointerfaces*, **144**: 118–124.

Chapter 5

Synthesis and Characterization of Copper–Ruthenium Composites

Rasidi Sule,[a] Iakovos Sigalas,[a] Joseph Kwaku Ofori Asante,[b] and Peter Apata Olubambi[c]

[a]*School of Chemical and Metallurgical Engineering, DST/NRF Centre of Excellence, University of the Witwatersrand, Johannesburg, South Africa*
[b]*Department of Physics, Tshwane University of Technology, Pretoria, South Africa*
[c]*Department of Chemical Engineering, University of Johannesburg, Johannesburg, South Africa*

rasidi.sule@gmail.com; rasidi.sule@wits.ac.za

The demand for significantly enhanced materials to keep copper as potential interconnection and thermal management materials in microelectronics packaging has led to widespread investigations. This chapter summarizes many of the research accomplishments in the area of powder metallurgy using ruthenium as reinforcement or additive to improve the hardness, oxidation, thermal conductivity, and specific capacitance of their matrix materials. The chapter also discusses different fabrication methods as well as the characterization techniques.

Ruthenium Chemistry
Edited by Ajay Kumar Mishra and Lallan Mishra

ISBN 978-981-4774-39-0 (Hardcover), 978-1-315-11058-5 (eBook)
www.panstanford.com

5.1 Introduction to Microelectronic Packaging Material

The demand for faster and cheaper integrated circuits prevalent in today's electronic and semiconductors technology continues to rise. Semiconductor manufacturers have been shrinking transistor size in integrated circuits (IC) to improve chip performance. This has been characterized by a continued miniaturization of electronic devices (Stangl et al., 2009). However, miniaturization of electronic devices is a demanding trend in current and future technologies for varying microelectronic and semiconductor applications. Thus, this has become a challenge to produce smaller and faster circuits in less time and for less cost (Cheng et al., 2003). In order to achieve this miniaturization, packaging technologies such as embedded active and passive components and 3D packaging, where new packaging designs and materials are required have emerged (Li et al., 2010). Electronic packaging in a variety of microelectronic applications however, involves interconnecting, powering, protecting, and cooling of semiconductor circuits (Schubert et al., 2008). Interconnect material purpose is to join two electrical terminals with low parasitics such as electrical resistance, inductance and capacitor to be reliable in field used (Pan et al., 2006). Meanwhile thermal management device in the electronic packaging is incorporated to cool the electronics components and systems and protect it from excessive heat during operation (Datsyuk et al., 2011). With the growing concern to achieve optimum performance of smart IC's at high operation temperature without reducing performance (Schubert et al., 2008); the requirements of both interconnect and thermal management materials properties are now enormous (Moore and Shi, 2014).

In earlier technology generation of microelectronic circuits and MEMS devices, metals such as aluminum, copper, gold, molybdenum and tungsten were commonly used for thermal management in electronic industries (Ekpu et al., 2011). Among all the metal materials used for the thermal management, Al has been widely used for heat sink materials due to its thermal conductivity (220 W/mK), electrical resistivity (2.8 μΩcm), light weight, low cost and manufacturability (Gallagher et al., 1998). However, Al reaches its performance limit as the demand for production of fast processors and miniaturization of microelectronic devices results in overheating of the computer central processing unit (CPU) (Ekpu et al., 2011). Therefore, aluminum was replaced by copper due to its higher thermal conductivity (400 W/mK) when compared with other metals such as gold, molybdenum, and tungsten (Gallagher et al., 1998; Yoshida and Morigami, 2004). Another metal that has been considered besides copper is silver (Ag). Silver is a promising material due to its high thermal and electrical conductivity values. However, the cost of Ag is too high from the economic point of view; this makes copper a preferred material.

With the increasing performance and packing density required in microelectronic devices, copper has shown to be a material of choice due to its attractive properties (Yi et al., 2008). Despite the attractive thermal and electrical properties of copper, its traditional properties however, can no longer satisfy the recent technological requirements and needs of the rapid development of the microelectronic technology.

To overcome these challenges, dispersion strengthening of copper grain boundary by carbon nanotube (CNT) and hard particles such as ruthenium is considered as a promising reinforcement in producing good Cu-based composite materials. Recently, ruthenium (Ru) has emerged as a suitable material to improve the non-protective surface scale of copper as well as its diffusion through the grain boundaries to the substrate (Au, 2010).

5.2 Ruthenium as Emerging Material

Ruthenium is a metal of a considerable importance in electrochemical science and large-scale integrated (VLSI) technology (Marinkovic et al., 2006; Gujar et al., 2007). It acts as a catalyst or co-catalyst material in Pt–Ru alloys for methanol- and reformate hydrogen-oxidation in fuel cells. Meanwhile, ruthenium oxide, a component in chlorine evolution catalysts, exhibits interesting material properties for electrochemical supercapacitors (Marinkovic et al., 2006). Recently, Ru has emerged as an alternative copper diffusion barrier due to its unique properties (Yang et al., 2006). Ruthenium has low bulk resistivity (7.1 μΩcm) and very low solubility with copper. Ruthenium does not easily oxidize and has good adhesion with copper (Yang et al., 2006).

5.2.1 Early History of Ruthenium

Ruthenium belongs to the family of platinum group of metals. However, its existence was not known until more than a century after the discovery of platinum (Weeks, 1932). According to *Philosophical Magazine* (New metals in the Uralian Platina, *Phil. Mag.*, 2, 391, November, 1827), cited by Weeks (1932), Berzelius and Osann, professors of chemistry at Dorpan, Russia, in 1828 examined the residues left after dissolving crude platinum from the Ural Mountains in aqua regia. Osann thought he had found new metals, but Berzelius did not see any unusual metals except palladium, rhodium, osmium, and iridium, which had already been reported. In 1844, Karl Karlovich Klaus, another Russian chemist, showed that Osann's ruthenium oxide was very impure but it did contain a small amount of a new metal, which he called ruthenium. Professor Karl was the first to publish a report on the residues from the platinum ores on the new metal ruthenium in German journals. Due to the convincing evidence he gave, in 1845 Berzelius announced his acceptance of ruthenium as a new element (Weeks, 1932).

Ruthenium is a noble transitional metal that has gained recognition in the field of science and technology due to its attractive properties. Ruthenium crystallizes in a hexagonal close-packed structure, (hcp) and has low bulk resistivity (7.1 μΩcm). Ruthenium has elastic and shear modulus values of 438 GPa and 170 GPa, respectively (Tsui et al., 2013). Well-annealed bulk ruthenium is said to have an ultimate tensile strength of 495 MPa (Tsui et al., 2013). Since the publication of Karl on the discovery of ruthenium in 1854 (Week, 1832), ruthenium has been incorporated into several materials to improve their properties.

Figure 5.1 shows the morphology of ruthenium powder particles. The SEM micrographs of the ruthenium powders revealed that the Ru particles are densely packed and had large agglomerates of small primary particles (Sule et al., 2015). Figure 5.2 shows the XRD pattern of ruthenium powder analyzed using monochromatic Cu$K\alpha$ radiation. The XRD result showed that the ruthenium powder was quite pure as only Ru peaks were obviously observed in the trace. The prominent peak of ruthenium powder was found at the 2θ angle of 44° with d-spacing of

2.05324 Å. The application of ruthenium is usually due to its structure and unique properties, which give the desire properties.

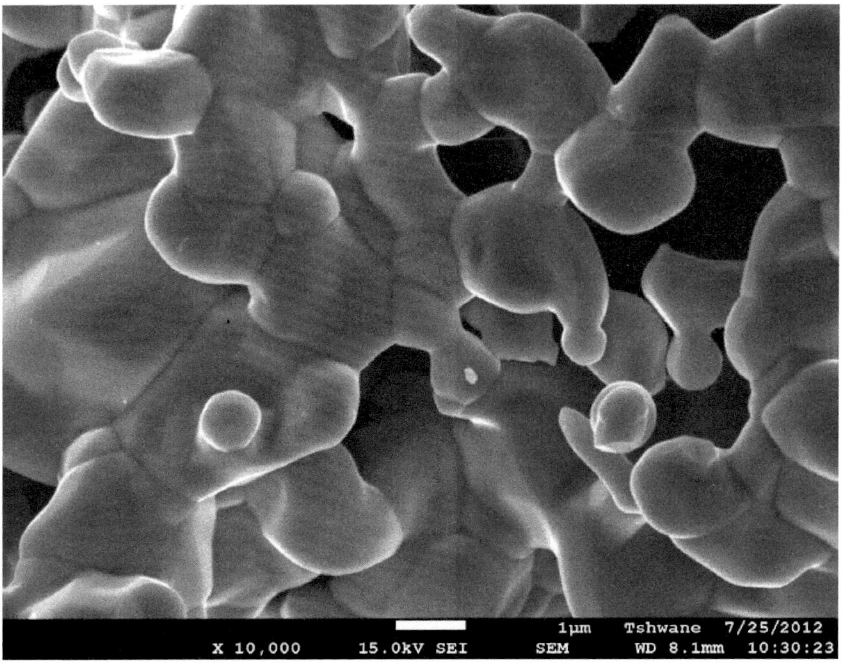

Figure 5.1 SEM image of Ru powder, showing its particle shape.

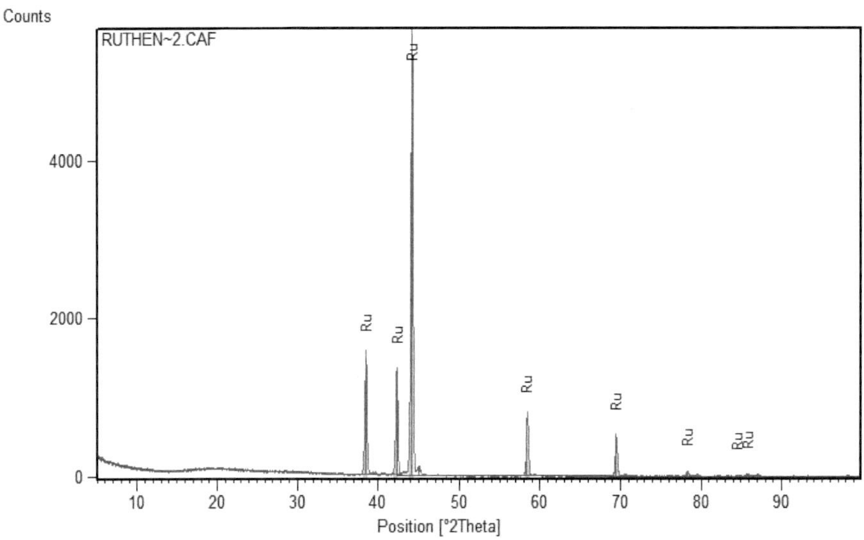

Figure 5.2 X-ray diffraction spectra of Ru powder.

5.2.2 Ruthenium–Aluminum Alloy

Nickel and titanium aluminide systems have been widely used for high-temperature structural applications due their strength, high oxidation and corrosion resistance, and high melting point (Bora et al., 2004). The application of these alloy systems at high temperature is limited by their poor room temperature ductility and fracture toughness (Bora et al., 2004). In order to overcome these challenges, a material with high strength at room and elevated temperatures as well as room temperature ductility and toughness could be incorporated.

Ruthenium has been considered as a promising material in producing good Al-based alloy system for high-temperature structural environment (Wolff, 1997). The Ru–Al alloy was reported to combine the high melting point (≈2050°C), high strength at room and elevated temperatures, high resistance to environmental corrosion in very severe chemical media as well as room temperature ductility and toughness in structural applications (Bora et al., 2004). The Ru–Al has shown good oxidation resistance up to at least 900°C. However, the formation of volatile oxides seriously curtails its resistance at higher temperature as severe internal oxidation occurs at 1300°C (Wolff, 1997). Thus, the application of Ru–Al without the further alloying in high-temperature environment is restricted by its oxidation behavior. However, the electrical resistivity and thermal conductivity of Ru–Al have suggested that the material has the potential to be used as electric contact as well as diffusion barrier in microelectronics (Wolff, 1997; Wang et al., 2011).

In earlier technology generation of microelectronic circuits and micro-electro-mechanical system (MEMS) devices, metal such as aluminum were commonly used as integrated circuit metallization element (Ekpu et al., 2011). Al has been widely used due to its thermal conductivity (220 W/mK), electrical resistivity (2.8 μΩcm), light weight, low cost, and manufacturability (Gallagher et al., 1998). However, Al reaches its performance limit as the demand for increasing chip performance and miniaturization of microelectronic devices to allow increase device density (McGahay, 2010). The performance and reliability problem encounter in continual scaling of aluminum-based material has led to the introduction of copper due to its attractive properties (Hau-Riege, 2004).

5.3 History of Copper

Historically, copper has remained one of the key prestigious materials chosen by mankind to improve its quality of life. In the Mississippian culture, which encompassed much of the present day Southeastern and Midwestern United States dated around AD1050, the use of copper is seen in items such as beads, repoussé plates and copper-clad personal adornments for ceremony and decoration (Chastain et al., 2011; Ehrhardt, 2009). From the early uses of copper as ornament up to its metallic material for the modern age, the choice of copper has largely been due to its availability, ease of fabrication and resistance to corrosion (not in an aggressive environment) and lately due its electrical and thermal conductivity. However, little knowledge of the electrical and thermal conductivity properties of copper was explored until the late 19th and early 20th centuries when the industrial revolution gave rise to electrical industries (Boschma, 1999).

The electrical engineering industrial sense grew following the Michael Faraday's (1831) discovery of electromagnetic induction, which was followed by the invention of the electric dynamo by Werner von Siemen in 1866 and Thomas Edison's invention of the electric light bulb in 1878; then the construction of an electric power plant in 1882 gave added impetus to the industrial revolution (Davis, 2001). These inventions have made copper a relevant material in the electrical industries up to the present age of semiconductor electronics industries. The choice of copper as the leading material in microelectronic can be attributed to its outstanding properties such as low resistivity and high thermal conductivity. Although copper offers high electrical and thermal conductivity that makes it a suitable material for both interconnect and thermal management. Pure copper with conventional grain size are soft materials (Mula et al., 2011; Yum et al., 2005). Apart from the fact that Cu is a soft material, the widespread application of copper technology has been limited by the oxidation of copper at elevated temperature by forming a non-protective surface scale (O'reilly et al. 1995). Recently, efforts have been focused on the improvement of copper for effective interconnection and thermal management in microelectronics packaging. Among the electrical and thermal conductive materials that can be used to augment copper for effective applications in advanced integrated circuit and power electronics is ruthenium (Sule et al., 2012).

5.3.1 Ruthenium as Diffusion Barrier in Copper Interconnect

Copper-based metallization has specific resistance of less than 2 μΩ-cm, compared to more than 3 μΩ-cm for aluminum metallization (Sharma and Teverovsky, 2001). Copper offer a reduced susceptibility to electromigration failure. The combination of these properties enables the design of highly scaled device with significantly higher performance and *RC* time delay. These features are mostly beneficial for devices such as high performance microprocessor and fast static RAMs (FSRAM) (Sharma and Teverovsky, 2001). Moreover, copper offers low resistivity and high electromigration resistance, its diffusion under electrical bias into surrounding dielectric layers, RC signal delay and electromigration failure has been a major challenge.

Diffusion is a major challenge in copper interconnects used in microelectronic devices. Usually, grain boundaries are the fast copper diffusion paths (Perng et al., 2008). The more electron scattering at the copper grain boundaries due to high current density, the higher will be the chance of copper atom to migration. Rapid diffusion along the grain boundaries increases the average copper concentration at that path. The relatively higher concentration gradient will in turn helps further diffusion and short circuit of the system. Copper diffusion barrier TaN/Ta has been commonly implemented in integrated circuit as a reliable barrier for copper interconnects (Damayanti et al., 2007). With continual scaling of integrated critical dimension, the cross-sectional area of the Tan/Ta barrier would contribute a significant increase in resistance and cause difficulty in copper filling (Perng et al., 2008). However, adding impurities to stuff grain boundaries is an effective way to improve barrier properties with reduce process step (Chang et al., 1998) cited by Perng et al. (2008).

Moreover, ruthenium (Ru) has emerged as impurities to stuff copper grain boundaries for effective barrier properties (Perng et al., 2008). The AES depth profiles of the Cu/Ru/PSG films with and without P doping in Ru after annealing at 400°C for 30 min in vacuum was investigated. The result showed that the diffusion of phosphorus into Ru film improves barrier properties as compared to the sample without P doping. Hsu et al. (2012) reported on ultrathin Cr added Ru film as a seedless Cu diffusion barrier for advanced Cu interconnects. Fourier transform-electron diffraction pattern reveal that a Cr contained Ru (RuCr) film has a glassy microstructure and is an amorphous-like film. The RuCr film was found to improve the thermal stability by over 200°C, which makes it more stable barrier than a pure Ru film against Cu diffusion.

5.3.2 Ruthenium as Alloying Element for Thermal Management Materials

Thermal management materials are manufactured into various consumer electronic products such as aerospace, servers, power supply subsystems, laptop and notebook computers, cellular phone, printed circuit board (PCB), optoelectronics packages, and plasma display (Zweben, 2005). Applications in which heat is generated within the electronic module itself, such as power and portable electronic can generally be addressed by improved thermal management techniques (Gallagher et al., 1998).

In order to meet the desired thermal management requirements, several materials have been added to Cu to improve its coefficient of thermal expansion. Among these materials are Mo, W, diamond, carbon or SiC particles (Qu et al., 2011). Copper reinforced with SiC would have been a good candidate for thermal management, owing to its low density and low coefficient of thermal expansion that matched up with that of the ceramic substrate (Bukhari et al., 2011). Copper/SiC offers high thermal conductivity ranges from 250–325 W/mK and the thermal expansion coefficients could be tailored to between 8.0 and

12.5 per Kelvin. However, Cu reacts chemically with SiC at temperatures necessary for Cu/SiC densification (850 to 1200°C). The diffusion of Si into Cu at that temperature tends to decrease the Cu thermal conductivity (Sundberg et al., 2006). Carbon nanotubes have been investigated as promising material for improving copper due to their high thermal conductivity and low coefficient of thermal expansion of approximately zero (Deng et al., 2008). Several authors have reported on the thermal conductivity of Cu-CNT composites for thermal management in microelectronics (Chai and Chen, 2010; Chu et al., 2010).

With the rapid development of electronic technologies, the requirements of thermal management properties at high operating temperature without reducing performance are becoming higher (Schubert et al., 2008); the requirements of thermal management materials properties are now involving combined properties such as oxidation behavior (Cocke et al., 2005; Orth and Wheat, 1997). Sule et al. (2015) reported on the effect of ruthenium on thermal conductivity as well as oxidation behavior of copper–carbon nanotube composites.

The surface oxidation behavior of copper–carbon nanotube containing ruthenium was investigated. X-ray photoelectron spectroscopy (XPS) analytical technique was used to identifying the oxidation state of the element. Figure 5.3a–f shows the deconvolutions of the Cu 2p3/2 spectrum of the various samples produced. The percentage area contribution of Cu(0), Cu(I) and Cu(II) was quantified by applying specific constraint on BE, FWHM and peak shape parameters. It was observed that the addition of 0.5 vol% Ru tends to reduce the contribution of Cu(I) oxide and Cu(II) oxide when comparing the composites without ruthenium content (Sule et al., 2015).

In addition, Ding et al. (2011) reported on the effect of ruthenium on severe oxidation of copper and galvanic corrosion during chemical mechanical polishing (CMP) process. It was found that addition of carbon (C) into Ru resulted in inhibition of Cu oxidation and promotion of Cu_2O reduction. The XPS analysis results showed that Cu_2O could be partially reduced to Cu on the RuC layer. It was concluded that the Cu loss and adhesion degradation due to oxidation can be improved by the RuC layer.

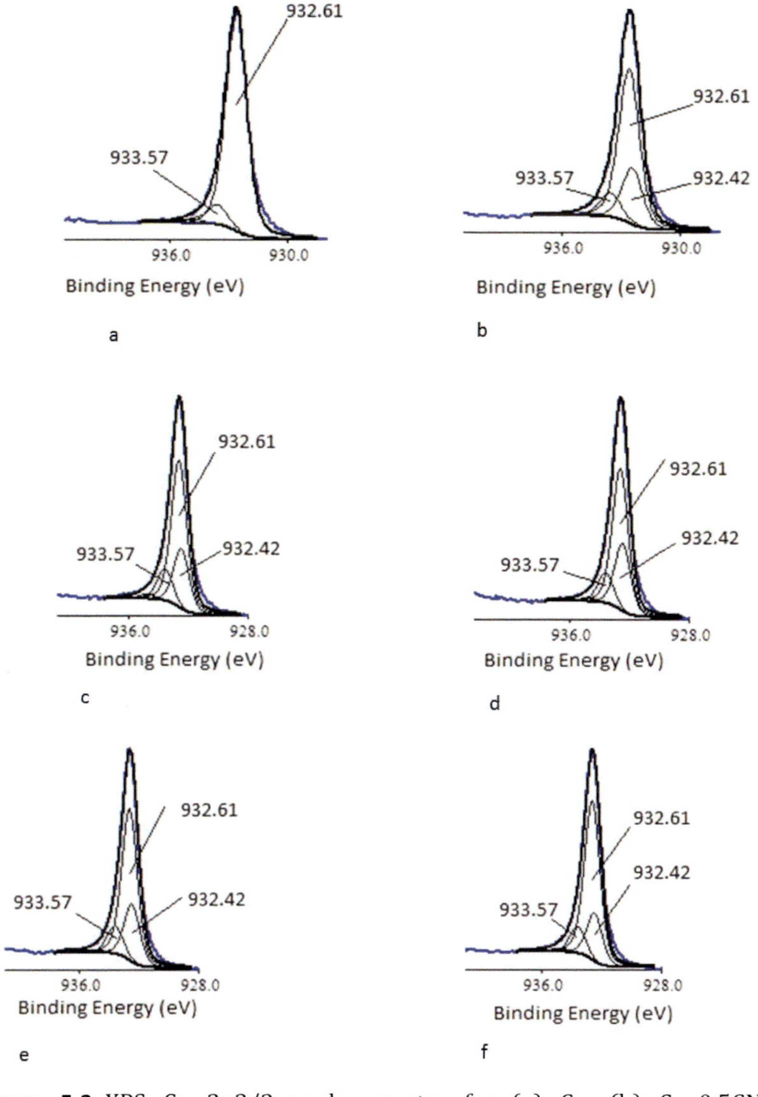

Figure 5.3 XPS Cu 2p3/2 peaks spectra for (a) Cu, (b) Cu–0.5CNT, (c) Cu–1CNT, (d) Cu–2CNT, (e) Cu–1CNT–0.5Ru, (f) Cu–2CNT–0.5Ru.

5.3.3 Ruthenium as Oxidation Resistance for Copper

Copper has been seen to be a material of choice for interconnection material in ultra-large-scale integrated (ULSI) technology due to its high conductivity and resistance to electromigration

(Gao et al., 2001). However, application of copper has been limited by the oxidation of copper at elevated temperature and humidity (Chuang et al., 2006). When copper is heated in air at low temperature around 200°C it forms Cu_2O film on the surface (Li and Marer, 1991). Further increase in temperature will result in formation of cupric oxide (CuO) over the copper surface (Bateni et al., 2001). The resultant effects of Cu oxidation are cracks, decrease in the interfacial shear strength and poor bonding between the component and the substrate which causes device failure (Chuang et al., 2006). Therefore, it is essential to understand and modify their surface properties in order to improve the performance of copper interconnect materials.

Sule et al. (2012) investigated the oxidation behavior of Cu and Cu–Ru composites synthesized by powder metallurgy route using X-ray photoelectron spectroscopy (XPS). Prior to XPS investigation, the samples were heated at temperature ranging from 20 to 800°C at the heating rate of 10°C/min inside the furnace. The XPS spectra measurements were carried out in a PHI 80–360 electron energy analyzer using Al Kα photons (1486.6 eV) in an ultrahigh vacuum chamber with a base pressure better than 10^{-9} Torr. The anode type X-ray power source was run at 15 KV, 300 W. The measurements were performed in sample surface at a takeoff angle 45° between the direction of the analyzer and specimen plane. The binding energy scale was calibrated using sputter clean, pure copper foil. Spectral analysis was performed using the peak fitting software (XPSPEAK version 4.1) after a Shirley background subtraction (Platzman et al., 2008). Peak fitting solutions were obtained for $x^2 < 2$, where x^2 is the standard deviation. A broad scan survey spectrum range from 0–1000 eV was first obtained to identify the elements present in the samples. Therefore, narrow detailed scans of selected peaks was carried out for more detailed analysis of the chemical composition of oxidized Cu, Cu+2.5 vol% Ru and Cu+5 vol% Ru samples.

Detailed scan was carried out on the oxidized pure copper from 928.0 eV to 939 eV. Figure 5.4a reveals two peaks at binding energy position of 932.44 eV and 934.27 eV after curve fitting. The bigger peak has a binding energy of 932.44 eV, which corresponds to Cu (Platzman et al., 2008). The smaller peak has a binding energy of 934.27 eV, which corresponds to CuO $2p_{3/2}$ (when examined from the XPS database spectra line). The results indicated that copper undergoes significant oxidation after heat treatment (Chuang et al., 2006).

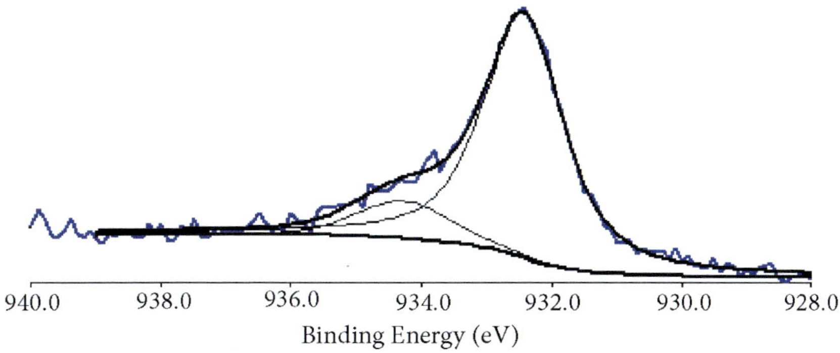

Figure 5.4a XPS Cu $2p_{3/2}$ peaks spectra for after heat treatment.

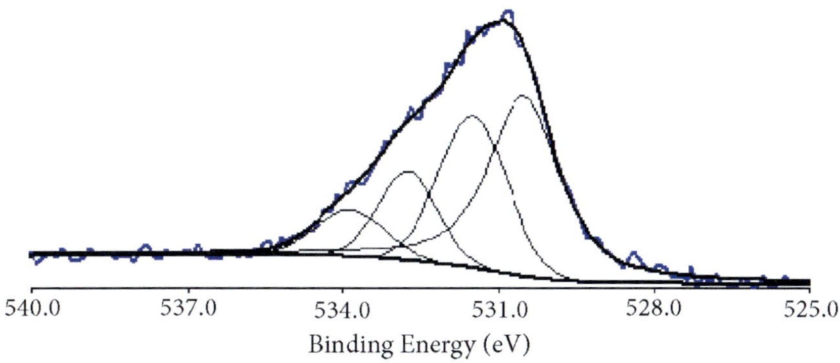

Figure 5.4b XPS O1s spectra for oxygen peaks on pure Cu after heat treatment.

A scan survey was carried out on the oxygen peaks of oxidized pure Cu sample as well. Figure 5.4b shows the O1s XPS spectrum observed at 530.97 eV binding energy position. After detailed analysis, four peaks were fitted with binding energy of 530.51 eV, 531.42 eV, 532.66 eV and 533.86 eV, respectively. Judging from the XPS standard spectra line, binding energy corresponding to 533.8 eV was found to be O_2/Cu. The atomic species with O1s binding energy of 533.8 eV showed the characteristic feature of molecular oxygen on copper surface (Rajumon et al., 1990). In addition, a unique O1s feature which could be O^{1-} species was found at 531.42 eV (Rajumon et al., 1990).

The influence of 2.5 vol% Ru on Cu composites samples was examined using XPS. The results revealed two peaks at binding energy position of 932.4 eV and 934.02 eV. The results from curve fitting indicate that, the major peak at 932.4 eV with FWHM of 1.45 corresponds to pure Cu $2p_{3/2}$ as observed by Platzman et al. (2008) and the minor peak at 934.02 eV with FWHM of 1.45 corresponds to CuO as reported by Kim and Winograd (1974) as shown in Fig. 5.5a. Figure 5.5b shows the binding energy spectra of O1s observed in Cu–2.5 vol% Ru sample. The O1s peak was observed at energy position 530.79 eV. The fitted peak values for O1s were 532.95 eV, 530.70 eV, 532.05 eV and 533.57 eV. The peak values for O1s binding energy 532.95 and 530.70 fall in H_2O and OH domain according to Nefedov et al., 1982). The binding energy corresponding to 533.8 eV was found to be O_2/Cu (Rajumon et al., 1990). However, O1s at binding energy position 532.05 eV could not be identified due to the compound formation with Ru in binding energy position. On the other hand, two peaks were observed at energy position of 280.33 eV and 284.66 eV in which the binding energy corresponds to RuO_2 and Ru, respectively (Riga et al., 1977; Al-Ajlonya et al., 2012). The results showed that 2.5 Ru volume percent resulted in the formation of RuO_2 layer on cupper surface after oxidation process as shown in Fig. 5.5c.

Furthermore, oxidized Cu+5 vol% Ru was also examined. Figure 5.6a shows the results from curve fitting of Cu+5 vol% Ru sample with two peaks at energy position of 932.72 eV and 934.14 eV which correspond to pure Cu and CuO $2_{p3/2}$, respectively, when examined from XPS database spectra line (Rajumon et al., 1990). Figure 5.6b shows the binding energy spectra of O1s observed in Cu+5 vol% Ru sample. The O1s main peak was observed at energy position 530.94 eV. The fitted peak values for O1s were 530.51 eV, 532.66 eV, 533.86 eV and 531.42 eV (Platzman et al., 2008; Yamamoto et al., 2008).

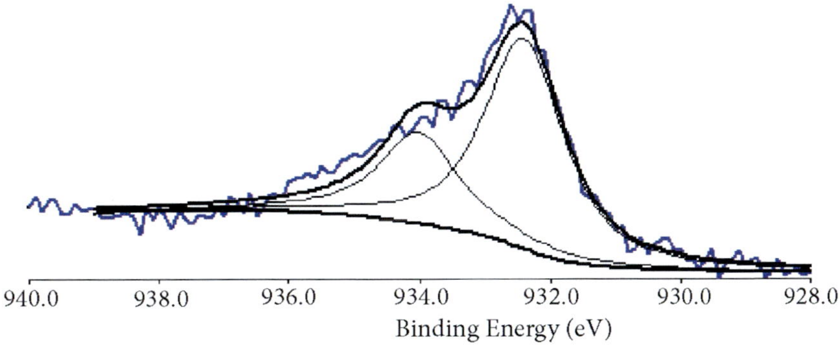

Figure 5.5a XPS spectra for Cu $2p_{3/2}$ in Cu + 2.5 vol% Ru.

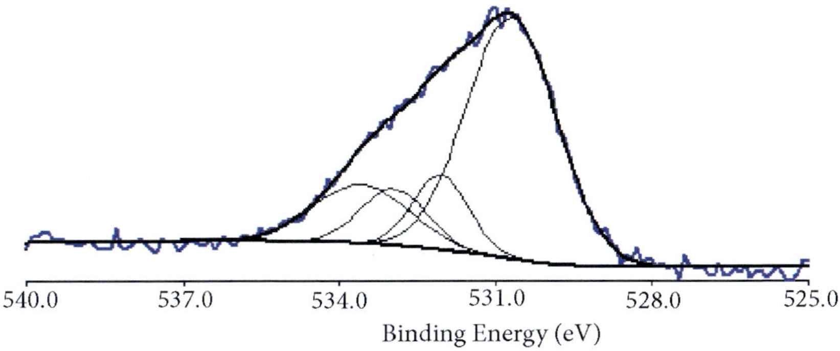

Figure 5.5b XPS spectra for O1s in Cu+2.5 vol% Ru.

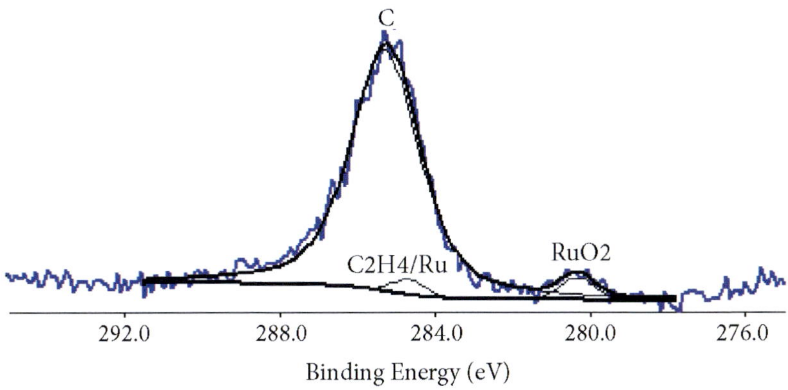

Figure 5.5c XPS spectra for RuO_2 $2p_{5/2}$ at BE of 280.34 eV in Cu+2.5 vol% Ru.

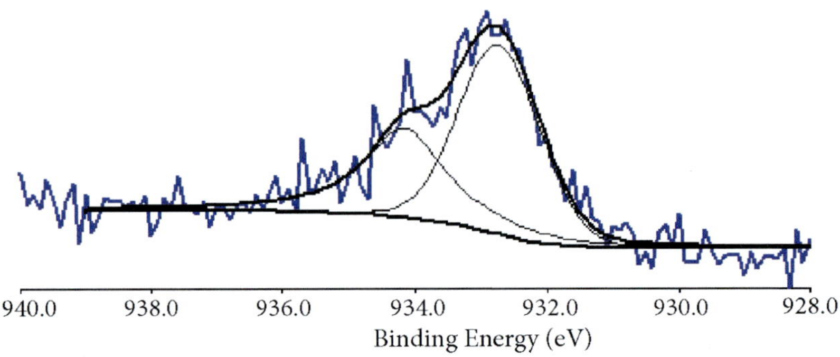

Figure 5.6a XPS spectra for Cu $2p_{3/2}$ in Cu+5 vol% Ru.

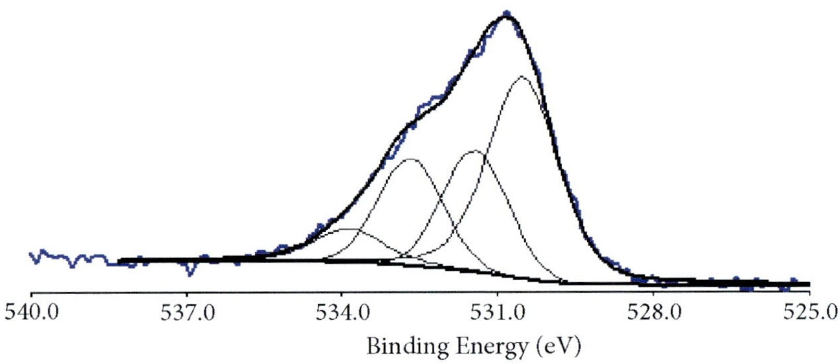

Figure 5.6b XPS spectra for O1s in Cu+5 vol% Ru.

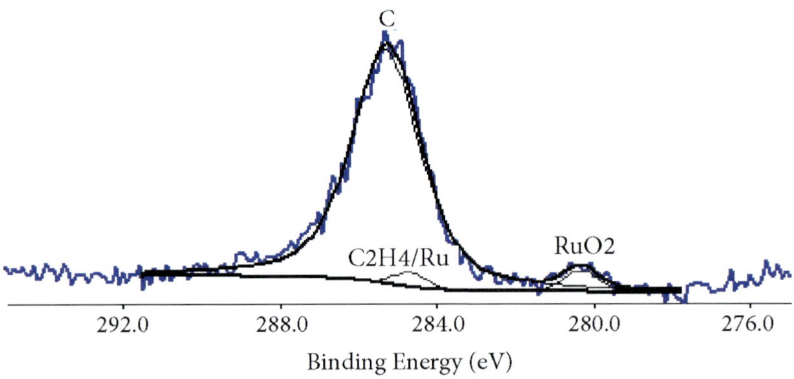

Figure 5.6c XPS spectra for RuO_2 $2p_{5/2}$ at BE of 280.34 eV in Cu+5 vol% Ru.

The peak of RuO_2 $3d_{5/2}$ was observed at energy position of 280.34 eV as shown in Fig. 5.6c. The observed energy position was reasonably consistent with previous results (Riga et al., 1977; Chen et al., 2004). Besides, a small peak was observed at energy position of 284.66 eV that corresponds to $5d_{3/2}RuO_3$ as observed by Kim and Winograd (1974). However, analyzing the Ru 3d peak is always a challenging task due to the presence of C 1s peak overlapping with Ru $3d_{3/2}$ peak (Al-Ajlonya et al., 2012). Moreover, Ernst and Sloof (2008) reported that RuO_2 was the only known stable oxide of Ru at room temperature up to 900 K and above this temperature the volatile RuO_4 was formed. Thus, binding energy position 284.66 eV was considered as Ru since volatile RuO_3 and RuO_4 could only form at a temperature above 900 K. In general, formation of RuO_2 was found at binding energy position 280.3 eV in Cu–2.5 vol% Ru and Cu–5 vol% Ru. The peak area of RuO_2 was relatively greater in Cu–5 vol% Ru indicating a thicker layer with improve oxidation resistance of copper matrix.

5.4 Processing Techniques of Metal Reinforced with Ruthenium

Series of efforts have been made toward studying the structure and properties of metal matrix reinforced with ruthenium using various processing techniques. The processing method utilized to consolidate or alloy ruthenium with metals largely depends on the temperature of the metallic matrix during processing. Ruthenium–aluminum alloy for high-temperature structural application has been fabricated by induction furnace, vacuum arc-melting, and powder metallurgy (Mucklich and Ilic, 2005).

Ingot metallurgy, which is commonly known as high-frequency induction furnace, has been used in the fabrication of Ru–Al alloyed with very small weight loss. However, this technique was abandoned. Vacuum arc-melting would have been an alternative for the above mentioned processing technique due to the unique ability of this process to produced material with good properties (Tryon et al., 2004). The main drawback of vacuum arc-melting was inhomogeneity, poor densification, and difficulty in controlling the desire composition. Powder metallurgy (P/M) is a promising method of producing homogenous and near fully dense Ru–Al in bulk form (Mucklich and Ilic, 2005).

The PM technique has the unique advantage of successfully processing several alloys and composites, which cannot be prepared by conventional melt techniques, especially in systems containing elements with remarkably different melting points (Bora et al., 2004). The material processed by P/M technique is free of coring and segregation which are observed in the products produced by melting technique. In addition, grain size distribution is narrow in materials prepared by P/M route. On the other hand, near net shaped components could be developed ensuring minimum material wastage and circumventing the need of machining process (Bora et al., 2004).

5.4.1 Powder Metallurgy

Powder metallurgy is recognized today as one of the important manufacturing processes for making several industrial products (Goudah et al., 2010). Powder metallurgy steps involve powder production, mixing of powders with or without additives and lubricants, compaction of the mixture and sintering of the compacts to bond the particles. The resulting sample is a solid body of material with sufficient strength and density for use in a diverse field of applications. The raw material powders are manufactured to precise specification to facilitate subsequent processing. These powders can be pure element, elemental blends or pre-alloyed. The choice of the starting material is influenced by the final product. Various methods have been used for powder production. These methods include atomization, reduction and electrolysis (Angelo and Subramanian, 2008).

Powder mixing is achieved through a combination of three mechanisms, which are diffusion, convection, and shearing (Van Scoik, 1992). Diffusion is the motion of a particle with respect to its neighbors, while convection is the motion of a group of particle in relation to their neighbors and shearing is the change of distribution layers of constituent materials in space. Blending is one of the techniques usually used to achieve a uniform distribution of all constituent material during powder mixing. In blending, each sample usually contains a large amount of the base materials. However, various factors influence mixing quality. These factors include mixing time, speed of mixing rotation, type of mixer and mixing media. Moreover, mixing time and speed of rotation are to be controlled to ensure uniform mixing. Improper mixing can cause non-uniform distribution of powders. At the same time, excessive mixing can harden the powders making the compaction difficult (Kung et al., 2009).

Sule et al. (2012) reported on the dry mixing of copper–ruthenium and copper–ruthenium–tantalum powder using Turbula T2F mixer shaker. The feedstock starting powder mixture was prepared by dry-mixing approximately 20 g per sample for 5 h in order to maximize homogeneity during mixing. The Turbula mixer speed was set at 101 revolutions per minute. The results showed that an increase in the mixing intensity increases shear force, which contributes to the powder agglomerate size reduction. The de-agglomeration during mixing improves the packing efficiency of the particles and promotes homogeneity of the mixture. Figure 5.7 shows the photograph of a Turbula mixer shaker.

Figure 5.7 Picture of Turbula mixer.

Another method of producing metal/ruthenium powder is the mechanical alloying process. Mechanical alloying process has been widely employed in the production of composite metal powders with controlled fine microstructure (Tjong and Ma, 2000;

Murty and Ranganathan, 2013). In mechanical alloying, elemental powders were milled under controlled atmosphere using attrition, planetary and centrifugal. Among these mechanical alloying process, attrition ball mill has been widely used because it is simple to operate and very effective in powder particles reduction. In an attrition mill, the elemental powders are repeatedly cold welded, fractured, and re-welded leading to mechanical alloying (Suryanarayana, 2001). However, it is critical to establish a balance between fracturing and cold welding in order to achieve and homogeneous alloyed powder.

Liu and Mucklich (2001) reported on the preparation of nano grain size single-phase Ru–Al using mechanical alloying method. Elemental powders of Ru and Al with average particle sizes of 40 and 200 μm, respectively, were used as starting materials. The powders were milled in a Spex 8000 for 35 h using hardened steel balls and a vial with a ball-to-powder weight ratio of 10:1 at room temperature. The structure of as-milled and annealed powders was examined by X-ray diffraction (XRD). The XRD result showed that no phase transformation of the as-milled Ru–Al was detected other than a variation of broadening and intensity for diffraction lines at different annealing temperatures and time.

Bora et al. (2004) reported on powder metallurgy processing of ruthenium aluminide by a sequence of attrition ball milling. Elemental powders of Ru and Al were used as starting materials. The average particle size of elemental Ru and Al powders were 4 and 7 μm, respectively. The ball to powder weight ratio was maintained at 10:1. The process control agent (PCA) used was 1 wt% stearic acid ($C_{18}H_{36}O_2$). Structural analysis of the as milled powders was carried out by X-ray diffraction technique.

Borah et al. (2007) reported on the mechanical alloying of ruthenium and aluminum powder in an attrition mill. A ball-to-powder weight ratio of 10:1 was maintained in the milling process. At the beginning of the milling process, 1 wt% stearic acid ($C_{18}H_{36}O_2$) was added to the powder mixture as the process control agent (PCA). Subsequently, 0.5 wt% stearic acid was added to the milling chamber after every 10 h of milling. The milling was performed at a constant impeller speed of 400 rpm. Structural analyses of the milled powder samples were performed using a powder X-ray diffraction. The peak broadening observed in the

XRD pattern during milling was attributed to the crystallite size and strain effects. On the other hand, mechanical alloying process was also employed to synthesize Ru–Al/ZrO_2 nanocomposites from elemental ruthenium, aluminum, and ZrO_2 powder. The ball milling was performed in a laboratory Spex 8000 shaker mill with a ball-to powder weight ratio of 10:1 (Liu and Mucklich, 2003). The starting average particle sizes of the elemental Ru, Al and ZrO_2 were 40, 200, and 100 μm, respectively. The XRD results showed that there is no diffraction line corresponding to Al and ZrO_2 after 2 h of milling. In addition, Green et al. (1984) investigated the physical metallurgy and electrical contact resistance of mechanically alloyed Cu–15 vol% Ru composites using high energy attrition milling. The starting powder particle sizes of the elemental Cu and Ru were –100 and –200 mesh, respectively. An approximately 25 g of powders were milled by batches with ratio of balls to powder 10 to 1. The powders were annealed at temperatures ranges from 450°C to 600°C for 30 min and 1 h in H_2. The annealed powders needed to be consolidated to obtain bulk composite material.

5.4.2 Overview of Sintering Process

Sintering is an art that has its origin thousands of years ago; from its early usage in the fabrication of pottery and bricks where clay bodies were heated in an open pit fire to the age of making jewelry and other materials (Munir et al., 2006). However, the science of sintering did not increase noticeably until October 1879 when tungsten filament was made from consolidated tungsten powder by the Coolidge process (Ramakrishnan, 1983).

Advancement in sintering took place in the 19th and early 20th centuries by using the principle of the liquid phase sintering technique to produce cemented carbides and heavy alloys which revolutionized the machining industry (Ramakrishnan, 1983). Hence, the knowledge on the theory and science of sintering has since been on the increase and more engineering parts are now produced by sintering process. Sintering of the powders can be carried out through traditional method.

Traditional sintering is the process in which powders are compacted to the desired shape and size. The green compact is heated to a temperature close to but not at the melting point of the major constituent powder in a controlled atmosphere furnace where the atoms can diffuse. Other sintering processes are pressure-assisted sintering methods such as hot pressing, high-temperature extrusion and high isostatic pressing which have been employed to understand the microstructural stability of composite materials at different temperatures (Chang et al., 1999). The hot pressing techniques combined the two stages of compaction and sintering in traditional sintering technique together in a single step. However, the conventional sintering methods are characterized by grain growth that affects the properties of the material produced (Suárez et al., 2013).

5.4.3 Spark Plasma Sintering Technique

Spark plasma sintering (SPS) also known as field-assisted sintering technology (FAST) is currently a widely used sintering technique employing current-activated consolidation. The spark plasma sintering process is based on the principle of the electrical spark discharge phenomenon. Spark plasma sintering uses high energy, low voltage, pulsed DC current, and uniaxial pressure to consolidate metallic/ceramic powders (Munir et al., 2006). The accurate control of sintering energy as well as high sintering speed makes SPS a promising technique of producing highly dense materials with controlled grain growth (Suárez et al., 2013). However, there is a similarity between hot pressing and spark plasma sintering as both techniques are usually involve a uniaxial press coupled with a heating devices. In hot press technique, the sample is heated by radiation from enclosed furnace through external heating elements. However, in SPS, the joule effect caused by a pulsed direct current lead to very rapid and efficient heating (Suárez et al., 2013). A schematic of the SPS the showing tooling set up is given in Fig. 5.8.

The SPS machine is assisted by a uniaxial press, punch electrode in a vacuum chamber which can also house a controlled atmosphere. The SPS consists of multiple measuring components that control the DC pulse generator, temperature, and pressure. However, voltage and current cannot be controlled directly;

instead the temperature profile is programmed by the user and the current is adjusted by the SPS unit accordingly. The temperature is measured by thermocouple inserted at the die wall as well as by optical pyrometer, typically viewing the bottom of a deep hole drilled into the top piston (Saheb et al., 2012). Sintering is carried out in SPS by pouring a powder inside a conductive die placed between the two punches. The punches are pushed close to each other to establish a minimum pressure for electrical contact. The powder is heated by the pulsed DC current that passes through the electrode and die in the presence of low voltage electric field (Munir et al., 2006). The heat dissipated by punches, die and powder resulted in joule heating which in turn offers the advantage of rapid heating rate and short holding time to obtain fully dense materials with superior properties (Tiwari et al., 2009). Munir et al. (2006) have published a comprehensive review on the features of SPS. It is reported that current distribution in the SPS process depends primarily on die geometry and thermal and electrical conductivity of the die and powder sample.

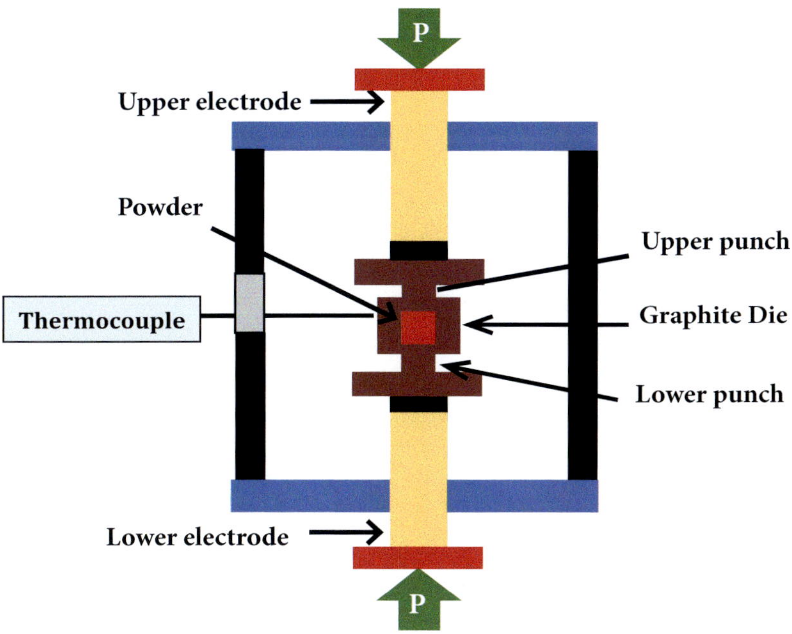

Figure 5.8 Schematic of SPS.

5.4.4 Sintering of Ruthenium Alloy

Sintering is a heat treatment method employed to understand the microstructural stability of composite materials at different temperature (Chang et al., 1999). Based on this understanding, optimal parameters can be selected and used to produce fully dense and high quality bulk metal matrix composite. For any material produced by a powder metallurgy route to be widely used in an engineering application it must be fully dense.

Several research groups have used conventional method to synthesize ruthenium materials composite (Bora et al., 2004; Gobran et al., 2004; Borah et al., 2007).

In a study by Bora et al., 2004), Ru and Al powders were cold compacted into 10 mm diameter green compacts under pressures ranging from 380 to 782 MPa. The cold compacted samples were sintered at 1400°C for 24 h under inert atmosphere. A relative density of 81% was obtained. The SEM/EDAX analysis of the sintered sample revealed the presence of a high percentage of oxygen in the dark phase (porous) while the gray region was found to consist of the Ru–Al phase that was free from oxygen.

Furthermore, Gobran et al. (2004) reported on the effect of particle size and pressure of the reactive sintering (SR) and reactive hot pressed (RHP) samples. A relative density of about 81% was obtained in reactive sintering product processed at 1480°C with Ru particle size of 5 μm. The density of a reactive sintering sample was found to depend on the particle size ratio. The high porosity observed in a reactive sintering method was attributed to molar volume changes between reactants and product, poor green compact and lack of interconnectivity of aluminum particles within the green compact. Meanwhile, reactive hot pressed samples resulted in the formation of multiphase structure even at low pressure (33 MPa). A maximum relative density of 94.4% was obtained. In addition, a relative density of 97% was obtained in the Ru–Al alloy produced by cold compaction at a pressure of 500 MPa and pre-sintered at 300°C for 3 h followed by sintering at 1450°C for 24 h in an argon atmosphere (Borah et al., 2007). The observed density value was attributed to the fact that 120 nm Ru crystallites size were obtained after milling; and the result was corroborated with the report of Gobran et al. (2004) that stated that density is strongly influenced by the Ru particle size ratio. Furthermore, Sule et al. (2012) reported on the synthesis and characterization of sub-micron-sized copper–ruthenium–tantalum composites for interconnection application using hot pressed. The mixed powders were consolidated at 850°C with pressure of 30 MPa and holding time of 30 min. A relative density of 99.89% was obtained in sample containing 97.5 vol% Cu–2.5 vol% Ru. It was observed that density improves when increase the Ru volume fraction in copper matrix by up to 2.5 vol%. The high density was attributed to good adhesion of copper with ruthenium. On the other hand, Ru content might aid in filling of the voids or pores which resulted in particle to particle interlocking. However, addition of Ru to Cu-Ta samples revealed a reduction in their density value. This could be due to insufficient sintering temperature, as both tantalum and ruthenium have high melting point of 3017°C and 2310°C, respectively.

Owing to the time and energy consumption involved in full densification of this class of material, there is a need for an efficient sintering technique.

Microwave sintering have been investigated as an alternative to conventional sintering method due to it attractive properties such as enhanced diffusion processes, reduced energy consumption, very rapid heating rates and considerably reduced processing times, decreased sintering temperatures, improved physical and mechanical properties, simplicity, unique properties, and lower environmental hazards (Oghbaei and Mirzaee, 2010).

Recently spark plasma sintering technique has emerged as a novel method sintering for metal, ceramic, and ceramic particulate-reinforced metal matrix composites capable of achieving full density at a relatively low temperature (Munir et al., 2006). Maekawa et al. (2005) reported on the thermoelectric properties of perovskite-type strontium ruthenium oxide ($SrRuO_3$) fabricated by SPS. A bulk density of about 97% of the theoretical density was obtained at a sintering temperature of 1673 K (1400°C). Angerer et al. (2009) reported on the residual stress of ruthenium powder samples compacted by spark-plasma-sintering (SPS). The powders were loaded into a graphite mould 12.5 mm. The consolidation was carried out under applied pressure of 20–30 MPa with a heating rate of 100°C/min and a holding time of 1 min. Sintering was done at 1200, 1400, and 1600°C. A relative density of 87%, 94%, and 92% was obtained at the three sintering temperatures, respectively. However, no reason was provided for the reduction in densification as temperature increased from 1400 to 1600°C.

Furthermore, Sule et al. (2015) investigated the spark plasma sintering of Cu–Ru–CNT composites for thermal management application. The mixed powders were annealed for 30 min under inert Ar of high-purity atmosphere, at 550°C with a heating rate of 5°C/min to reduce the oxygen content that could exist as CuO layers on the surface of Cu powder particles, by its thermal decomposition. The powder mixtures were consolidated by SPS (HPD5, FCT System GmbH) at 650°C under a pressure of 50 MPa with heating rate of 80°C/min and holding time of 5 min. A relative density of 97.41%, 98.37%, and 98.03% was obtained for Cu–1 vol% Ru, Cu–1 vol% CNT–0.5 vol% Ru, and Cu–2 vol% CNT–0.5 vol% Ru, respectively.

5.5 Properties of Ruthenium Alloy

Metals reinforced or alloyed with ruthenium particles have found applications in several consumer electronic products such as cell phones, microprocessors, fast static RAM, supercapacitors, home appliances and structural materials (Li and Wong, 2006). Hence, it is important to examine the unique properties of metal ruthenium alloys.

5.5.1 Properties of Copper with Ruthenium Additive

Copper composites have been well developed and applied in the electronics industry for several decades due to their high electrical and thermal conductivity. These materials have been used for electrical contact such as make-break contacts and brush (Shen et al., 1990; Nadkarni et al., 1984; Hoyer, 1984). However, the traditional engineering properties of copper interconnections suffer major material compatibility limitation because pure copper and copper-based composites with conventional grain size (micron size) are soft materials (Mula et al., 2011; Yum et al., 2005; Karwan-Baczewska et al., 2005). The softness of copper needs to be improved in order to achieve the optimum performance of smart ICs in the electronic devices. This limiting property of copper could therefore be improved through the incorporation of suitable materials having the desired properties into the matrixes of copper using appropriate technology.

Sule et al. (2012) investigate the effect of ruthenium on the hardness value of copper fabricated by powder metallurgy route. Copper powder was reinforced with 0.28 wt%, 0.69 wt%, and 1.37 wt% ruthenium particles. The maximum microhardness of the composites was 127.8HV in Cu+0.69 wt% Ru sample. With the addition of 0.28 wt% Ru, the microhardness of the composite was found to be 123.2HV, which was 1.03 times higher than that of pure copper. The increased in copper hardness with Ru addition could be due to the fact that copper matrix was first deformed upon applying the load, whilst Ru acted as obstacles to the moving dislocations in the copper matrix, leading to an increase in the microhardness of the composite. Studies on metal matrix composite materials have shown that reinforcement particulate prevents plastic deformation by blocking the dislocation movement in the matrix phase under mechanical loading (Long et al., 2010). Figure 5.9 shows the influence of Ru content on hardness of hot pressed Cu–Ru composites with various Ru volume fractions (Sule et al. 2012).

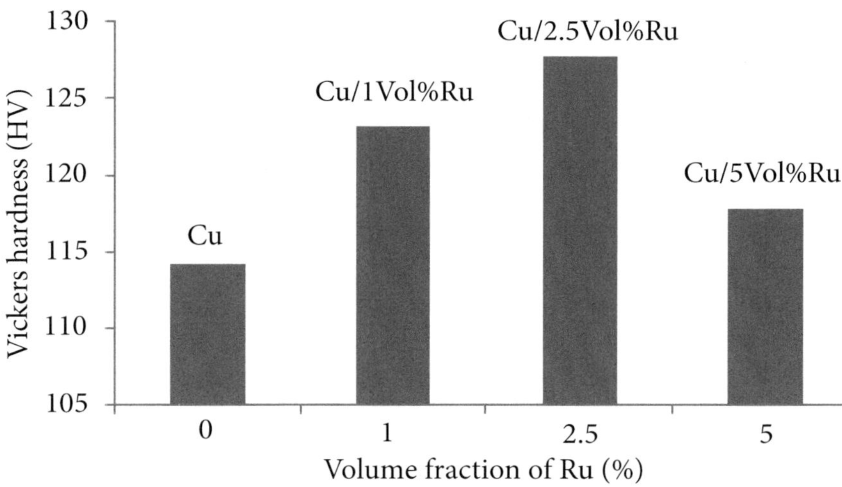

Figure 5.9 Effect of ruthenium on Vickers hardness of Cu–Ru composites.

On the other hand, 1 vol% Ru was found to decreased the electrical conductivity of pure Cu from 2.6 × 10^6 to 1.06 × 10^6 S/m due to interfacial resistance and much higher resistivity of Ru (7.1 × 10^{-6} Ωcm) versus (1.68 × 10^{-6} Ωcm) for copper.

Furthermore, the effect of ruthenium on copper–carbon nanotube composites was studied by the spark plasma sintering technique and their thermal conductivity property was investigated (Sule et al., 2015). It was observed that the addition of ruthenium increases the thermal conductivity of copper/CNTs composites as shown in Fig. 5.10.

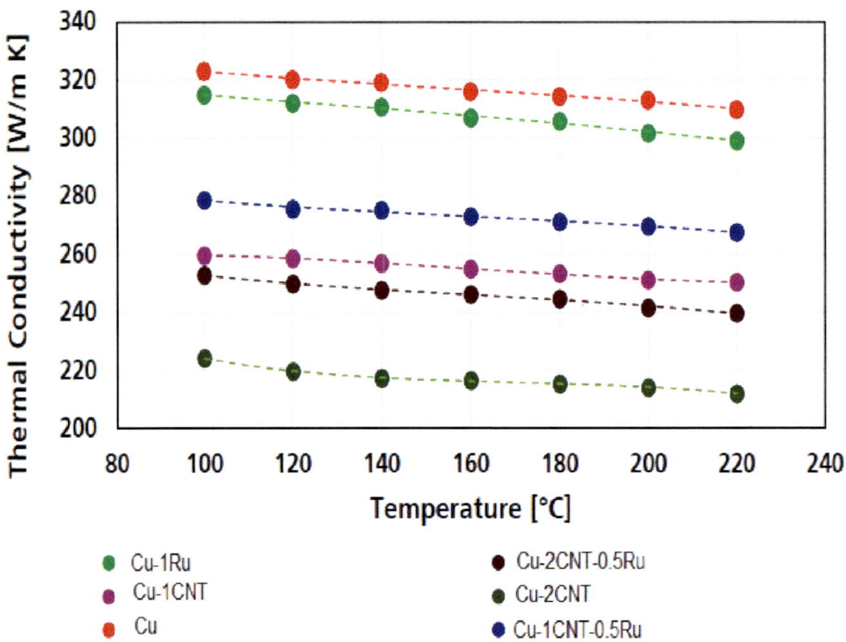

Figure 5.10 Calculated thermal conductivity and the regression curve.

5.5.2 Properties of WC-Co with Ruthenium Additive

Cemented carbides are composite materials consisting of hard carbide and a ductile binder. They are metallurgically manufactured, where liquid phase sintering is one of the main steps. The most common cemented carbide consists of WC and Co, and it is widely used for cutting tools. Two of the most important factors controlling the mechanical properties are the

WC grain size and the grain size distribution, and thus it is of great interest to understand the grain growth behavior. Shing et al. (2001) reported on the effect of Ru addition to WC-Co. The results show that the hardness of WC-Co increases and the toughness decreases with increasing ruthenium content. The WC-Co hardness was found to increase from 1420 HV to 1520 HV with 3 wt% Ru content, while the toughness of the material reduces. It should be noted that the increase in hardness and decrease in toughness became pronounced only when the ruthenium content in the cobalt is higher than about 15%. In addition, the mean grain size of the WC grain decreases with increasing Ru content in WC-Co-Ru sample. Luyckx (2002) also investigated the high-temperature hardness of WC-Co-Ru. Although studies have shown that the Ru content increases the hardness of WC-Co at room temperature measurement, it is important to determine their hardness at higher temperature since WC-Co materials used in cutting tools could reach up to around 800–900°C depending on the area of application (Luyckx, 2002). It was observed that the hardness of WC–Co–3Ru sample is higher than that of the WC-Co sample at about 600°C. However, above 600°C its hardness value decreases rapidly compared to that of the WC-Co sample. The reason for the decrease in hardness above 600°C was attributed to the fact that WC-Co boundaries present less serious obstacle to the dislocation compared to when temperature was below 600°C, which resulted in high contribution of WC grain to plastic deformation. Since the composition and properties of WC grain are not affected by the Ru contents, the hardness of this material at above 600°C became inert to ruthenium contents.

Moreover, Potgieter et al. (2011) reported on the influence of Ru additions on the corrosion behavior of WC-Co cemented carbide alloys in sulfuric acid. The corrosion behavior was investigated using electrochemical polarization test and chronoamperometry. The results show that ruthenium addition up to 3 wt % increased the corrosion resistance of the WC-Co alloys. It was also found that ruthenium affects the cathodic Tafel constant (B_c) in 1M H_2SO_4. This indicates that Ru influences the cathodic part of the corrosion reaction. In general, it can be deduced that Ru is more effective in improving corrosion resistance than vanadium carbide.

5.6 Properties of Ruthenium Oxide

There is a growing demand for metal oxide thin films for many important technological applications such as smart windows, gas sensors, solar cells, and microelectronic packaging (Patake and Lokhande, 2008). Literature surveys have revealed the used of electrically conductive metal oxide such as PtO_2, IrO_2, RuO_2, TiO_2, OsO_2, RhO_2, and Ta_2O_5 as active components in solid-state potentiometric electrodes for pH sensing (Liao and Chou, 2008). On the other hand, noble metal oxides have been identified as ideal electrode materials for supercapacitors due to their great specific capacitance, wide potential windows, and high electrochemical stability. Therefore, materials associated with fast Faradaic redox processes on a high surface area are recognized as the most promising candidates for supercapacitors (Zheng et al., 2008). Among the metal oxide materials, ruthenium oxide (RuO_2) have shown to be a material of choice in very large-scale integration (VLSI), thick-film resistors, and supercapacitors due to its very high specific capacitance, perfect reversibility, and excellent stability in H_2SO_4 (Hu and Chen, 2004). Ruthenium oxide thin films have been prepared using various methods such as reactive sputtering, sol–gel, electrodeposition, spray pyrolysis (Patake and Lokhande, 2008). The amorphous RuO_2 prepared by a sol–gel method exhibit a high specific capacitance of 720 F g^{-1}, which results from the surface Faradaic redox reactions between Ru ions and H ions in the H_2SO_4 electrolyte.

5.6.1 Properties of RuO_2 for Supercapacitor Applications

Supercapacitors are electrochemical energy storage devices that have attracted global research interest due to its usage in various applications. Since they are characterized by fast discharge rates and low maintenance, the demand for them is predicted to be most extensive in frequency regulation applications (Jampani et al., 2010). The mechanisms of energy storage for supercapacitors could be regarded as charge/discharge of non-Faradaic charge separation across the electrical double-layer (Sarangapani et al., 1996) and fast Faradaic redox reactions within the electrode materials (Miller et al., 1997).

High electrochemical reversibility and high specific capacitance are the two most important characteristics for supercapacitors. These properties exist in ruthenium oxide (RuO_2), and it has been widely used as an electrode material in supercapacitors (Patil et al., 2011; Patake et al., 2009). Patake et al. (2009) reported on the effect of surface treatment on the supercapacitive performance of as-deposited RuO_2 films using cyclic voltammetric (CV) in 0.5M H_2SO_4 electrolyte. The potential range between −100 and +600 mV/SCE was used with a scanning rate of 20 mV s^{-1}. It was found that RuO_2 electrode with 0.349 mg cm^{-2} film thickness possesses high specific capacitance of 650 F g^{-1}. This high specific capacitance was achieved on a stainless steel electrode due to the amorphous and porous nature of deposited material (Subramanian, 2004). However, the CV performance of annealed electrode shows that specific capacitance decreases from 650 F g^{-1} to 111 F g^{-1} after annealing. The reduction in specific capacitance was attributed to the loss in hydrous content of RuO_2 electrode (Hu et al., 2002). Similarly, the performance of RuO_2 electrode was found to reduce after anodization and ultrasonic weltering. It was observed that specific capacitance decreases from 650 F g^{-1} to 70 F g^{-1} after anodization. The SEM cross-sectional image revealed more pores after surface treatment, which could result in a decrease contact angle. Hu et al. (2007) also reported on the capacitive behavior of sol–gel-derived and co-annealed ruthenium–tin oxide composites. A relatively high specific capacitance of 690 F g^{-1} was obtained when 0.2 weight fraction of SnO_2 was added to the RuO_2. However, the total specific capacitance of mixed Ru–Sn oxides is highly composition-dependent and a precise composition control for this series of oxides is complicated. Notwithstanding, the specific capacitance and power characteristics of mixed Ru–Sn oxides can be enhanced by a careful control of annealing temperature and time due to a compromise between electronic conductivity and active sites. Wu et al. (2010) investigated the potential of graphene-based composites in the development of high-performance energy-storage systems. Hydrous ruthenium oxide (RuO_2)/graphene sheet composites (ROGSCs) with varying Ru content were prepared by combining sol–gel and low-temperature annealing processes. The specific capacitance of ROGSC-based supercapacitors shows a monotonic increment with increasing Ru content. A maximum specific capacitance of 570 F g^{-1} is obtained for ROGSCs with 38.3 wt% Ru. It was observed that the specific capacitance of pure GSs (148 F g^{-1}) increases tremendously with Ru addition.

5.7 Conclusion

In the past decades, the study of ruthenium has progressed from primarily a laboratory enterprise with a narrow commercial significance to a diverse and robust class of materials with numerous important applications. This transformation has been accomplished as a result of extensive material research and development followed by innovations in engineering processes.

Various techniques have been employed to synthesize Ru in the metal matrix such as low-energy ball milling and mechanical milling followed by conventional sintering and spark plasma sintering. However, the dense packing of the Ru atom is one of its unique features. The rich properties of Ru could be seen in its surface electrochemical and catalytic properties as well as its mechanical and thermal properties. Therefore, Ru and its alloys or composites remain a field for discovery in micro-electronics packaging, energy generation, and sensors using powder metallurgy process.

References

Al-Ajlony A, Kanjilal A, Catalfano M, Harilal S, Hassanein A, Rice B. (2012). *Low Energy Electron Bombardment Induced Surface Contamination of Ru Mirrors.* Paper presented at the SPIE Advanced Lithography.

Angelo P, Subramanian R. (2008). *Powder Metallurgy: Science, Technology and Applications*: PHI Learning Pvt. Ltd.

Angerer P, Wosik J, Neubauer E, Yu L, Nauer G, Khor KA. (2009). Residual stress of ruthenium powder samples compacted by spark-plasma-sintering (SPS) determined by X-ray diffraction. *International Journal of Refractory Metals and Hard Materials*, 27(1), 105–110.

Au Y, Lin Y, Kim H, Beh E, Liu Y, Gordon RG. (2010). Selective chemical vapor deposition of manganese self-aligned capping layer for Cu interconnections in microelectronics. *Journal of the Electrochemical Society*, 157(6), D341–D345.

Bateni M, Mirdamadi S, Ashrafizadeh F, Szpunar J, Drew R. (2001). Oxidation behaviour of titanium coated copper substrate. *Surface and Coatings Technology*, 139(2), 192–199.

Bora A, Singha P, Robi P, Srinivasan A. (2004). Powder metallurgy processing of ruthenium aluminum alloys. *Journal of Materials Processing Technology*, 153, 952–957.

Borah A, Robi P, Srinivasan A. (2007). Synthesis of nano-crystalline RuAl by mechanical alloying. *Metals and Materials International*, 13(4), 293–302.

Boschma RA. (1999). The rise of clusters of innovative industries in Belgium during the industrial epoch. *Research Policy*, 28(8), 853–871.

Chai G, Chen Q. (2010). Characterization study of the thermal conductivity of carbon nanotube copper nanocomposites. *Journal of Composite Materials*, 44(24), 2863–2873.

Chang S-Y, Lin J-H, Lin S-J, Kattamis TZ. (1999). Processing copper and silver matrix composites by electroless plating and hot pressing. *Metallurgical and Materials Transactions A*, 30(4), 1119–1136.

Chastain ML, Deymier-Black AC, Kelly JE, Brown JA, Dunand DC. (2011). Metallurgical analysis of copper artifacts from Cahokia. *Journal of Archaeological Science*, 38(7), 1727–1736.

Chen C, Chen R, Tsai T, Huang Y, Tsai D, Tiong K. (2004). The growth and characterization of well aligned RuO_2 nanorods on sapphire substrates. *Journal of Physics: Condensed Matter*, 16(47), 8475.

Cheng C, Arunagiri T, Chyan O. (2003). Electrodeposition of silver and copper/silver multilayer on ruthenium substrate as a potential new metal interconnect for integrated circuits. *American Journal of Undergraduate Research*, 2(1), 11–18.

Chu K, Guo H, Jia C, Yin F, Zhang X, Liang X, Chen H. (2010). Thermal properties of carbon nanotube–copper composites for thermal management applications. *Nanoscale Research Letters*, 5(5), 868–874.

Chuang C-L, Aoh J-N, Din R-F. (2006). Oxidation of copper pads and its influence on the quality of Au/Cu bonds during thermosonic wire bonding process. *Microelectronics Reliability*, 46(2), 449–458.

Cocke D, Schennach R, Hossain M, Mencer D, McWhinney H, Parga J, Mollah M. (2005). The low-temperature thermal oxidation of copper,

Cu_3O_2, and its influence on past and future studies. *Vacuum,* 79(1), 71–83.

Damayantia M, Sritharan T, Mhaisalkar SG, Phoon E, Chan L. (2007). Study of Ru barrier failure in the Cu/Ru/Si system. *Journal of Materials Research,* 22(9), 2505–2511.

Datsyuk V, Firkowska I, Gharagozloo-Hubmann K, Lisunova M, Vogt A, Boden A, Reich S. (2011). *Carbon Nanotubes Based Engineering Materials for Thermal Management Applications.* Paper presented at the Semiconductor Thermal Measurement and Management Symposium (SEMI-THERM), 2011 27th Annual IEEE.

Davis JR, (ed) (2001). *ASM Specialty Handbook: Copper and Copper Alloys,* ASM International, pp. 3–9.

Deng C, Ma Y, Zhang P, Zhang X, Wang D. (2008). Thermal expansion behaviors of aluminum composite reinforced with carbon nanotubes. *Materials Letters,* 62(15), 2301–2303.

Ding S-F, Xie Q, Müeller S, Waechtler T, Lu H-S, Schulz SE, Gessner T. (2011). The inhibition of enhanced Cu oxidation on ruthenium/diffusion barrier layers for Cu interconnects by carbon alloying into Ru. *Journal of The Electrochemical Society,* 158(12), H1228–H1232.

Ehrhardt KL (2009). Copper working technologies, contexts of use, and social complexity in the Eastern Woodlands of Native North America. *Journal of World Prehistory,* 22(3), 213–235.

Ekpu M, Bhatti R, Ekere N, and Mallik S. (2011). *Advanced Thermal Management Materials for Heat Sinks Used in Microelectronics.* Microelectronics and Packaging Conference (EMPC), 18th European, Brighton.

Ernst M, Sloof W. (2008). Unraveling the oxidation of Ru using XPS. *Surface and Interface Analysis,* 40(3–4), 334–337.

Gallagher C, Shearer B, Matijasevic G. (1998). Materials selection issues for high operating temperature (HOP) electronic packaging. *IEE,* 1(1), 180–189.

Gao W, Gong H, He J, Thomas A, Chan L, Li, S. (2001). Oxidation behaviour of Cu thin films on Si wafer at 175–400 C. *Materials Letters,* 51(1), 78–84.

Gobran H, Ilić N, Mücklich F. (2004). Effects of particle size and pressure on the reactive sintering of RuAl intermetallic compound. *Intermetallics,* 12(5), 555–562.

Goudah G, Ahmad F, Mamat O. (2010). Microstructural studies of sintered carbon nanotubes reinforced copper matrix composite. *Journal of Engineering Science and Technology,* 5(3), 272–283.

Green M, Coleman E, Bader F, Sproles E. (1984). The physical metallurgy and electrical contact resistance of mechanically alloyed Cu–Ru composites. *Materials Science and Engineering,* 62(2), 231–239.

Gujar T, Shinde V, Lokhande C, Kim WY, Jung KD, Joo OS. (2007). Spray deposited amorphous RuO_2 for an effective use in electrochemical supercapacitor. *Electrochemistry Communications,* 9(3), 504–510.

Hau-Riege CS. (2004). An introduction to Cu electromigration. *Microelectronics Reliability,* 44(2), 195–205.

Hoyer N. (1984). Electrical-contact materials. In Metals Handbook on Powder Metallurgy, Davis JR and Refsnes SK (eds), American Society for Metals, Metals Park, OH, pp. 662–695.

Hsu K-C, Perng D-C, Yeh J-B, Wang Y-C. (2012). Ultrathin Cr added Ru film as a seedless Cu diffusion barrier for advanced Cu interconnects. *Applied Surface Science,* 258(18), 7225–7230.

Hu C-C, Chen W-C. (2004). Effects of substrates on the capacitive performance of $RuO_x \cdot nH_2O$ and activated carbon–RuO_x electrodes for supercapacitors. *Electrochimica Acta,* 49(21), 3469–3477.

Hu C-C, Huang Y-H, Chang K-H. (2002). Annealing effects on the physicochemical characteristics of hydrous ruthenium and ruthenium–iridium oxides for electrochemical supercapacitors. *Journal of Power Sources,* 108(1), 117–127.

Hu C-C, Wang C-C, Chang K-H. (2007). A comparison study of the capacitive behavior for sol–gel-derived and co-annealed ruthenium–tin oxide composites. *Electrochimica Acta,* 52(7), 2691–2700.

Jampani P, Manivannan A, Kumta PN. (2010). *The Electrochemical Society Interface, Fall* 2010, 57–61.

Karwan-Baczewska J, Gotman I, Gutmanas E, Shapiro M. (2005). Small particles with better contacts make nanocomposites kings of conductivity. *Metal Powder Report,* 60(6), 28–34.

Kim K, Winograd, N. (1974). X-ray photoelectron spectroscopic studies of ruthenium-oxygen surfaces. *Journal of Catalysis,* 35(1), 66–72.

Kung C, Liao T-T, Tseng K-H, Chen K-Y, Chuang M-S. (2009). The influences of powder mixing process on the quality of W-Cu composites. *Transactions of the Canadian Society for Mechanical Engineering,* 33(3), 361.

Li J, Mayer J, Colgan E. (1991). Oxidation and protection in copper and copper alloy thin films. *Journal of applied physics,* 70(5), 2820–2827.

Li Y, Moon KJ, Wong C. (2010). Nano-conductive adhesives for nano-electronics interconnection *Nano-Bio-Electronic, Photonic and MEMS Packaging* (pp. 19–45).

Li Y, Wong C. (2006). Recent advances of conductive adhesives as a lead-free alternative in electronic packaging: Materials, processing, reliability and applications. *Materials Science and Engineering: R: Reports,* 51(1), 1–35.

Liao Y-H, Chou J-C. (2008). Preparation and characteristics of ruthenium dioxide for pH array sensors with real-time measurement system. *Sensors and Actuators B: Chemical,* 128(2), 603–612.

Liu K, Mücklich F. (2001). Thermal stability of nano-RuAl produced by mechanical alloying. *Acta Materialia,* 49(3), 395–403.

Liu K, Mücklich F. (2003). Synthesis of RuAl/ZrO_2 nanocomposite by mechanical alloying and subsequent annealing. *Scripta Materialia,* 49(3), 207–212.

Long B, Othman R, Umemoto M, Zuhailawati H. (2010). Spark plasma sintering of mechanically alloyed in situ copper–niobium carbide composite. *Journal of Alloys and Compounds,* 505(2), 510–515.

Luyckx S. (2002). High temperature hardness of WC-Co-Ru. *Journal of Materials Science Letters,* 21(21), 1681–1682.

Maekawa T, Kurosaki K, Muta H, Uno M, Yamanaka S. (2005). Thermoelectric properties of perovskite type strontium ruthenium oxide. *Journal of Alloys and Compounds,* 387(1), 56–59.

Marinkovic NS, Vukmirovic MB, Adzic RR. (2006). Some recent studies in Ruthenium Electrochemical and Electrocatalysis (pp. 1–28).

McGahay V. (2010). Porous dielectrics in microelectronic wiring applications. *Materials,* 3(1), 536–562.

Miller J, Dunn B, Tran T, Pekala R. (1997). Deposition of ruthenium nanoparticles on carbon aerogels for high energy density supercapacitor electrodes. *Journal of The Electrochemical Society,* 144(12), L309–L311.

Moore AL, Shi L. (2014). Emerging challenges and materials for thermal management of electronics. *Materials Today,* 17(4), 163–174.

Mücklich F, Ilić N. (2005). RuAl and its alloys. Part I. Structure, physical properties, microstructure and processing. *Intermetallics,* 13(1), 5–21.

Mula S, Sahani P, Pratihar S, Mal S, Koch C. (2011). Mechanical properties and electrical conductivity of Cu-Cr and Cu-Cr-4%SiC nanocomposites for thermoelectric applications. *Materials Science and Engineering A,* 528, 39–44.

Munir Z, Anselmi-Tamburini U, Ohyanagi M. (2006). The effect of electric field and pressure on the synthesis and consolidation of materials:

A review of the spark plasma sintering method. *Journal of Materials Science,* 41(3), 763–777.

Murty B, Ranganathan S. (1998). Novel materials synthesis by mechanical alloying/milling. *International Materials Reviews,* 43(3), 101–141.

Nadkarni AV, Samal PK. (1984). Metal modified dispersion strengthened copper: Google Patents.

Nefedov V, Firsov M, Shaplygin I. (1982). Electronic structures of $MRhO_2$, MRh_2O_4, $RhMO_4$ and Rh_2 MO_6 on the basis of X-ray spectroscopy and ESCA data. *Journal of Electron Spectroscopy and Related Phenomena,* 26(1), 65–78.

Oghbaei M, Mirzaee O. (2010). Microwave versus conventional sintering: A review of fundamentals, advantages and applications. *Journal of Alloys and Compounds,* 494(1), 175–189.

O'reilly M, Jiang X, Beechinor J, Lynch S, NiDheasuna C, Patterson J, Crean G. (1995). Investigation of the oxidation behaviour of thin film and bulk copper. *Applied Surface Science,* 91(1), 152–156.

Orth J, Wheat H. (1997). Corrosion behavior of high energy-high rate consolidated graphite/copper metal matrix composites in chloride media. *Applied Composite Materials,* 4(5), 305–320.

Pan M, Gupta M, Tay A, Vaidyanathan K. (2006). Development of bulk nanostructured copper with superior hardness for use as an interconnect material in electronic packaging. *Microelectronics Reliability,* 46(5), 763–767.

Patake V, Lokhande C. (2008). Chemical synthesis of nano-porous ruthenium oxide (RuO_2) thin films for supercapacitor application. *Applied Surface Science,* 254(9), 2820–2824.

Patake V, Lokhande C, Joo OS. (2009). Electrodeposited ruthenium oxide thin films for supercapacitor: Effect of surface treatments. *Applied Surface Science,* 255(7), 4192–4196.

Patil U, Kulkarni S, Jamadade V, Lokhande C. (2011). Chemically synthesized hydrous RuO_2 thin films for supercapacitor application. *Journal of Alloys and Compounds,* 509(5), 1677–1682.

Perng D-C, Yeh J-B, Hsu K-C. (2008). Phosphorous doped Ru film for advanced Cu diffusion barriers. *Applied Surface Science,* 254(19), 6059–6062.

Platzman I, Brener R, Haick H, Tannenbaum R. (2008). Oxidation of polycrystalline copper thin films at ambient conditions. *The Journal of Physical Chemistry C,* 112(4), 1101–1108.

Potgieter JH, Thanjekwayo NP, Olubambi PA, Maledi N, SSP. (2011). Influence of Ru additions on the corrosion behaviour of WC-Co cemented carbide alloys in sulphuric acid. *International Journal of Refractory Metals and Hard Materials,* 29(4), 478–487.

Qu X-H, Zhang L, Wu M, Ren S-B. (2011). Review of metal matrix composites with high thermal conductivity for thermal management applications. *Progress in Natural Science: Materials International,* 21(3), 189–197.

Rajumon M, Prabhakaran K, Rao C. (1990). Adsorption of oxygen on (100), (110) and (111) surfaces of Ag, Cu and Ni: An electron spectroscopic study. *Surface Science,* 233(1), L237–L242.

Ramakrishnan P. (1983). History of powder metallurgy. *Induan Journal of History of Science,* 18(1), 109–114.

Riga J, Tenret-Noel C, Pireaux J, Caudano R, Verbist J, Gobillon Y. (1977). Electronic structure of rutile oxides TiO_2, RuO_2 and IrO_2 studied by X-ray photoelectron spectroscopy. *Physica Scripta,* 16(5–6), 351.

Saheb N, Iqbal Z, Khalil A, Hakeem AS, Al Aqeeli N, Laoui T, Kirchner R. (2012). Spark plasma sintering of metals and metal matrix nanocomposites: A review. *Journal of Nanomaterials,* 2012, 18.

Sarangapani S, Tilak B, Chen CP. (1996). Materials for electrochemical capacitors theoretical and experimental constraints. *Journal of The Electrochemical Society,* 143(11), 3791–3799.

Schubert T, Trindade B, Weißgärber T., Kieback B. (2008). Interfacial design of Cu-based composites prepared by powder metallurgy for heat sink applications. *Materials Science and Engineering A,* 475, 39–44.

Sharma AK, Teverovsky A. (2001). Preliminary reliability evaluation of copper-interconnect metallization technology. *http://nepp. nasa. gov/ imd/eee_links,* 32.

Shen Y, Lattari P, Gardner J, Wiegard H. (1990). Electrical contact materials. *ASM International, Metals Handbook,* 10th Edition, 2, 840–868.

Shing T, Luyckx S, Northrop I, Wolff I. (2001). The effect of ruthenium additions on the hardness, toughness and grain size of WC–Co. *International Journal of Refractory Metals and Hard Materials,* 19(1), 41–44.

Stangl M, Lipták M, Acker J, Hoffmann V, Baunack S, Wetzig K. (2009). Influence of incorporated non-metallic impurities on electromigration in copper damascene interconnect lines. *Thin Solid Films,* 517(8), 2687–2690.

Suárez M, Fernández A, Menéndcz J, Torrecillas R, Kessel H, Hennicke J, Kessel T. (2013). Challenges and opportunities for spark plasma sintering: A key technology for a new generation of materials, in: Ertuğ B (ed), Sintering Application, pp. 320–338.

Subramanian V, Hall SC, Smith PH, Rambabu B. (2004). Mesoporous anhydrous RuO_2 as a supercapacitor electrode material. *Solid State Ionics*, 175(1), 511–515.

Sule R, Olubambi PA, Abe BT, Johnson OT. (2012). Synthesis and characterization of sub-micron sized copper-ruthenium-tantalum composites for interconnection application. *Microelectronics Reliability*, 52, 1690–1698.

Sule R, Olubambi PA, Asante JKO, Roos WD. (2012). Effect of ruthenium on oxidation behaviour of copper interconnect. Book of Abstract, South African Institute of Physics, *SAIP* 2012, 9–13.

Sule R, Olubambi PA, Sigalas I, Asante JKO, Garrett JC, Roos WD. (2015). Spark plasma sintering of sub-micron copper reinforced with ruthenium–carbon nanotube composites for thermal management applications. *Synthetic Metals*, 202 123–132.

Sundberg G, Paul P, Sung C, Vasilos T. (2006). Fabrication of CuSiC metal matrix composites. *Journal of Materials Science*, 41(2), 485–504.

Suryanarayana C. (2001). Mechanical alloying and milling. *Progress in Materials Science*, 46(1), 1–184.

Tiwari D, Basu B, Biswas K. (2009). Simulation of thermal and electric field evolution during spark plasma sintering. *Ceramics International*, 35(2), 699–708.

Tjong SC, Ma Z. (2000). Microstructural and mechanical characteristics of in situ metal matrix composites. *Materials Science and Engineering: R: Reports*, 29(3), 49–113.

Tryon B, Feng Q, Pollock T. (2004). Intermetallic phases formed by ruthenium–nickel alloy interdiffusion. *Intermetallics*, 12(7), 957–962.

Tsui TY, Jahed Z, Jin S, Burek MJ. (2013). Suppression of size-dependent softening effects in sub-micron nanocrystalline ruthenium columnar structures. *Materials Science and Engineering: A*, 565, 219–227.

Van Scoik KG. (1992). Solid pharmaceutical dosage in tablet triturate form and method of producing same: Google Patents.

Wang Y, Cao F, Zhang ML, Liu YT. (2011). Comparative study of Cu-Zr and Cu-Ru alloy films for barrier-free Cu metallization. *Thin Solid Film*, 519(10), 3169–3172.

Weeks ME. (1932). The discovery of the elements. VIII. The platinum metals. *Journal of Chemical Education*, 9(6), 1017.

Wolff I. (1997). Toward a better understanding of ruthenium aluminide. *Journal of Minerals, Metals and Materials Society,* 49(1), 34–39.

Wu ZS, Wang DW, Ren W, Zhao J, Zhou G, Li F, Cheng HM. (2010). Anchoring hydrous RuO_2 on graphene sheets for high-performance electrochemical capacitors. *Advanced Functional Materials,* 20(20), 3595–3602.

Yamamoto S, Bluhm H, Andersson K, Ketteler G, Ogasawara H, Salmeron M, Nilsson A. (2008). In situ x-ray photoelectron spectroscopy studies of water on metals and oxides at ambient conditions. *Journal of Physics: Condensed Matter*, 20(18), 184025.

Yang C-C, Spooner T, Ponoth S, Chanda K, Simon A, Lavoie C, Gignac L. (2006). *Physical, Electrical, and Reliability Characterization of Ru for Cu Interconnects.* Paper presented at the Interconnect Technology Conference, 2006 International.

Yi S-M, Jang K-H, An J-U, Hwang S-S, Joo Y-C. (2008). The self-formatting characteristics of Cu-Mg/SiO_2 and Cu-Ru/SiO_2 films for Cu interconnects. *Microelectronics Reliability*, 48, 744–748.

Yoshida K, Morigami, H. (2004). Thermal properties of diamond/copper composite material *Microelectronics Reliability,* 44, 303–308.

Yum Y-J, Lee K-S, Kim J-S, Kwon Y-S, Chang M-G, Kwon D-H, Sung C-H. (2005). *Mechanical Properties of Cu-TiB_2 Nanocomposite by MA/SPS.* Paper presented at the Proceedings. The 9th Russian-Korean International Symposium on Science and Technology, 2005. KORUS 2005.

Zheng Y-Z, Ding H-Y, Zhang M-L. (2008). Hydrous–ruthenium–oxide thin film electrodes prepared by cathodic electrodeposition for supercapacitors. *Thin Solid Films*, 516(21), 7381–7385.

Zweben, C. (2005). *Ultrahigh-Thermal-Conductivity Packaging Materials.* Paper presented at the Semiconductor Thermal Measurement and Management Symposium, 2005 IEEE Twenty First Annual IEEE.

Chapter 6

Orbital Tuning of Ruthenium Polyimine Complexes by Ligand Design: From Basic Principles to Applications

Joe Otsuki, Guohua Wu, Ryuji Kaneko, and Yayoi Ebata

College of Science and Technology, Nihon University, Tokyo, Japan

otsuki.joe@nihon-u.ac.jp

Ruthenium polyimine complexes have been one of the central subjects of inorganic electrochemistry and photochemistry. In this chapter, attempts are organized to tune the orbitals of ruthenium polyimine complexes making use of interactions between the ruthenium and the ligands orbitals. Starting from the basic principles, selected examples of design of non-pyridine-based ligands and "active tuning" are discussed for the tuning of orbital energies of the ruthenium complexes, followed by the discussion of applications to dye-sensitized solar cells and organic light-emitting diodes from the orbital viewpoints.

Ruthenium Chemistry
Edited by Ajay Kumar Mishra and Lallan Mishra

ISBN 978-981-4774-39-0 (Hardcover), 978-1-315-11058-5 (eBook)
www.panstanford.com

6.1 Basic Principles of Orbital Tuning for Ruthenium Polyimine Complexes

6.1.1 Ruthenium Polyimine Complexes

Ruthenium polyimine complexes have been one of the central subjects of inorganic electrochemistry and photochemistry, exemplified by a ruthenium tris(bipyridine) complex $[Ru(bpy)_3]^{2+}$ and a ruthenium bis(terpyridine) complex $[Ru(tpy)_2]^{2+}$ (bpy = 2,2′-bipyridine; tpy = 2,2′:6′,2′-terpyridine) (Garino et al., 2016; Juris et al., 1988; Meyer, 1986; Prier et al., 2013; Thompson et al., 2013; Wang et al., 2004). The $[Ru(bpy)_3]^{2+}$ has been occupying the special position in the chemistry of ruthenium polyimine complexes. The first oxidation occurs at +1.26 V vs. NHE in aqueous solution (Bock et al., 1975) corresponding to the oxidation of the metal center from Ru(II) to Ru(III). The first reduction occurs at −1.26 V vs. NHE in aqueous solution (Bock et al., 1979) corresponding to the reduction of one of the bpy ligands, rendering $[Ru(bpy)_3]^{2+}$ into $[Ru(bpy)_2(bpy)^{\bullet-}]^{+}$. The complex $[Ru(bpy)_3]^{2+}$ has a broad metal-to-ligand charge-transfer (MLCT) absorption band peaking at 450 nm in the visible region, in which a d electron on the ruthenium is promoted to a bpy π^* orbital, so that visible light can be utilized to effect photoreaction. The initially formed singlet MLCT excited state (^{1}MLCT) is quickly relaxed through intersystem crossing to the triplet MLCT excited state (^{3}MLCT). The lifetime of the ^{3}MLCT state is hundreds of nanoseconds, long enough to collide with other molecules to make something happen before decaying back to the ground state. The oxidation potential in the excited state, −0.62 V vs. NHE in H_2O (Diamantis et al., 2014) is more negative than the potential for the reduction of proton to hydrogen (−0.414 V vs. NHE at pH 7) and the reduction potential in the excited state, +0.84 V vs. NHE in H_2O (Creutz et al., 1976) is more positive than the oxidization of water to molecular oxygen (+0.816 V vs. NHE at pH 7).

The $[Ru(bpy)_3]^{2+}$, despite its excellent photophysical properties, has some structural complications. The six nitrogen atoms from the bpy ligand coordinates to the metal ion along the axes of the Cartesian coordinates (x, y, z). As such, the complex is chiral depending on the screw sense of the bpy ligands (Δ and Λ enantiomers). Further, the introduction of a substituent to each bpy ligand necessarily leads to a mixture of meridional and facial isomers. These complications may not cause any problem in many cases but are detrimental if the $[Ru(bpy)_3]^{2+}$ unit is to be incorporated into an extended supramolecular system.

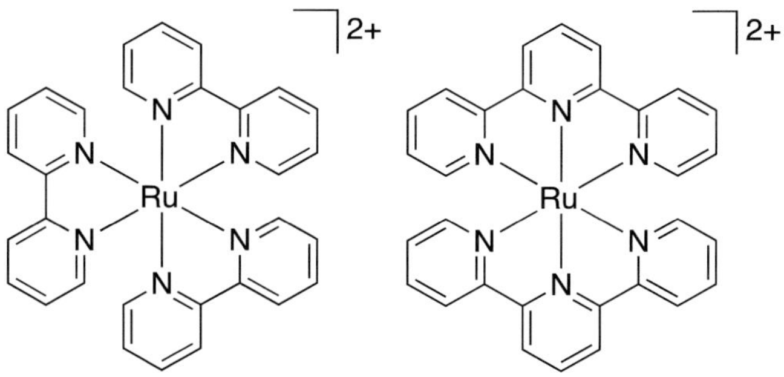

$[Ru(bpy)_3]^{2+}$ $[Ru(tpy)_2]^{2+}$

The $[Ru(tpy)_2]^{2+}$ is another basic motif of ruthenium polyimine complexes. This complex with two tridentate ligands has no chirality. Also, the introduction of substituent at the 4′ position (carbon para to the nitrogen in the central pyridine ring) will lead to a linear single component species with no isomer. These geometrical properties are suitable as a building block for a larger extended supramolecular system. However, this complex has a severe problem. The excited-state lifetime is very short due to its electronic structure, which we will delineate in this chapter.

The bpy and tpy ligands can be modified for further tuning of optical and redox properties for specific purposes. While a vast number of substituents have been introduced on the bpy and tpy skeletons and exploited for various purposes, we do not concern ourselves with substituents effects in this chapter. Another strategy to tune the properties of the ligand may, instead putting substituents, modify the bpy or tpy framework itself. Incorporating more nitrogen atoms in the ring will pull the electron affording an electron-poorer heterocyle. Replacing a nitrogen atom by a carbon atom (i.e., cyclometallic ligands) will make the ring electron-richer. Changing the number of atoms constituting the rings (i.e., hexagonal vs. pentagonal rings) is yet another strategy to alter the electronic structure of the ligand.

In this chapter, attempts are organized to tune the orbitals of ruthenium polyimine complexes making use of σ and π interactions between the ruthenium 4d orbitals and the ligand n and π/π^* orbitals. We start with a brief background knowledge of simplified molecular orbital theory and ligand field theory. Specific examples of orbital tuning of ruthenium complexes are given on cyclometalated complexes and complexes with azole/azolate ligands, among others. We briefly introduce "active tuning" of the ligand properties by redox reactions. The examples are not comprehensive but rather selective. Our focus is on the redox and optical properties of these complexes and we do not deal with ruthenium complexes for catalysis purposes. We conclude this chapter by seeing how the orbital tuning described in the former parts is applied to ruthenium complexes used for working devices, i.e., dye-sensitized solar cells and organic light-emitting diodes.

6.1.2 Orbital Tuning of Metal Complexes: A Crash Course on Molecular Orbitals and Ligand Field Theory

Theoretical interpretation or prediction of properties of metal complexes is done by density functional theory (DFT) computation most of the cases at the research level. Still, further interpretation of the results of the theoretical calculations by a human brain requires basic understanding of molecular orbital theory. Ligand field theory is molecular orbital theory as applied to metal complexes. Before going into the ligand field theory, we briefly summarize the basics of molecular orbital theory. The Hartree–Fock method is the standard theory in molecular orbital theory. This method incorporates electron–electron repulsion terms as an electron interacting with the average field the other electrons produce. As this theory is already complicated enough for back-of-the-envelope treatments, we restrict ourselves more or less on the extended Hückel theory level of molecular orbital theory. As the extended Hückel theory does not include electron–electron repulsion terms explicitly, which are hidden in the parameters used, the treatment is greatly simplified and the qualitative understanding becomes easier.

An atom has atomic orbitals in which electrons are accommodated. When atoms approach each other, the atomic orbitals are not suitable description of orbitals anymore. Instead, molecular orbitals are used in which electrons are accommodated similarly. However, it is still convenient to express the molecular orbitals in terms of atomic orbitals, or more specifically, as a linear combination of atomic orbitals.

The molecular orbitals ψ_+ and ψ_- resulting from two atomic orbitals ϕ_1 and ϕ_2 are represented approximately as

$$\psi_- \approx c_1\phi_1 - c_2\phi_2$$

$$\psi_- \approx c_2\phi_1 - c_2\phi_2,$$

where c_1 and c_2 are positive constants. Let us assume that the energies of ϕ_1 is higher than or equal to that of ϕ_2: $E_1 \geq E_2$. Then it turns out that $c_1 \geq c_2$. There are several characteristics worth remembering how the resulting molecular orbital is related to component atomic orbitals, referring to Fig. 6.1.

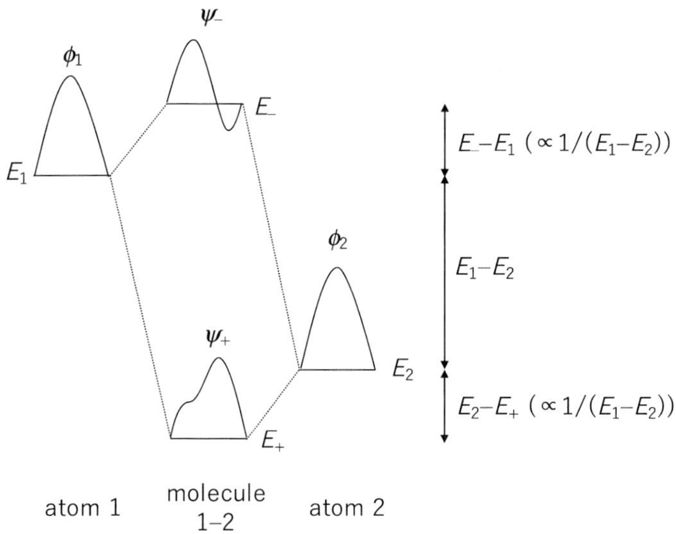

Figure 6.1 Molecular orbitals.

- The lower molecular orbital is similar to the lower atomic orbital. Because $c_1 \geq c_2$, ψ_+ is similar to ϕ_2 rather than ϕ_1.

- The higher molecular orbital is similar to the higher atomic orbital. Because $c_1 \geq c_2$, ψ_- is similar to ϕ_1 rather than ϕ_2.
- The energy of the lower atomic orbital gets lower: $E_+ < E_2$.
- The energy of the higher atomic orbital gets higher: $E_- < E_1$.
- The lower molecular orbital is bonding: coefficients of ϕ_1 and ϕ_2 have the same sign.
- The higher molecular orbital is antibonding: coefficients of ϕ_1 and ϕ_2 have different signs.
- The smaller the difference in energy of the two atomic orbitals, the larger the energy changes: $E_2 - E_+$ and $E_- - E_1$ are proportional to $(E_1 - E_2)^{-1}$.
- The smaller the difference in energy of the two atomic orbitals, the larger the mixing of the two atomic orbitals: c_1 and c_2 becomes closer to each other.
- When the energy of the two atomic orbitals are the same $E_1 = E_2$, the resulting molecular orbitals consists of a 1:1 mixture of the atomic orbitals, $c_1 = c_2$.
- The largest energy changes ($E_- - E_1$ and $E_2 - E_+$) result when the energy of the two atomic orbitals are the same: $E_1 = E_2$.

The above "rules" has been derived from a simple treatment for a one-electron system. Thus electron–electron repulsion will modify the resulting energy levels.

The same treatment can also be applied to metal complexes. There is a complication, however, because a metal complex is a three-dimensional object and several ligands form metal–ligand bonds at the same time. Nonetheless, basic ideas discussed above are valid for each metal–ligand bond. Here we take a six-coordinated octahedral metal complex as an example to see how metal–ligand bonds are constructed by separate metal and ligand orbitals. When a metal ion is isolated, the 5d orbitals have the same energy, which are said to be degenerate. In the case of ruthenium, which is a 5th period element, the relevant d orbitals are 4d orbitals. Ligands, with a lone pair in a nonbonding (n) orbital pointing toward the metal ion, approach to the metal ion along the six Cartesian coordinate axes. Three of the d orbitals (d_{xy}, d_{yz}, and d_{zx}) are orthogonal in symmetry with the ligand n orbitals and hence do not mix with the ligand orbitals. On the other hand, the remaining two d orbitals ($d_{x^2-y^2}$ and

d_{z^2}), which are oriented along the Cartesian axes, have the right symmetry to mix with the ligand n orbitals to form σ bonds. As the ligand n orbitals are lower in energy than the metal d orbitals in general, the ligand n orbitals are stabilized and the metal $d_{x^2-y^2}$ and d_{z^2} orbitals are raised in energy, lifting the degeneracy of the d orbitals into a set of higher-energy e_g orbitals ($d_{x^2-y^2}$ and d_{z^2}) and lower-energy t_{2g} orbitals (d_{xy}, d_{yz}, and d_{zx}), as shown in Fig. 6.2. The generated energy splitting is termed Δ_O, where the subscript o stands for octahedral. In general, more basic the ligand is, the higher the energy of the n orbital. Higher n orbital means that it is closer to the metal d orbitals in energy, mixing better with $d_{x^2-y^2}$ and d_{z^2} and making the energy level of these orbitals higher. The interaction of the n orbital of the ligands and $d_{x^2-y^2}$ and d_{z^2} of the metal is called the σ interaction.

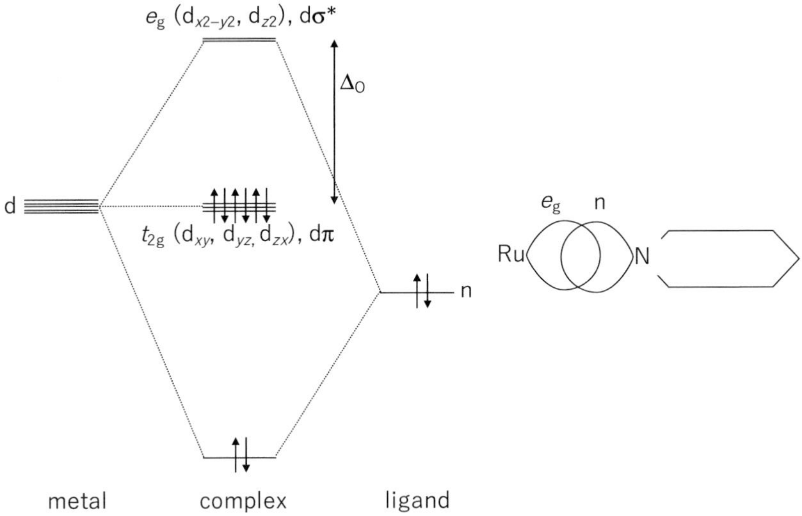

Figure 6.2 Orbital diagram for a σ bond between a metal and a ligand.

The σ interaction is not the whole story. The three orbitals (d_{xy}, d_{yz}, and d_{zx}) which are not influenced by the *n* orbitals of the ligands may be affected by the π-type interactions (Fig. 6.3). When the ligand has additional lone pairs (e.g., halide ions, a coordinating oxygen atom, and so on) after making a σ bond with the metal ion, it could donate the electron pairs to the t_{2g} orbitals as a π base. The energy of π-basic orbitals is relatively high for a

π orbital. Therefore, the energy of the π-basic orbital is close to the t_{2g} orbital from below, which makes it possible to interact effectively.

When the ligand has unsaturated bonds such as a double or a triple bond or the ligand is an aromatic molecule (e.g., bpy, etc.), it may also interact with the t_{2g} orbitals through the empty, antibonding π^* orbital of the ligand as a π acid. The energy of π-acid orbitals is relatively low for a π^* orbital. Therefore, the energy of the π-acid orbital is close to the t_{2g} orbital from above, which makes it possible to interact effectively. The resulting filled orbital has a partial character of ligand π^* orbital, which means that the d electrons are partially given to the ligand. This bonding is called a "back-bonding." The back-bonding plays an important role to determine the t_{2g} orbital level.

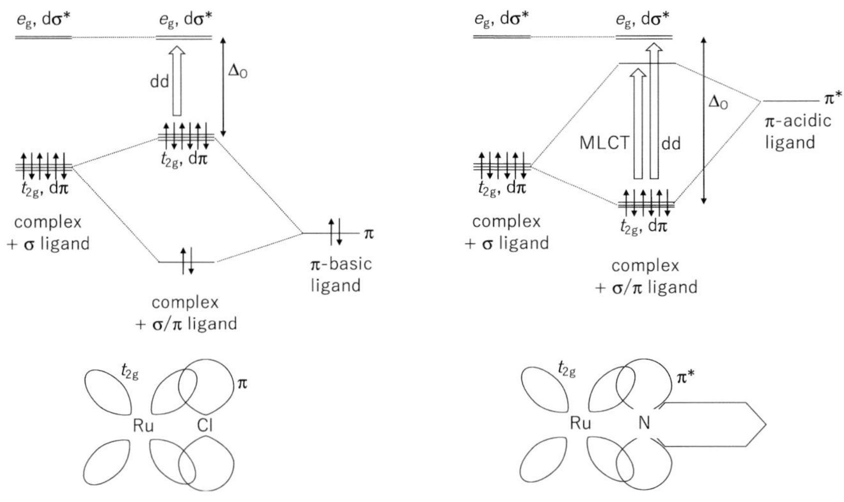

Figure 6.3 Orbital diagram for a π bond between a metal and a ligand. (Left) Metal–π-donor bond. (Right) Metal–π-acceptor bond ("back-bonding").

In summary, ligand n orbitals raise the energy of $d_{x^2-y^2}$ and d_{z^2} orbitals (e_g) and ligand π orbitals raise or lower the energy of d_{xy}, d_{yz}, and d_{zx} orbitals (t_{2g}) depending on the ligand has a relatively high occupied π orbital (π-basic) or a relatively low empty π^* orbital (π-acidic), respectively. The difference Δ_O is determined by the relative positions of the e_g and t_{2g} orbitals and gets smaller with a π-basic ligand and larger with a π-acidic ligand. The sequence of ligands according to the magnitude of Δ_0 they produce are known to be the *spectrochemical series* (Cotton et al., 1976):

$I^- < Br^- < Cl^- < F^- < OH^- < C_2O_4^{2-} < H_2O < -NCS^- < py < NH_3 < en < bpy < phen < NO_2^- < CN^-$,

where py = pyridine, en = ethylenediamine, phen = phenanthroline.

Krause and co-workers proposed a use of 2-(phenylazo) pyridine (Azpy) as a π-bonding probe (Fig. 6.4) (Krause et al., 1980; Krause et al., 1982). They prepared a series of ruthenium complexes [Ru(Azpy)$_2$LL′] and found that the N=N stretching wavenumber was a good measure of the π-acidity of ligands LL′. The more π-acidic LL′ is, the more electron density is taken away from the ruthenium ion by the LL′. This in turn reduces the degree of back-bonding to Azpy's, resulting in the stronger N=N bonding within the ligand, hence the increased N=N stretching wavenumber.

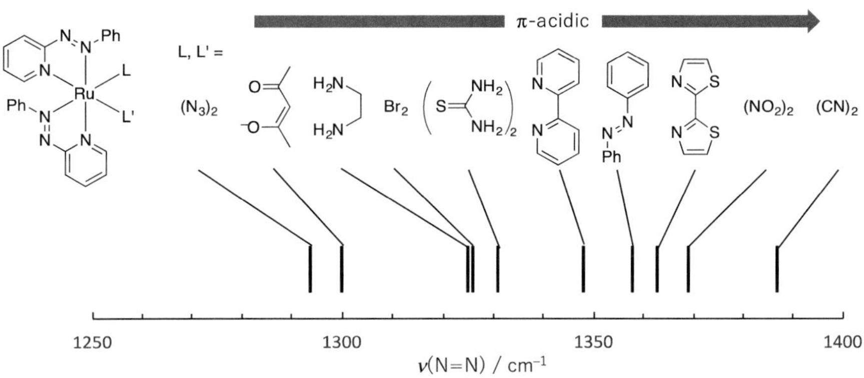

Figure 6.4 Stretching wavenumbers of N=N in the 2-phenylazopyridine ligand.

6.1.3 Orbitals and Redox and Photophysical Properties of Ruthenium Polyimine Complexes

The divalent Ru(II) ion has six electrons in 4d orbitals. All the six electrons are accommodated in the three t_{2g} orbitals in the case of $[Ru(bpy)_3]^{2+}$-type complexes. The spin of these electrons are all paired and hence there is no net spin. The t_{2g} orbitals are the highest-occupied molecular orbitals (HOMOs) of the complex. In electrochemistry the first oxidation takes an electron from the t_{2g} HOMO, thus can be represented as Ru(III)/Ru(II). The tuning of the HOMO level can be done by the π-acidity or π-basicity according to the simple ligand field theory. However, a σ-donating ligand also raise the HOMO level through electrostatic interactions. Overall, the oxidation potential of Ru(III)/Ru(II) couple is a good measure of the electron density on the metal center. A table of a large number of ligands listed in the order of Ru(III)/Ru(II) potentials is available constituting the *electrochemical series* (Lever, 1990). The lowest-unoccupied molecular orbitals (LUMOs) are the π^* orbitals of the bpy ligand. Thus the electro-chemical first reduction corresponds to adding electron on the bpy ligand.

The longest-wavelength, i.e., lowest-energy, optical transition occurs such that an electron in the t_{2g} orbital of ruthenium is excited into the π^* orbital of bpy ligand. Thus the visible light absorption corresponds to a MLCT transition (Garino et al., 2016). The processes that follow the excitation is schematically illustrated in Fig. 6.5. The singlet excited state ^{1}MLCT initially formed upon photoexcitation undergoes the intersystem crossing, facilitated by the mixing of the singlet and triplet states owing to the heavy atom effect of ruthenium, to be the triplet MLCT state (^{3}MLCT). This intersystem crossing occurs within 300 fs from excitation (Damrauer et al., 1997). The triplet excited state has a relatively long lifetime of several hundreds of nanoseconds, which is one of the major reasons that make the ruthenium complex useful as a photosensitizer. In one pathway, the ^{3}MLCT state decays either radiatively or nonradiatively back to the ground state. In the other pathway, ^{3}MLCT state can change into a ^{3}dd state, in which the electronic configuration is $t_{2g}{}^5e_g{}^1$. A dd transition is an optically forbidden process. Further, an electron is in a nonbonding d orbital and hence the metal–ligand bond is lengthened and weakened. This large geometry changes facilitates nonradiative crossing from the dd excited-state potential curve to the ground-state potential curve. Thus the dd state is quickly thermally decays back to the ground state. For this reason, the dd state is detrimental to the utility of the excited state and luminescence from the excited state. The transition from the ^{3}MLCT state to the ^{3}dd state is known to be temperature dependent. With this temperature dependence, the overall rate constant of the excited state decay k_{obs} is also temperature dependent, which is approximately expressed by (Sun et al., 2015)

$$k_{obs} = k_{MLCT} + k_{dd}e^{-\Delta E/k_BT},$$

where k_B is the Boltzmann constant and T is the temperature. Here k_{MLCT} is the rate constant for the processes, both radiative and nonradiative, from the ^{3}MLCT state to the ground sate. k_{dd} is the rate constant for the process from the ^{3}dd state to the ground state. The ΔE corresponds to the energy difference between the ^{3}MLCT state and the ^{3}dd state, although detailed analysis indicates that ΔE is actually a composite parameter that contains rate constants for several processes (Sun et al., 2015). It has been estimated that ΔE = ~4000 cm^{-1} in the case of $[Ru(bpy)_3]^{2+}$ (Barigelletti et al., 1983; Caspar et al., 1983).

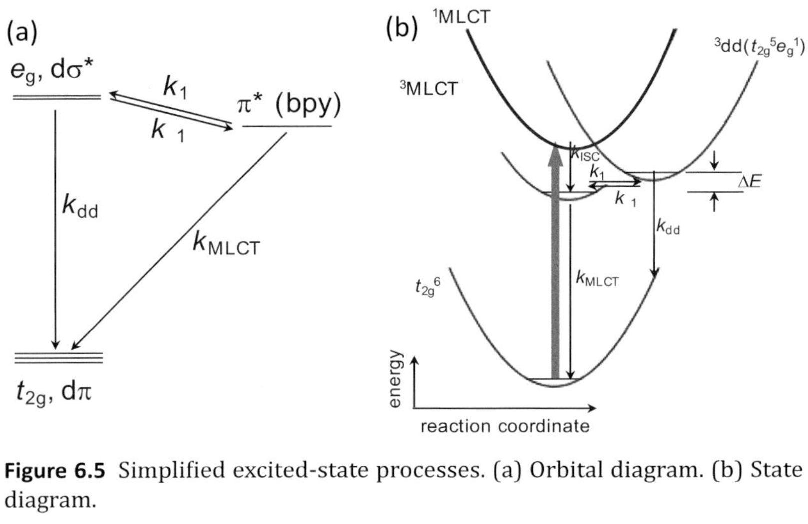

Figure 6.5 Simplified excited-state processes. (a) Orbital diagram. (b) State diagram.

The geometry of the bis(tpy) complex of ruthenium is attractive because the introduction of 4-substituent on the para position of the central pyridine moiety will lead to a linear complex, which can be a building block for extended linear arrays of metal complexes. Such extended arrays may be envisaged as components of molecular wire for molecular electronics (Otsuki et al., 2008a; Sauvage et al., 1994). Unfortunately, however, the excited-state lifetime of $[Ru(tpy)_2]^{2+}$-type complexes limits its utility as a photoactive component. The short excited-state lifetime (250 ps at room temperature (Winkler et al., 1987) is attributed to a small gap between the emissive ^{3}MLCT state and nonemissive ^{3}dd excited state. The N–Ru–N bite angles deviate from an ideal 90° for an octahedral complex due to the constraint imposed by the ligand structure. This leads to a weakened ligand field, resulting in a small Δ_O. As a result, the energy of the ^{3}dd is above the ^{3}MLCT only by ΔE = 1500 cm^{-1} for $[Ru(tpy)_2]^{2+}$ (Hecker et al., 1991).

Therefore, the strategy to increase the excited-excited lifetime is to lower the ^{3}MLCT state energy relative to the ^{3}dd state energy. To decrease the ^{3}MLCT state energy, electron-withdrawing groups are introduced into the ligands to lower the π^* orbital energies. To extend the π-system of the ligand is another effective strategy to lower the π^* orbital energies. In this case, the nuclear displacement associated with the MLCT excitation is small that further helps reduce the nonradiative decay rate. On the other hand, raising the energy of dd state is a complementary method to enlarge ΔE. Introducing a stronger σ-donating ligand, which is not necessarily the ligand that provides the π^* orbital for the MLCT state, is a straightforward strategy. Even a π-basic ligand would help, which increases the electron density on the metal that in turn raises the dσ^* orbital energy not only the dπ orbital energy, giving rise to a larger ΔE. A few examples of emissive ruthenium complexes with a bis(tridentate) ligand will be given in the next section.

6.2 Ligand Variation for Orbital Tuning

We will see selected ligand sets which were used in attempts to tune the redox potentials of ruthenium complexes. First we describe cyclometalated ruthenium complexes followed by azole ligands. We will then discuss some examples of ruthenium complexes with nonstandard geometries. Finally, we will discuss about the "active tuning" of properties of ruthenium complexes by redox reactions.

6.2.1 Cyclometalated Ruthenium Complexes

In place of ubiquitous metal–nitrogen bonds, metal–carbon bonds may be used to generate metal complexes. Often the metal–carbon bond formation is assisted by the coordination of another donor atom within the same ligand, such as a nitrogen atom, resulting in a metallacyclic structure, which is therefore called cyclometalation (Albrecht, 2010; Djukic et al., 2009). The cyclometalating ligand with the formally anionic carbon atom is a stronger σ-donor and a weaker π-acceptor than nitrogen heterocycles (Constable et al., 1990).

Figure 6.6 displays the first oxidation and reduction potentials for tris(bidentate) complexes and compares N-ligands and cyclometalated C^--ligands (Bomben et al., 2009). The Ru(III)/Ru(II) potential for $[Ru(bpy)_2(ppy)]^+$ is negatively shifted from that for $[Ru(bpy)_3]^{2+}$ by ca. 0.8 V (Bomben et al., 2009; Constable et al., 1986; Reveco et al., 1986), where ppy is 2-phenylpyridine in which the ortho proton on the phenyl group is deprotonated. The first reduction potential is also negatively shifted but by a smaller extent by 0.3–0.4 V. The first reduction corresponds to the reduction of a bpy ligand common to both $[Ru(bpy)_2(ppy)]^+$ and $[Ru(bpy)_3]^{2+}$. Therefore, the effect of the introduction of ppy on the LUMO level is indirect. Increase in the electron density on the ruthenium ion by the electron-donating ppy ligand makes the back-donation stronger, raising the π^* orbital energy of the bpy ligands. The decreased HOMO–LUMO gap brings about a decrease in energy of transition leading to a longer wavelength absorption in the visible spectra. The magnitude of the red shift upon replacing bpy ppy is roughly 100 nm. Another feature of $[Ru(bpy)_2(ppy)]^+$ is a broader absorption than $[Ru(bpy)_3]^{2+}$ due to less symmetry. To put it simply, there are transitions Ru→ppy as well as Ru→bpy. The effect of replacing a phen ligand by benzo[h]quinoline (bhq) is nearly the same as the effect of replacing a bpy ligand by a ppy ligand (Bomben et al., 2009; Reveco et al., 1986).

The bis-cyclometalated ruthenium complex, $[Ru(bpy)(ppy)_2]$, is known and the Ru(III)/Ru(II) potential is as low as −0.2 V vs. SCE (Le Lagadec et al., 2006). The preparation of the tris-cyclometalated ruthenium complex, $[Ru(ppy)_3]$ was attempted (Saavedra-Díaz et al., 2008), but this complex has not been isolated so far.

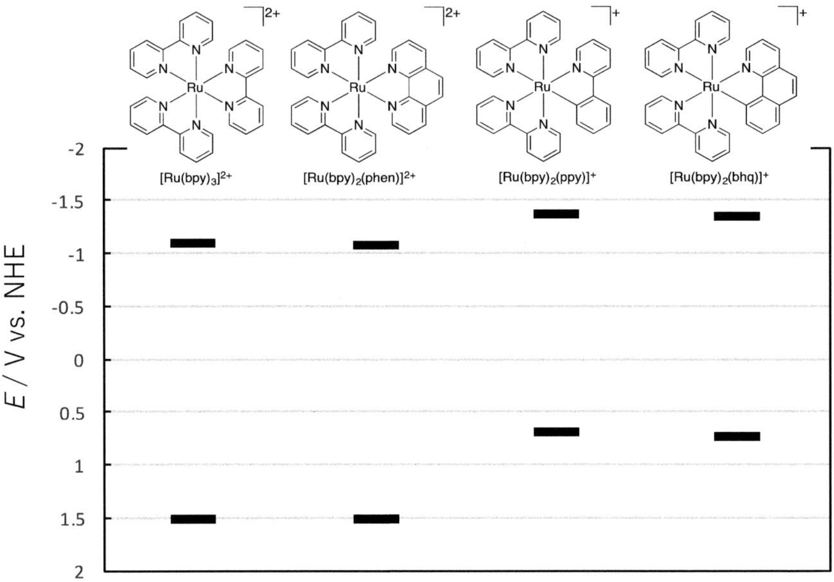

Figure 6.6 First oxidation and reduction potentials of cyclometalated complexes of bidentate ligands.

Figure 6.7 displays the first oxidation and reduction potentials of ruthenium complexes containing a tpy and a noncylcometalating or cyclometalating ligand (Bomben et al., 2009). The effect of cyclometalation is to increase the energy of the t_{2g} orbitals, manifested by a 0.8 V negative shift of the first oxidation potential (Beley et al., 1993; Bomben et al., 2009; Collin et al., 1991; Wadman et al., 2009). The magnitude of the effect of replacement by the phenyl group of terminal pyridine (phenylbipyridine; pbpy) and the central pyridine (benzo[h]quinoline; bhq) are nearly the same.

Figure 6.7 also includes complexes with a chloride ligand. Comparing $[Ru(bpy)_3]^{2+}$ and $[Ru(tpy)(bpy)Cl]^+$ reveals that the Ru(III)/Ru(II) potential is negatively shifted by 0.5 V. The halide ion is positioned at the smallest Δ_O end of the spectrochemical series because halide ions are π-bases and raise the energy of the t_{2g} orbitals, although the effect of a chloride ion is smaller than that of cyclometalated ligand. Complexes having both a Cl^- ion and a cyclometalated ligand raises the t_{2g} orbital energies by as much as 1.0 eV.

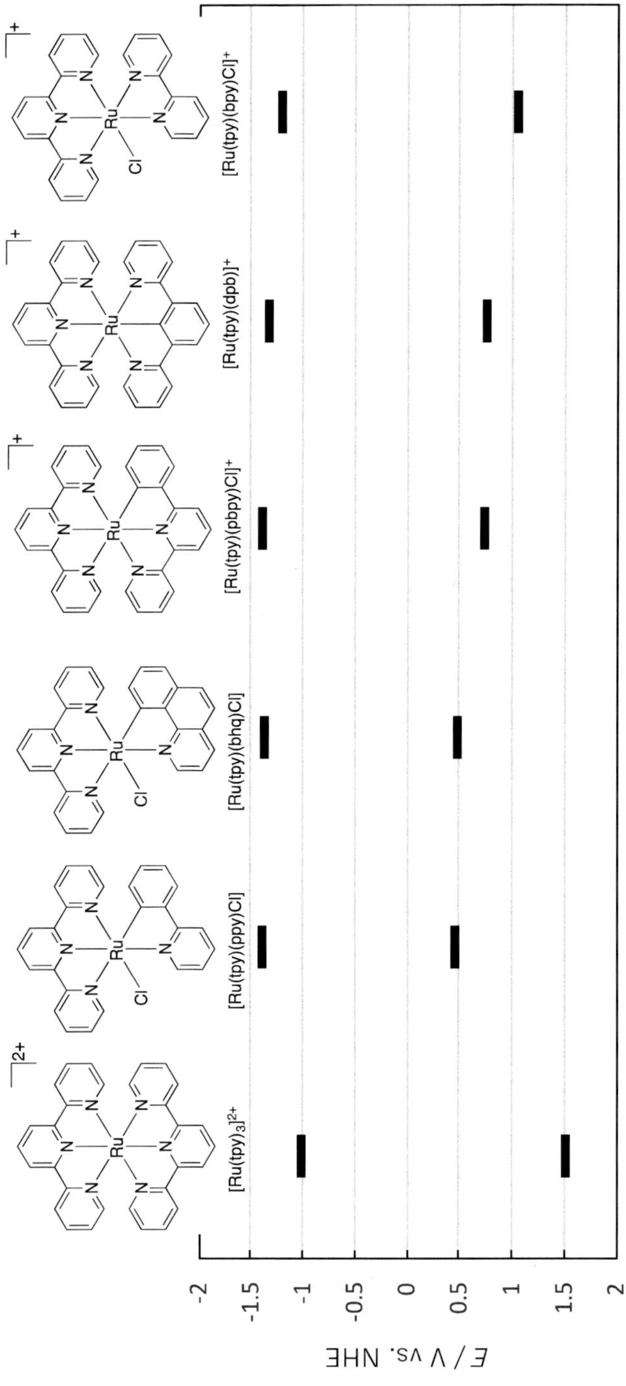

Figure 6.7 First oxidation and reduction potentials of cyclometalated complexes containing tpy.

6.2.2 Azole Ligands

An aromatic nitrogen-containing heterocycle commonly used as a ligand can either be a six-membered or five-membered ring. The six-membered rings are called azines; pyridine is the prominent example of azines. Azines are generally π-deficient and have relatively low π^* orbitals which allow good metal-to-ligand back-bonding by accepting t_{2g} electrons from the metal. On the other hand, the five-membered rings are called azoles; pyrrole is an example which contains one nitrogen atom as a ring-constituting member. Azoles are generally more π-electron-rich aromatic systems. Further, azoles contain a NH unit that can release a proton if proton acceptor (base) is present. Thus, pH-responsive ruthenium complexes may be prepared using an azole ligand.

The high π and π^* levels of azoles compared to those of azine can be understood from a simple Hückel theory calculations. According to the Hückel theory, the energy levels E_n of molecular π orbitals of a N-membered carbon ring is

$$E_n = \alpha + 2\beta \cos\left(2\pi \frac{n}{N}\right); \quad n = 0, \ldots, N-1,$$

where α is the Coulomb integral and β (< 0) is the resonance integral. The energy levels are conveniently visualized by the corners of regular polygons. For example, the orbital energy levels of benzene (N = 6) are expressed by a hexagon as shown in Fig. 6.8, left. On the other hand, cyclopentadienyl anion is a pentagonal aromatic system having 6π electrons. As it is a five-membered ring, the energy diagram is like the one shown in Fig. 6.8, right. As the diagram shows, both the HOMO and LUMO of a cyclopentadienyl anion are higher in energy than those of benzene. Note that this argument neglects the electron–electron interactions, which would further raise the energy of anionic species.

The argument about the six-membered vs. five-membered rings may apply to heterocyclic systems containing nitrogen atoms for example, although the values of α and β involving nitrogen atom are different from those only involving carbon atoms. This is corroborated by more accurate DFT calculations. The energy of the lowest π^* orbital of pyridine and pyrrole are –0.03967 eV and +0.02254 eV, respectively; the highest π orbital of pyridine and pyrrole are −0.27247 eV and −0.21611 eV, respectively, according to the B3LYP/6-31++G(d,p) calculations, as shown in Fig. 6.9.

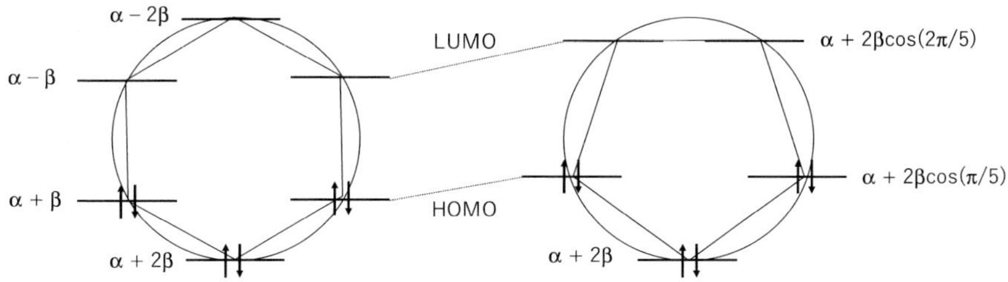

Figure 6.8 Orbital energy levels of benzene (left) and cyclopentadienyl anion (right) according to the Hückel theory.

Another way to tune the orbital levels of the ligand is to change the number of nitrogen atoms in the ring. Figure 6.9a displays the lowest π^* orbital energies and the highest π orbital energies of five-membered *N*-containing heterocycles, with those of pyridine being included for reference, calculated by a DFT method. The terms "the highest π^* orbital" and "the lowest π orbital," not the HOMO and LUMO, are used because we are only concerned with π (or π^*) orbitals here and these might not be the HOMO and LUMO, in which case an n orbital or even a σ orbitals may be the HOMO or LUMO. As pointed out above, five-membered pyrrole has higher π^* and π levels than those of six-membered pyridine, both containing a single nitrogen atom. As the diagram clearly shows, the orbitals are more and more stabilized as the number of electronegative nitrogen atom is increased. It is seen from the diagram that the energy of the π orbital is more influenced by the number of nitrogen atoms than that of the π^* orbital.

When these five-membered *N*-heterocycles make a coordination bond to a metal (ruthenium), deprotonation may occur. We also calculated the lowest π^* orbital energies and the highest π orbital energies of deprotonated anionic species by the same DFT method and the results are shown in Fig. 6.9b. It is seen from the diagram that as the number of nitrogen atom increases the highest π orbitals are more and more stabilized, while the lowest π^* level is rather insensitive to the number of nitrogen atoms in these anionic species.

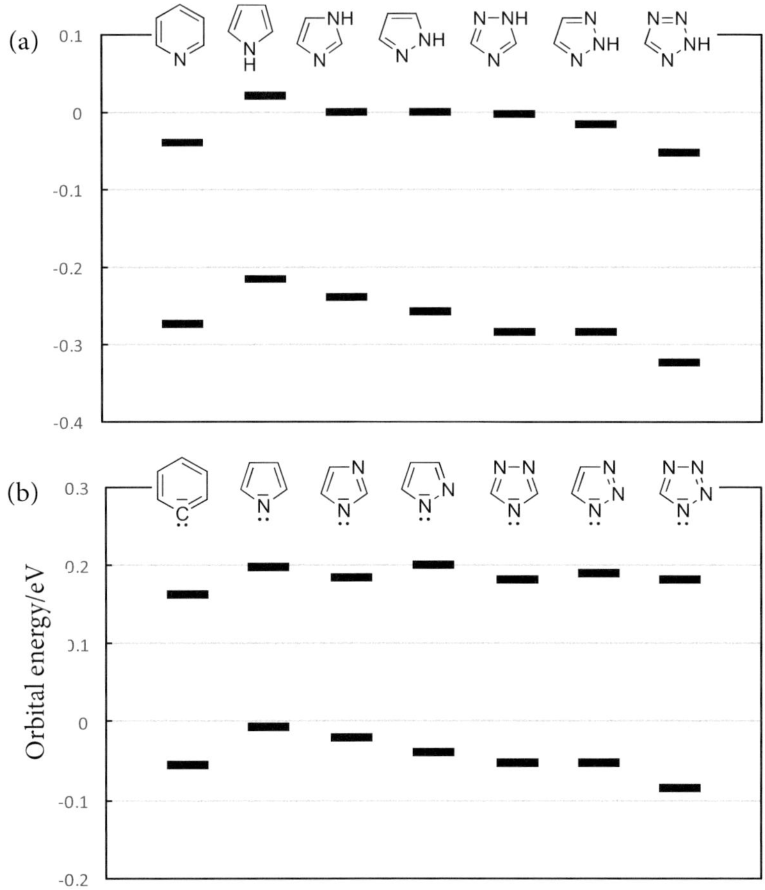

Figure 6.9 The lowest π^* orbitals and the highest π orbitals (B3LYP/6-31++G(d,p)). From left to right: pyridine, 1H-pyrrole, 1H-imidazole, 1H-pyrazole, 1H-1,2,4-triazole, 2H-1,2,3-triazole, 2H-1,2,3,4-tetrazole. (a) Neutral *N*-containing heterocycles. (b) Deprotonated anionic *N*-containing heterocycles and benzene.

Steel and co-workers investigated physical properties of ruthenium complexes for the formula $[Ru(bpy)_2L]$, which is summarized in Fig. 6.10 (Downard et al., 1995). The oxidation reactions observed at around +1 V represent the process of Ru(III)/Ru(II), a measure of the position of t_{2g} orbitals. The t_{2g} orbital energy are raised by the increased electron density on the metal atom as well as π-donating ligand. The reduction reactions observed at around −1.5 V represent the process of the ligand reduction. In the case of mixed ligand complexes as in the present case, the most electron-deficient ligand is reduced first. The involved orbitals are the π^* orbitals of the ligand, which is mixed with metal dπ orbitals through π interactions. Comparing the first three entries in Fig. 6.10 with bpy, we can conclude that alkylated, neutral tetrazole ligands are more electron-withdrawing than bpy. Tetrazole without substituent coordinates to the ruthenium ion as the tetrazolate anion accompanied by deprotonation. Thus pyridyltetrazolate is more electron-rich than bpy. The pyridyltetrazolate in the ruthenium complex has noncoordinating nitrogen atoms that can be protonated. Addition of acid to [Ru(bpy)$_2$(pyridyltetrazolate)] protonate the ligand, resulting in a positive shift of the oxidation potential by 0.15 V.

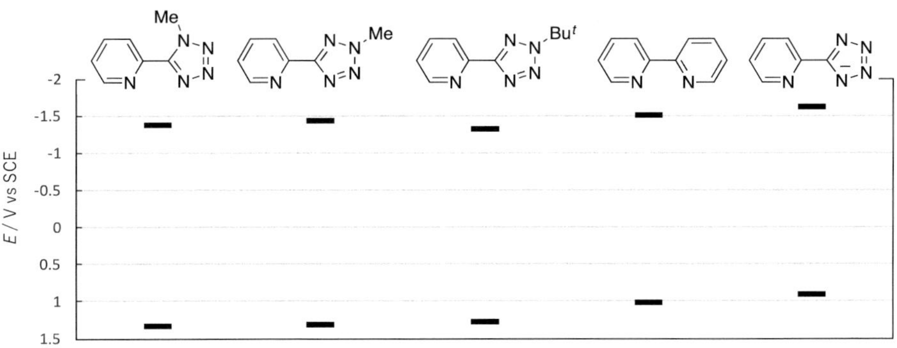

Figure 6.10 First oxidation and reduction potentials of complexes [Ru(bpy)$_2$L], where the structure of L is indicated on the diagram.

Unlike [Ru(tpy)$_2$]$^{2+}$, which is nonemissive (luminescence quantum yield $\phi < 5 \times 10^{-6}$), ruthenium complexes with a bis(triazolate) and a bis(tetrazolate) pyridine shown in Fig. 6.11 are emissive (Duati et al., 2003). The stronger σ-donor and less π-acidic properties of the anionic triazolate and tetrazolate ligands widen the Δ_O. As a result, the thermally activated transition from the emissive ^{3}MLCT to the nonemissive ^{3}dd state is retarded. The lifetime of these complexes are in a range of 24–77 ns at 298 K, which are two-orders of magnitudes longer than that of [Ru(tpy)$_2$]$^{2+}$. The overall geometry of tpy is maintained in these complexes, which implies that these photoactive complexes may be used as building blocks for well-defined, extended supramolecular structures.

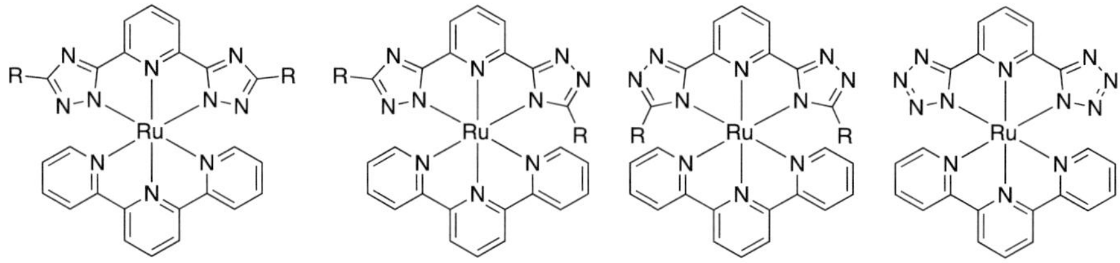

Figure 6.11 Emissive ruthenium bis(tridentate) complexes with triazolate and tetrazole ligands where ϕ is the luminescence quantum yield.

6.2.3 Out-of-Ring Coordination

Ruthenium complexes with coordination motifs in which both aromatic ring sites and out-of-the ring sites coordinate have also been studied. Ward and co-workers prepared ruthenium complexes in which pyridylphenolate or bipyridylphenolate ligands are contained (Holligan et al., 1992). The phenolate group is characterized by the negative charge, which electrostatically destabilizes the metal orbitals, a strong σ donation, which raises e_g orbital energies, and no π-accepting ability, which leads to a higher t_{2g} orbital energies compared to the π-accepting bpy ligand. With these effects, the Ru(III)/Ru(II) potential of $[Ru(bpy)_2(PyPhO)]^+$ is negatively shifted by 0.86 V compared to $[Ru(bpy)_3]^{2+}$. The Ru(III)/Ru(II) potential is negatively sifted further by 0.57 V by additional introduction of a negative phenolate group in $[Ru(BpyPhO)_2]^+$. The Ru(III)/Ru(II) potential is now so negative that the oxidation state of Ru(III) is the stable state for $[Ru(BpyPhO)_2]^+$ under ambient conditions.

Ward and co-workers also prepared ruthenium complexes carrying a ligand containing a dimethylamino group, $BpyPhNMe_2$ (Bardwell et al., 1995). The Ru(III)/Ru(II) potential of $[Ru(BpyPhNMe_2)_2]^{2+}$ is negatively shifted by 0.1 V compared to that of $[Ru(tpy)_2]^{2+}$, because of the poorer π-accepting characteristics of the coordinating dimethylamino group than that of pyridines. The ligand $BpyPhNMe_2$ is ambidentate in that it can make bonds to the ruthenium metal in a N_3 fashion as well as in a cyclometalated N_2C fashion, in which case the dimethylamino group points outwards. A ruthenium complex of the $BpyPhNMe_2$ ligand in which one of the ligand coordinates in a cyclometalated fashion, $[Ru(BpyPhNMe_2)(cm\text{-}BpyPhNMe_2)]^{2+}$, was obtained, where *cm* represents a cyclometalated ligand. In the cyclometalated ligand, the coordinating carbon is formally anionic, which results in the Ru(III)/Ru(II) potential shifted negatively by as much as 0.84 V compared to the N_3 type coordination.

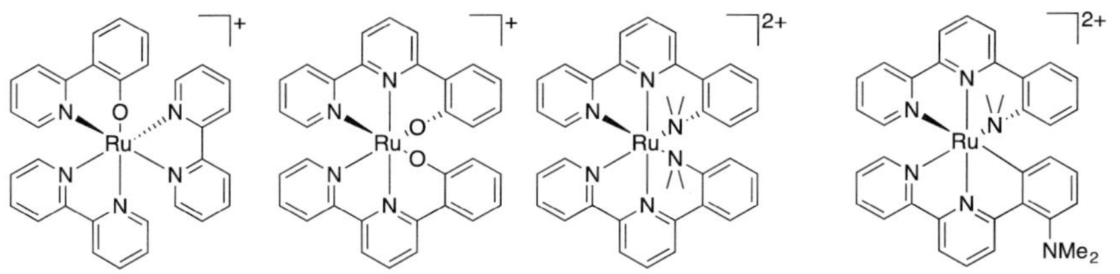

[Ru(bpy)$_2$(PyPhO)]$^+$ [Ru^{III}(BpyPhO)$_2$]$^+$ [Ru(BpyPhNMe$_2$)$_2$]$^{2+}$ [Ru(BpyPhNMe$_2$)(*cm*-BpyPhNMe$_2$)]$^{2+}$

We have examined heteroleptic ruthenium complexes involving above-mentioned "out-of-ring" coordination chemistry to investigate their use in DSSCs. At least one of the ligands of ruthenium complexes for use in DSSCs must have anchoring groups, such as a carboxyl group. In the case of ruthenium tpy complex dyes, 4,4′,4′-tricarboxy-2,2′:6′,2′-terpyridine is most commonly used as the anchoring ligand. As a model of dyes for DSSCs we have prepared heteroleptic complexes of ruthenium having 4′-(4-methylphenyl)-2,2′:6′,2′-terpyridine (ttpy) as one of the ligand and 6-(*o*-methoxyphenyl)-2,2′-bipyridine (BpyPhOMe) or 6-(*o-N,N*-dimethylaminophenyl)-2,2′-bipyridine ($BpyPhNMe_2$) as the other ligand (Otsuki et al., 2014).

We have isolated three ruthenium complexes from the reactions of $[Ru(ttpy)Cl_3]$ and these ligands. In the case of $BpyPhNMe_2$, the only isolated complex was a ruthenium complex in which $BpyPhNMe_2$ with N_3 coordination, i.e., $[Ru(ttpy)(BpyPhNMe_2)]^{2+}$. In the case of BpyPhOMe, however, ruthenium complexes both of the N_2O-type coordination ($[Ru(ttpy)(BpyPhO)]^{2+}$) and the N_2C-type coordination ($[Ru(ttpy)(cm\text{-}BpyPhOMe)]^{2+}$) were obtained (see Fig. 6.12). Interestingly, in the complex in which the O atom coordinates to ruthenium, the methyl group was missing. Demethylation occurred during the complex forming reaction.

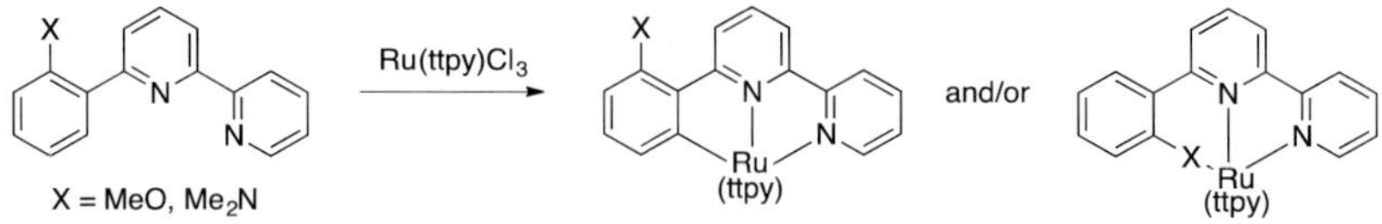

Figure 6.12 Reaction of ambidentate ligands with a metal complex fragment [Ru(ttpy)Cl_3].

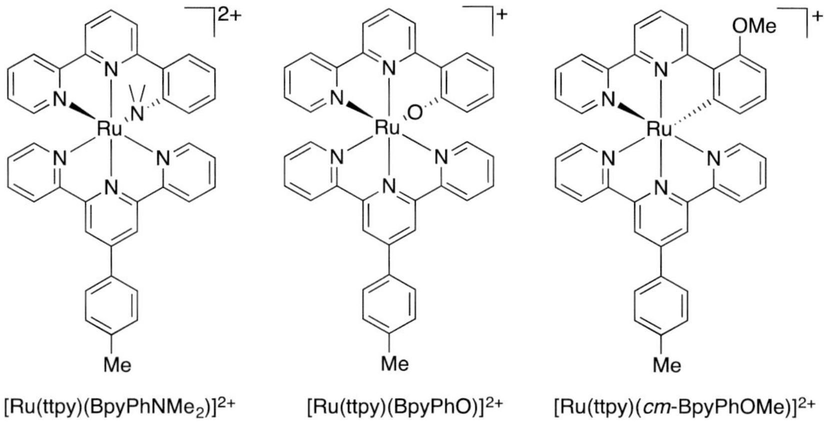

Figure 6.13 shows the cyclic voltammograms of these ruthenium complexes. We can see that these complexes are classified into two groups at a glance of the cyclic voltammograms; the upper two complexes have positively shifted redox waves than the lower two. The upper two complexes have neutral ligands while each of the lower two complexes has an anionic ligand. Apparently, the charge on the coordinating atoms has a large impact on the redox property of the complex. Compared to the charge difference, the change in the coordinating group from a pyridiyl (weaker σ-donor and good π-acceptor) group to a dimethylamino group (stronger σ-donor and not π-acceptor) has smaller effect on the redox potentials. The positive waves are for the Ru(III)/Ru(II) couple, while the reduction waves are for the ligand reduction. Even though every complex has a ttpy ligand, the reduction potentials differ depending on whether the other ligand is neutral or anionic. This observation suggests that the anionic ligand increases the electron density on the ruthenium atom, which in turn raises the orbital level of the other ligand (ttpy).

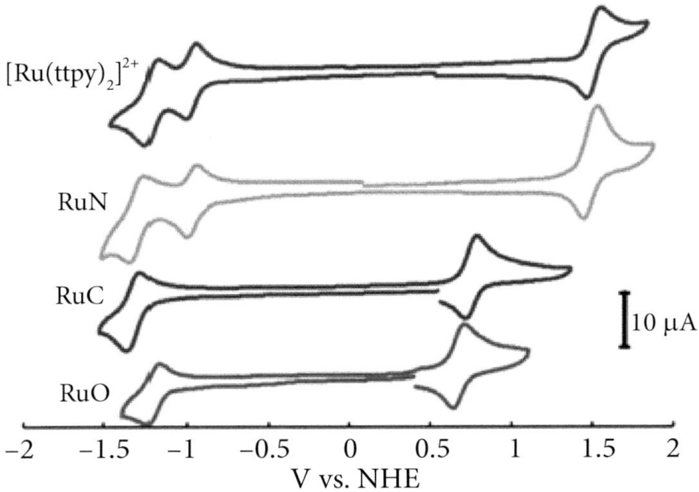

Figure 6.13 Cyclic voltammograms (0.1 M NBu_4PF_6, MeCN). RuN = $[Ru(ttpy)(BpyPhNMe_2)]^{2+}$, RuC = $[Ru(ttpy)(cm\text{-}BpyPhOMe)]^{2+}$, RuO = $[Ru(ttpy)(BpyPhO)]^{2+}$. Reprinted from *J. Organomet. Chem.* **2014**, 749, 312 with permission from Elsevier.

6.2.4 Active Orbital Tuning: Redox Reactions on Noninnocent Ligands

Orbital tuning can also be achieved in a post synthetic way. If the ligand itself is redox-active, the properties of the metal complex may be drastically changed depending on the redox state of the ligand. Here we describe examples of redox-responsive metal complexes taken from our laboratory.

The ruthenium complexes having an azobisbpy, bpyNNbpy, ligand exhibit a long wavelength absorption band, which is centered around 560 nm in addition to a common 450 nm band in the case of $[Ru(bpy)_2(bpyNNbpy)Ru(bpy)_2]^{4+}$ (Otsuki et al., 1996). The long-wavelength absorption is assigned to a MLCT transition, in which an electron is transferred from the ruthenium t_{2g} orbital to the π^*-orbital of the bpyNNbpy ligand. The long wavelength absorption reflects the low-lying π^* level of the bpyNNbpy ligand due to the extended π-system with electronegative nitrogen atoms. Indeed, the bpyNNbpy ligand in the complexes are reduced at −0.75 V and −0.59 V vs. Ag^+/Ag, which are much less negative compared to the third reduction potentials corresponding to the reduction of bpy ligands which occurs at −1.69 V and −1.74 V vs. Ag^+/Ag. Another important observation is that these complexes are almost nonluminescent in sharp contrast to $[Ru(bpy)_3]^{2+}$. This may partly be due to the energy gap law associated with the low-lying π^* level and partly due to vibrational modes associated with the azo group of the bpyNNbpy ligand. The nonluminescence also indicates that even the excited ^{3}MLCT state involving the bpy π^* orbital quickly relaxes into that involving the bpyNNbpy π^* orbital, which decays nonradiatively.

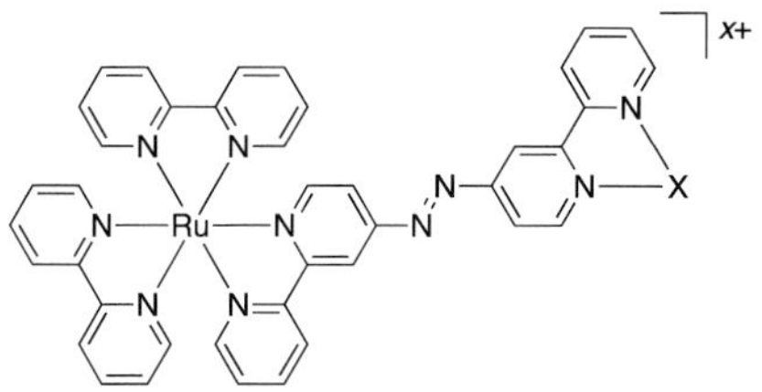

X = none ; $x = 2$ $[Ru(bpy)_2(bpyNNbpy)]^{2+}$

X = $Ru(bpy)_2$; $x = 4$ $[(bpy)_2Ru(bpyNNbpy)Ru(bpy)_2]^{4+}$

X = $Os(bpy)_2$; $x = 4$ $[[(bpy)_2Ru(bpyNNbpy)Os(bpy)_2]^{4+}$

Due to the least negative reduction potential of the bpyNNbpy ligand by a large margin, it is possible to selectively reduce the bpyNNbpy ligand electrochemically. Spectroelectrochemistry revealed that the long-wavelength absorption band disappeared upon reduction as shown in Fig. 6.14. A simplified interpretation of this phenomenon is as follows. The electron introduced in the π^* orbital of the bpyNNbpy ligand raises the energy of the π^* orbital of bpyNNbpy ligand for the next electron on account of the electronic repulsion. Then the π^* orbitals available for the MLCT of a t_{2g} electron are either the π^* orbital of bpy or the raised π^* orbital of bpyNNbpy. Interestingly, the reduced complexes are luminescent like $[Ru(bpy)]^{2+}$. The wavelength of the emission (640 nm) indicates that the luminescence comes not from the ^{3}MLCT involving the π^* orbital of pristine bpyNNbpy ligand, but from the ^{3}MLCT involving the π^* orbitals of bpy or the π^* orbital of one-electron reduced bpyNNbpy ligand.

This is a prototypical "molecular switch," in which the electronic structure of the excited state, manifested in the absorption and luminescence spectra, is controlled in response to redox input, the mechanism of which is schematically illustrated in Fig. 6.15. Controlling an excited state makes it possible to control processes that follow the excited state, e.g. energy transfer, electron transfer, and photoinduced chemical reactions. These aspects have been experimentally demonstrated with custom-made ruthenium complexes for energy-transfer switching in elaborate supramolecular systems (Akasaka et al., 2003; Akasaka et al., 2002; Otsuki et al., 2008b; Otsuki et al., 2003; Otsuki et al., 1997).

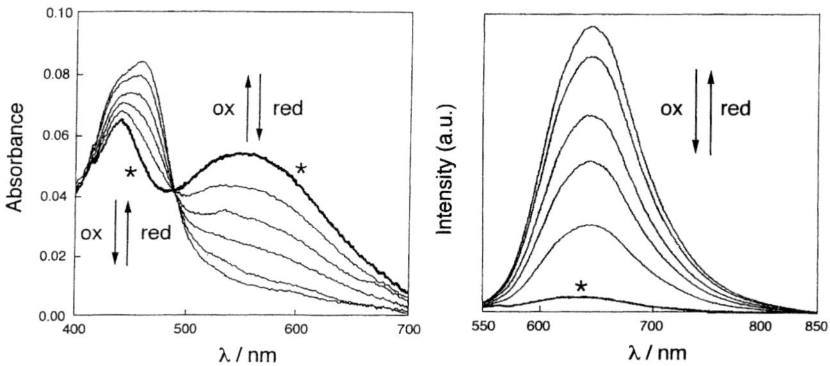

Figure 6.14 Changes in electronic absorption (left) and luminescence (right) spectra of $[(bpy)_2Ru(bpyNNbpy)Ru(bpy)_2]^{4+}$ upon one-electron reduction and reoxidation. The asterisks indicate the spectra for the as-prepared state. Reprinted from *Chem. Lett.*, **1996**, *25*, 847 with permission from the Chemical Society of Japan.

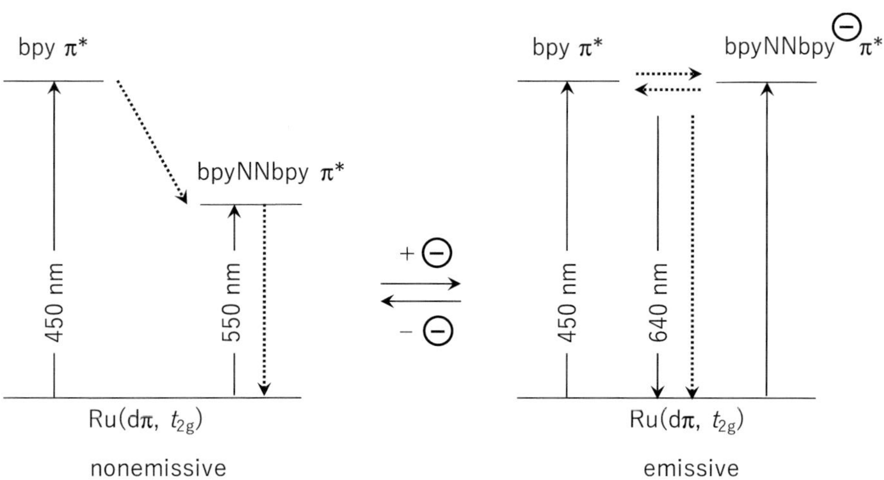

Figure 6.15 Redox-responsive molecular switch. Schematic mechanism of redox control between emissive and nonemissive states by redox reactions for $[(bpy)_2Ru(bpyNNbpy)Ru(bpy)_2]^{2+}$. The dotted arrows indicate nonradiative processes.

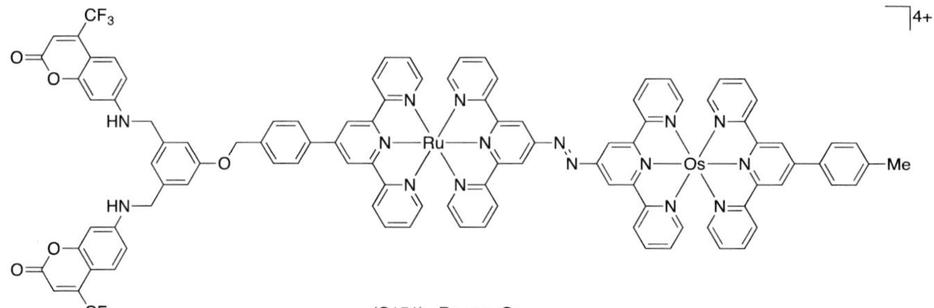

$(C151)_2$-Ru-azo-Os

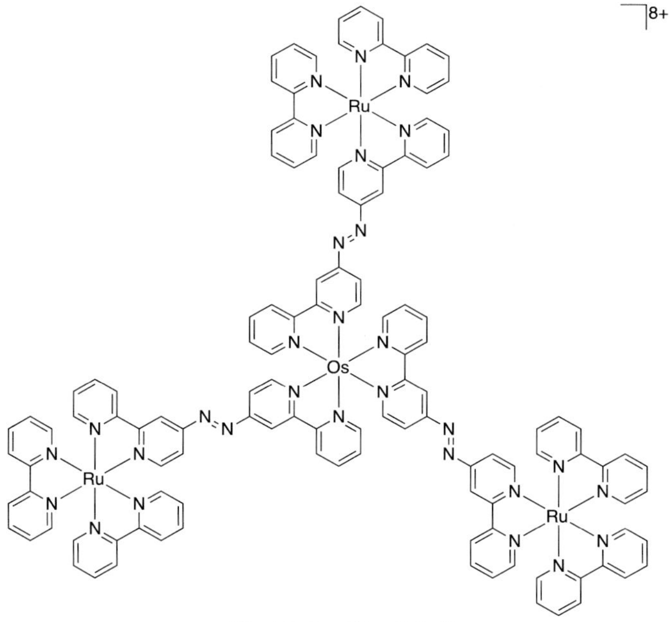

$[\{(bpy)_2Ru(bpyNNbpy)\}_3Os]^{8+}$

An example for molecular switches of energy transfer constructed from a ruthenium azobis(bpy) complex is obtained just by replacing one of the ruthenium atoms by an osmium atom, $[(bpy)_2Ru(bpyNNbpy)Os(bpy)_2]^{4+}$ (Otsuki et al., 1997). Only a residual luminescence was observed from the Ru/Os heterometallic complex in the as-prepared state. The reduction of the complex did not increase the luminescence from the ruthenium center as in homometallic $[(bpy)_2Ru(bpyNNbpy)Ru(bpy)_2]^{4+}$ but increase the luminescence from the osmium center. This observation indicates that efficient ruthenium to osmium energy transfer occurs in the reduced state. Effective quenching of the excited state occurs and no luminescence output was observed when the complex is in the as-prepared state, while efficient energy transfer occurs when the complex is electrochemically reduced. Thus it qualifies as "a molecular switch" that control the energy transfer process by the redox input to the molecule as shown schematically in Fig. 6.16. These switching components were incorporated in larger photoactive multiunit molecules, such as a linearly extended energy-transfer array, $(C151)_2$-Ru-azo-Os, (Akasaka et al., 2003) and a star-shaped tetrametallic complex, $[((bpy)_2Ru(azobpy))_3Os]^{8+}$ (Otsuki et al., 2008b).

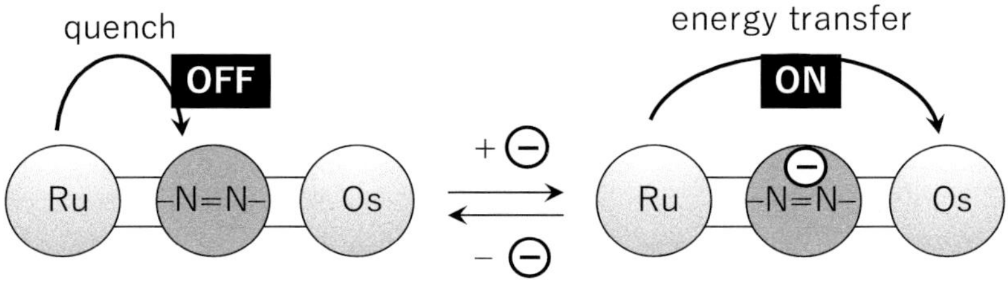

Figure 6.16 Redox-controlled switch for energy transfer.

6.3 Applications of Ruthenium Polyimine Complexes through Orbital Tuning

6.3.1 Dye-Sensitized Solar Cells

In 1991, O'Regan and Grätzel reported dye-sensitized solar cells (DSSCs) (O'Regan et al., 1991). Various attempts were being made to use molecule-based sensitizer to convert light energy into electricity before this report. However, the efficiency of the 1991 paper far exceeded those of previous attempts. The working principle is depicted by the schematic in Fig. 6.17 (Grätzel, 2005). The light is absorbed by the dye molecules, which are adsorbed onto the surface of TiO_2 nanoparticles. The excited electron is injected into the conduction band of TiO_2. The injected electrons are collected by the transparent electrode, go out from the cell to the external circuit and come back to the counter electrode. At the counter electrode, I_2, which is solubilized in the form of I_3^-, is reduced into I^-. The I^- then reduces the oxidized dye to regenerate the original dye, completing the cycle.

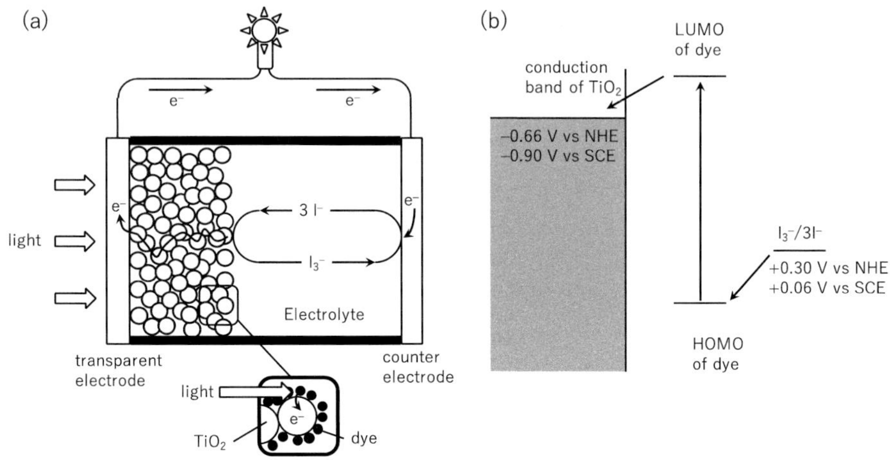

Figure 6.17 Dye-sensitized solar cells. (a) Schematic device. (b) Energy diagram. The values of potentials are taken from the literature (Redmond et al., 1993; Wang et al., 2006).

While the initial report used a trimetallic ruthenium complex, ruthenium complexes which are named N3 (N719 is a salt analogue in which two out of the four carboxyl moieties is deprotonated) (Nazeeruddin et al., 1993) and N749 ("black dye") (Nazeeruddin et al., 2001) emerged as standard dyes that make efficient DSSCs. N3 is bpy-based and N749 is tpy-based ruthenium complexes. The bpy and tpy ligands are decorated with carboxyl groups which play a role of an anchor to immobilize these complexes onto the surface of colloidal TiO_2 nanoparticles. These complexes also carry monodentate thiocyanate groups in contrast to the tris(bpy) or bis(tpy) type complexes which are most often used in demonstrating photoinduced charge separation in solutions.

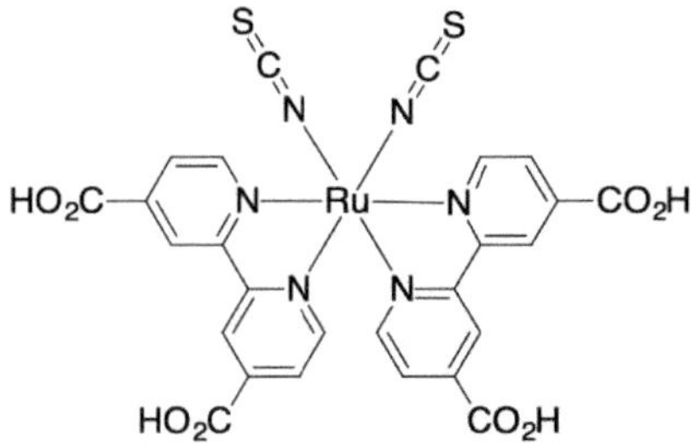

N3 (N719)

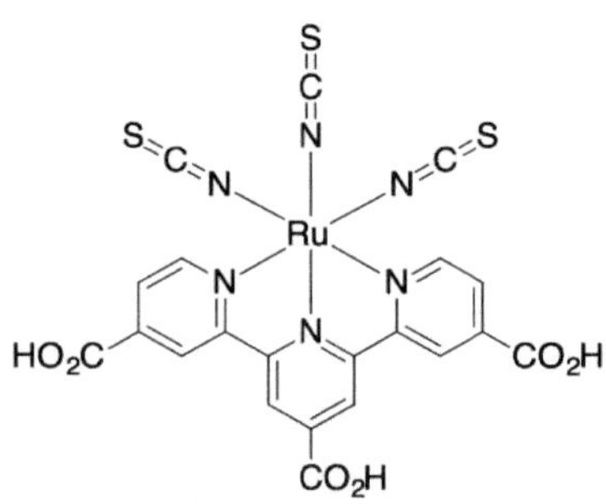

N749 "black dye"

Figure 6.18 shows the energy diagram for complexes $[Ru(dcbpy)_2L_2]$ where dcbpy is 4,4′-dicarboxy-2,2′-bipyridine (Kalyanasundaram et al., 1992). The oxidation potential represents the t_{2g} level, while the π^*(dcbpy) level was roughly estimated by subtracting the energy of emission maximum from the t_{2g} level. The energy of the t_{2g} orbitals with the NCS^- ligand is higher than that with the bpy-based ligand (Me_2bpy = 4,4′-dimethyl-2,2′-bipyridine). This is because bpy is higher in spectrochemical series than NCS^-, manifesting its better π-accepting ability and thus stabilizing the t_{2g} orbitals. The energy of π^* orbital of dcbpy is also higher with NCS^- than that with Me_2bpy. The increased electron density on the metal is partially transferred to the acceptor (dcbpy) ligand via mixing of the dπ and the ligand π^* orbitals. The raised energy of the π^* of dcbpy seems critically important as a sensitizer for DSSCs, because this is the orbital from which efficient electron injection to the conduction band of TiO_2 should occur. The driving force for the electron injection may be too low for $[Ru(dcbpy)_2(Me_2bpy)]^{2+}$, which is corrected for $[Ru(dcbpy)_2(NCS)_2]$. Another point is that the HOMO-LUMO gap of tris(bpy)-type complex is too wide: the λ_{max} of $[Ru(bpy)_3]^{2+}$ is 450 nm. More wide-ranging absorption is obviously desirable. On the other hand, t_{2g} must be more positive than +0.06 V vs. SCE (I_3^-/I^-) but this is satisfied by a large margin for both complexes.

One of the disadvantages of N3 and N749 is that the NCS^-, as a monodentate ligand, is prone to substitution reactions (Nguyen et al., 2011; Tuyet Nguyen et al., 2009). Bidentate ligands are less vulnerable to substitution reactions. In this sense, cyclometalated ligands have been explored (Bomben et al., 2012; Robson et al., 2012). Another attracting aspect in using cyclometalated ligands in place of NCS^- is that the former ligands can be easily structurally modified while the latter cannot. The phenylpyridine ligand with deprotonated phenyl group and the formally anionic carbon atom in the cyclometalated complex is a strongly electron-donating ligand. The cyclometalated ruthenium complex has a less positive oxidation potential than that of complexes with pyridine-based ligands, which indicates that t_{2g} orbital level is very high, even higher than NCS^- containing complexes.

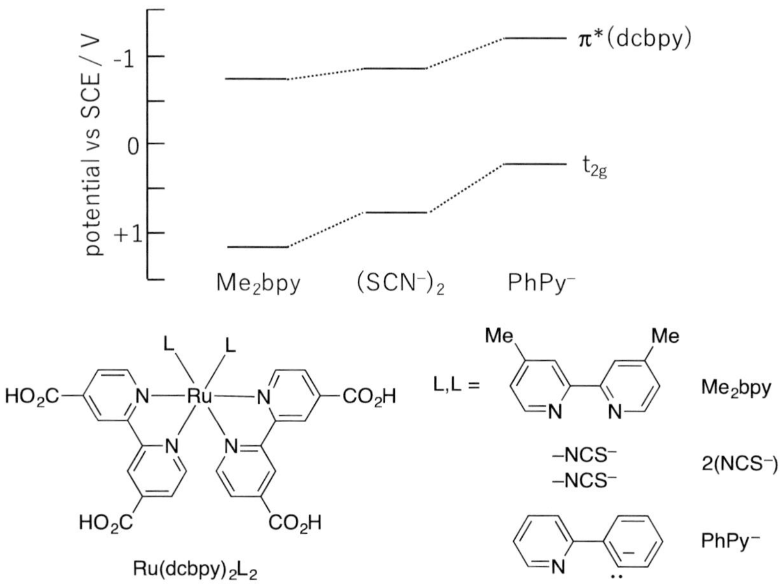

Figure 6.18 Orbital energy diagram of HOMO and LUMO $[Ru(dcbpy)_2L_2]$ complexes.

In 2007, van Koten and co-workers reported the first attempt to use cyclometalated complexes as dyes for DSSCs (Wadman et al., 2007) and subsequently reported the characterization of related complexes (Wadman et al., 2010). Figure 6.19 shows the first oxidation and reduction potentials of these complexes. Compared to the all *N*-coordinated bis(tpy) complex, $[Ru(tpy)(tpyCO_2Et)]^{2+}$, the cyclometalated complexes have negatively-shifted Ru(III)/Ru(II) potentials by as large as 0.7 V owing to the negatively charged electron donor nature of the cyclometalated ligand. The reduction potentials are also negatively shifted but not as large as those for the oxidation potentials. Hence the HOMO–LUMO gap get narrower by cyclometalation, bringing the absorption to longer wavelengths, which is beneficial for covering a wider visible range of the solar spectrum.

Good photocurrents comparable to that with N719 were obtained for $[Ru(tpy)(pbpyCO_2H)]^+$ and $[Ru(tpy)(pbpy(CO_2H)_2)]^+$ but poor photocurrents resulted for $[Ru(tpy)(dpybCO_2H)]^+$ (also poor for $[Ru(tpy)(tpyCO_2H)]^{2+}$). The former complexes have a C^N^N-type coordination and the latter complex has a N^C^N-type coordination. These coordination patterns give an impact on the position of the LUMO orbitals. For the complexes with the C^N^N pattern, the LUMO is located on the cyclometalated ligand, as a result of opposing effects of the electron-donating cyclometalation and the electron-withdrawing carboxyl group. On the other hand, for the complexes with the N^C^N pattern, the LUMO is located on the tpy ligand. The photoexcited electron is brought into the LUMO, which then is injected into the TiO_2. Therefore, the requirement for efficient electron injection is that the LUMO should be localized on a ligand with an anchoring group, which is satisfied only by the complexes with the C^N^N pattern.

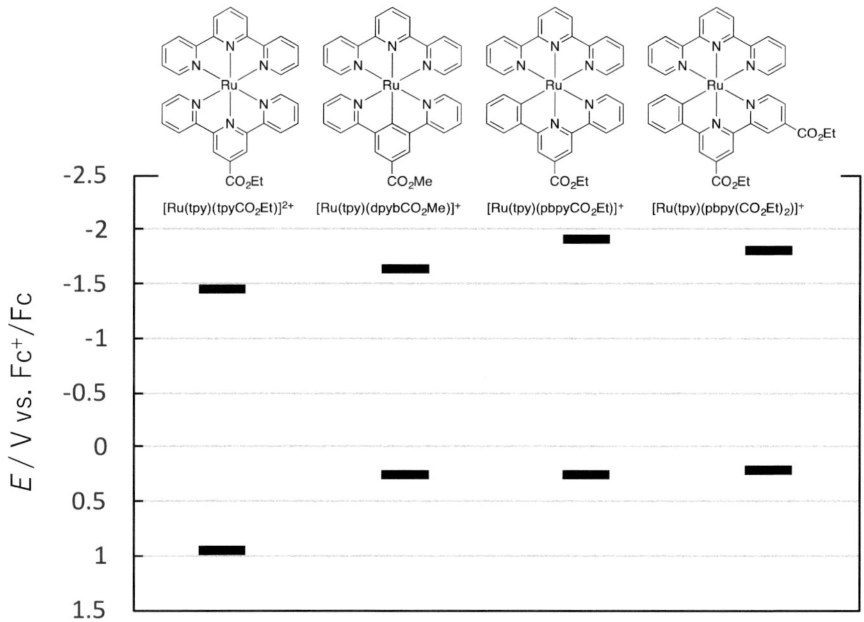

Figure 6.19 First oxidation and reduction potentials of [Ru(tpy)L] where L is a tridentate cyclometalated ligand with carboxyl group(s).

Only two years after this first attempt, a cyclometalated dye recording an efficiency as high as 10.1% was reported by Grätzel and co-workers (Bessho et al., 2009). The complex, $[Ru(dcbpy)_2(dfppy)]^+$, contains difluorophenylpyridine as the cyclometalating ligand. The Ru(III)/Ru(II) potential is +1.08 V vs. NHE, which is close to that of N719 ($[Ru(dcbpy)(NCS)_2]$ with one carboxyl group deprotonated) (+1.12 V vs. NHE). Apparently, electron-donating ability of the deprotonated anionic phenylpyridine ligand is judiciously attenuated with the introduction of electronegative fluorine atoms on the phenyl moiety.

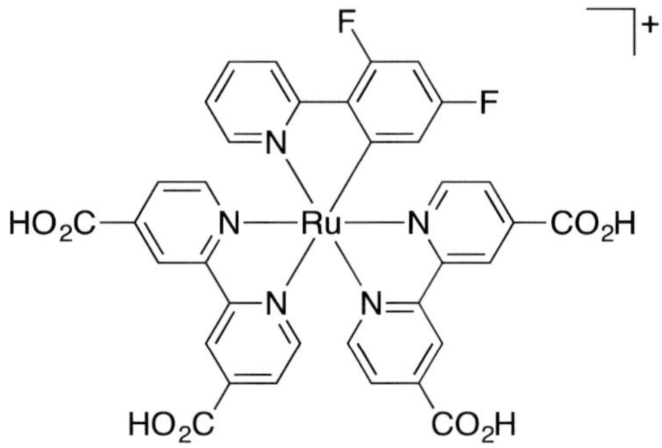

YE05

Berlinguette and co-workers systematically investigated the structure variation of cyclometalated complexes (Alpeeva et al., 2003; Bomben et al., 2010). Among them, TPA-decorated (TPA = triphenylamine) cyclometalated complexes were applied to DSSCs (Robson et al., 2011). The TPA unit was introduced for optical as well as electrochemical reasons. Optically TPA enhances the absorption coefficient. Electrochemically, TPA, as an electron-donating group, is redox active, i.e., oxidizable at a mildly positive potential. It can then be anticipated that charge transfer occurs from the TPA unit to the oxidized ruthenium complex part generated after injection of electron into the TiO_2, if the redox potentials of respective parts are aligned appropriately. The vectorial electron transfer scheme is very attractive to suppress the reverse unwanted electron flows, enhancing the energy conversion efficiency. This ordering of energy levels can be achieved by using N^N^C type cyclometalating ligand rather than N^C^N type ligand. The latter is so effective in raising the t_{2g} level of ruthenium that the Ru(III)/Ru(II) potential is shifted very close to or even to less positive than that of TPA oxidation. A moderately electron-donating N^N^C type ligand would be suitable to position the Ru(III)/Ru(II) potential somewhat more positive than that of TPA oxidation so that the vectorial electron flow from TPA to ruthenium to TiO_2 would be expected. Indeed, the highest efficiency DSSC (8.02%) among these series of complexes was realized by one of the N^N^C type ligand with R = Me.

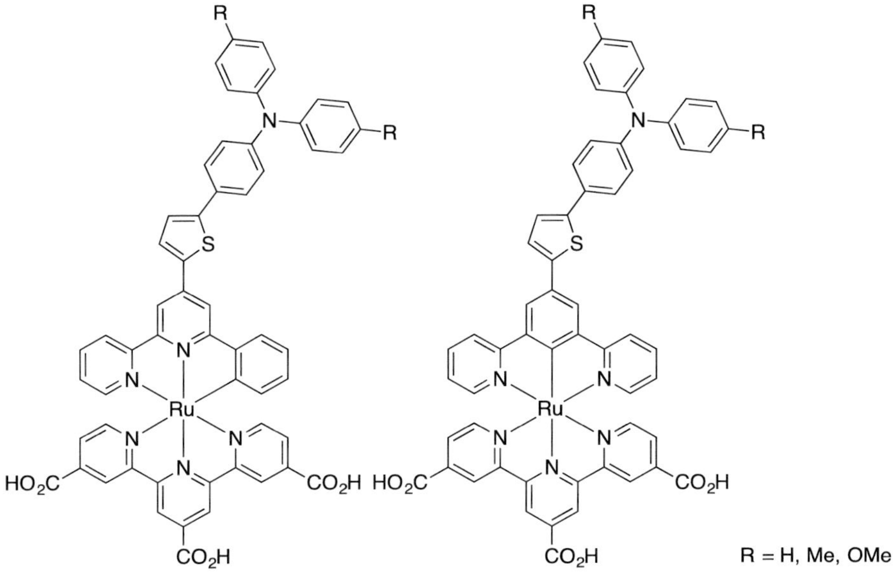

[Ru(tctpy)(pbpy-TPA)]$^{+}$ [Ru(tctpy)(dpb-TPA)]$^{+}$

Instead of hexagonal pyridine-based ligands, pentagonal azole-based ligands are potential candidates for ruthenium complexes for use as a DSSC dye (Chi et al., 2015). Chi and co-workers investigated dyes TF-1 to TF-4 as a dye for DSSCs (Chou et al., 2011). The oxidation potentials corresponding to Ru(III)/Ru(II) for these complexes are in a range of 0.94–0.97 V vs. NHE, slightly more positive than that of N749, which is 0.88 V vs. NHE. This is a result of various elementary contributions of the molecular design. If we start from a tpy complex, the HOMO and LUMO energies go up on changing from pyridine to pyrazole. Deprotonation further raises the orbital energies. However, the introduction of electronegative CF_3 lower the orbital energies. Finally, various R groups were compared but these groups seemed to have only small influences on the π-acceptor characteristics of the bis(pyrazolyl)pyridine ligand. These dyes recorded the power-conversion efficiency as high as 9.1–10.7% under the standard AM1.5 light irradiation.

Other series of complexes having pyrazolylpyridine ligand(s), PRT1–PRT4 (Chen et al., 2009) and TFRS1–TFRS3 (Wu et al., 2010) also exhibit good efficiencies that are comparable or even exceeding the efficiencies of N749 or N719. Similar complexes with triazolylpyridine ligands, TFRS21–TFRS24, also recorded comparable efficiencies (Wang et al., 2013).

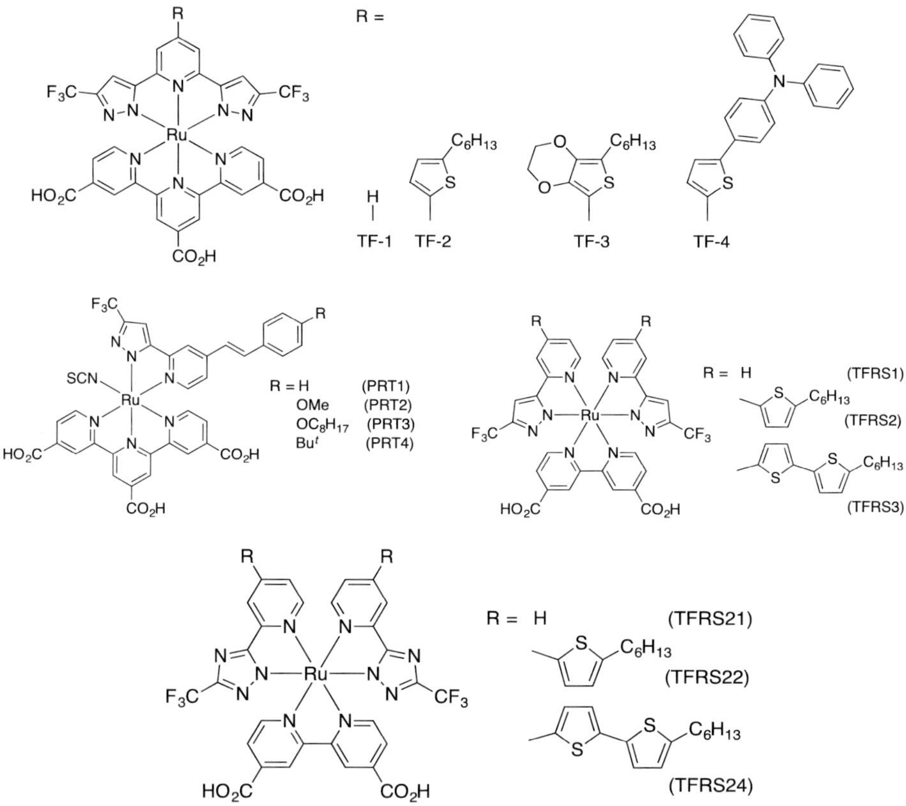

They also combined a pyrazole unit and cyclometalating unit in the same ligand (Hsu et al., 2012). The Ru(III)/Ru(II) potential of TF-21 is +0.84 vs. NHE, which is more negative than that of TF-1 by 0.10 V. Thus replacing the pyrazole group by cyclometalating phenyl group, the t_{2g} level is raised in energy by 0.10 eV. Replacing the pyrazole unit by the 1,2,4-triazolate (TF-23) unit, the Ru(III)/Ru(II) is shifted positively indicating the stabilization of the t_{2g} orbitals by 0.08 eV with the introduction of the electronegative nitrogen atom in the ligand. The Ru(III)/Ru(II) potentials for TF-22 to TF-24 are similar within a range from +0.92 to +0.96 V vs. NHE. Within the series of TF-21 to TF-24, the light energy conversion efficiency was as high as 8.6–9.0% except TF-21 (4.8%). The inferior performance of TF-21 in DSSCs was attributed to smaller driving force for dye regeneration (t_{2g} too high) compared to the other dyes.

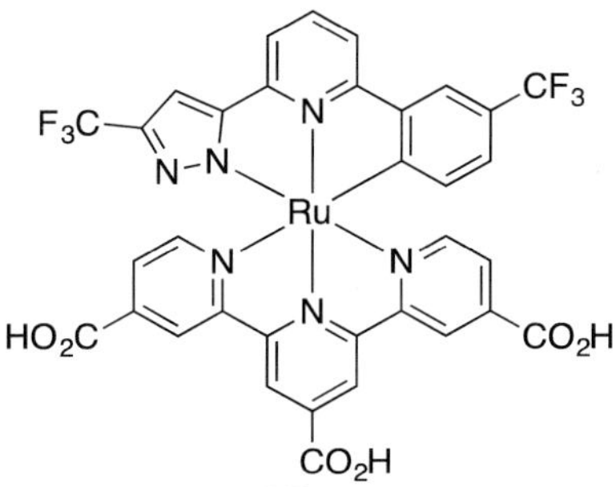

TF-21

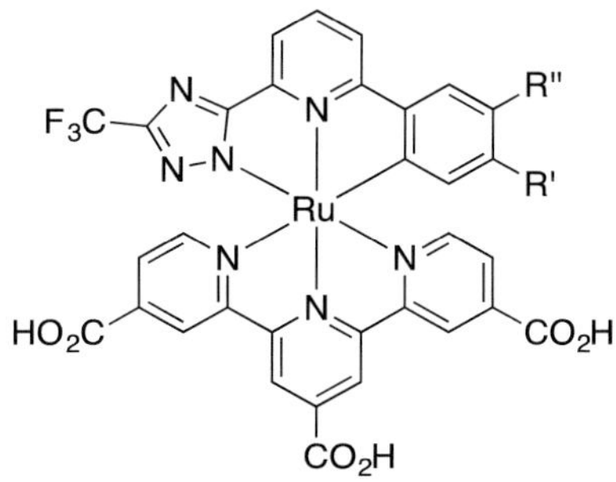

TF-22: R' = CF_3, R'' = H
TF-23: R' = H, R'' = CF_3
TF-24: R' = F, R'' = CF_3

We have prepared a ruthenium complex with a bis(tetrazolate)pyridine ligand (BTP dye) (Wu et al., 2016b). We also prepared a similar complex but in which the tetrazole units are methylated (BMTP dye). In BTP dye, each tetrazole unit is deprotonated and coordinates to ruthenium ion as an anionic ligand. The overall charge of BTP dye is neutral. On the other hand, the complex BMTP dye has a +2 overall charge because dimethylated bis(tetrazolyl)pyridine coordinates to the ruthenium ion as a neutral ligand. Surprisingly, the difference between Ru(III)/Ru(II) potentials is only 0.07 V for BTP and BMTP dyes with potentials of +1.29 and +1.23 V vs. NHE, respectively. The difference is unexpectedly small, considering the difference in charge between the two ligands involved, These and other data for the complexes indicated that the HOMO as well as LUMO levels are similar for these two complexes. Despite energetic similarity, the performances of these complexes as a dye for DSSCs was completely different. The light-to-electricity energy conversion efficiencies for BTP and BMTP dyes were 6.1% and 1.5%, respectively (N719: 7.6% under the same conditions). We surmise that difference in the spatial distribution of HOMO and

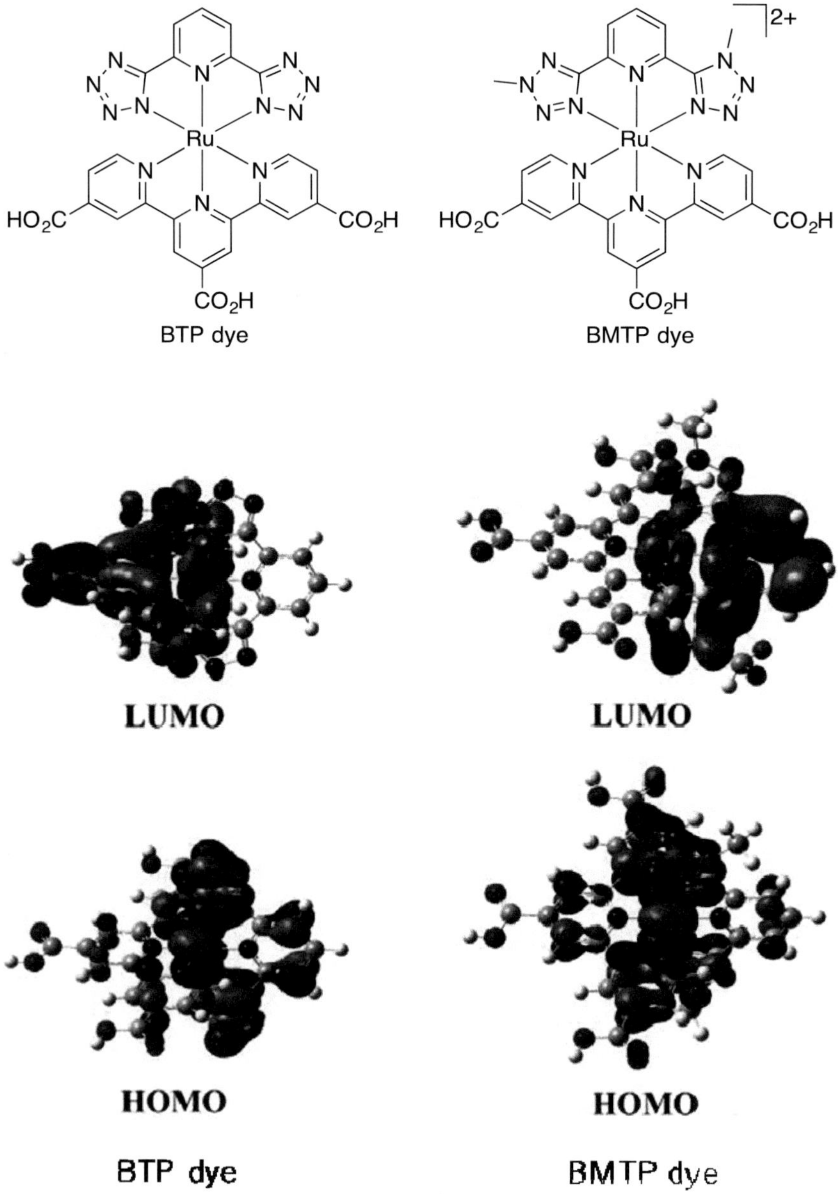

Figure 6.20 HOMO and LUMO of BTP and BMTP dyes as calculated with B3LYP/LanL2DZ. The ligand on the left side is tctpy while that on the right side is the bis(tetrazolyl)pyridine (BTP dye) or dimethylated bis(tetrazolyl)pyridine ligands (BMTP dye). Adopted from *J. Power Sources* **2016**, 307, 416 with permission from Elsevier.

LUMO is the major factor that led to the different performance between these complexes. As shown in Fig. 6.20, the HOMO is distributed over the ruthenium metal and bis(tetrazolyl) pyridine ligand in both complexes. The LUMO is over the tricarboxyterpyridine ligand and the ruthenium metal in the case of BTP dye. This is favorable for electron injection because the complex adsorbs onto the surface of the TiO_2 nanoparticles on the photoelectrode through the carboxyl groups of the anchoring ligand. The desirable vectorial electron transfer and injection is illustrated in Fig. 6.21. However, for BMTP dye, the LUMO is localized over the dimethylated bis(tetrazolyl)pyridine ligand, which is unfavorable for electron injection because the dimethylated bis(tetrazolyl)pyridine ligand is located away from the surface.

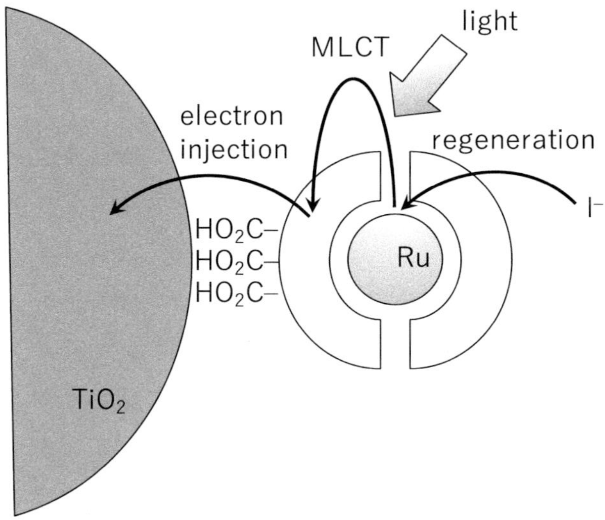

Figure 6.21 Directional electron transfer operative in DSSCs.

We prepared a set of three complexes, PYZ, TRZ, and TEZ dyes (Wu et al., 2016a). The tridentate anchoring ligand of these complexes are common terpyridine tricarboxylic acid. The tridentate ancillary ligand has a pyridine and an azole as well as a cyclometalating moieties. The azole group is deprotonated in the complexes, coordinating to ruthenium as an azolate anion. The ancillary ligand is thus a dianionic and the overall charge of the complex is neutral. The asymmetric nature of the ancillary ligand lifts the degeneracy of transitions and would help broaden the absorption range in the spectrum. On going from TRZ dye to TEZ dye, the number of an electron-withdrawing nitrogen atom increases but at the same time an electron-withdrawing CF_3 group is removed. It is not obvious from inspection of the structure which one is more electron withdrawing. Electrochemical and spectroscopic investigation revealed the HOMO and LUMO levels for these complexes, which are shown in Fig. 6.22. The HOMO and LUMO of TEZ are slightly more stabilized than those of TRZ dye. This observation reveals that the additional nitrogen atom in the azole ring is slightly more electron withdrawing than the CF_3 group. These dyes exhibit similar absorption feature and panchromatic absorption covering the visible and near-IR region. The efficiencies of DSSCs using PYZ, TRZ, and TEZ dyes as the sensitizer dye recorded 3.09%, 4.16%, and 6.44%, respectively, compared to 8.46% by N719 under the same experimental set up. TEZ dye has the smallest driving force for electron injection to the conduction band of TiO_2, but it has the largest driving force for regeneration by the iodide ions in the electrolyte solution. Time-resolved measurement indicated that the rates of charge injection from these complexes are similarly fast in tens of picoseconds, while the regeneration rates are in the order of TEZ dye > TRZ dye > PYZ dye. We concluded from these observations that the regeneration of the dye is the determining factor for the performance of the set of dyes. Finally, the efficiency of TEZ DSSCs remained to be over 90% of the initial value over 1000 h light irradiation. The tridentate ligands are apparently beneficial for the long term stability of the complexes.

The visible absorption spectra for ruthenium complexes are dominated by singlet–singlet MLCT transitions. The initially formed singlet ^{1}MLCT state undergoes intersystem crossing into the triplet ^{3}MLCT state. In this process, a large portion of energy (~5,000 cm^{-1}, 0.6 eV) is lost. If direct singlet-to-triplet excitation is possible, one can broaden the absorption range without compromising the HOMO and LUMO levels. Segawa and co-workers have developed a new ruthenium complex DX1 (Kinoshita et al., 2013). Two chloride ions and a phenyldimethoxyphosphine are coordinated to the ruthenium ion. The ^{31}P NMR analysis indicated that the phosphine atom is positively polarized on the Ru–P bond. The authors attributed the result to the relatively small electronegativity of the P atom compared with that of the ruthenium atom. On the other hand, ^{15}N NMR analysis indicated that the ruthenium atom is positively polarized on the Ru–N bond with NCS^-. Thus the electron density on the ruthenium atom is increased for the ligand set of (Cl^-, Cl^-, $Ph(MeO)_2P$) from the ligand set of (NCS^-, NCS^-, NCS^-) in which N is the coordinating atom, if one compares this complex with the black dye.

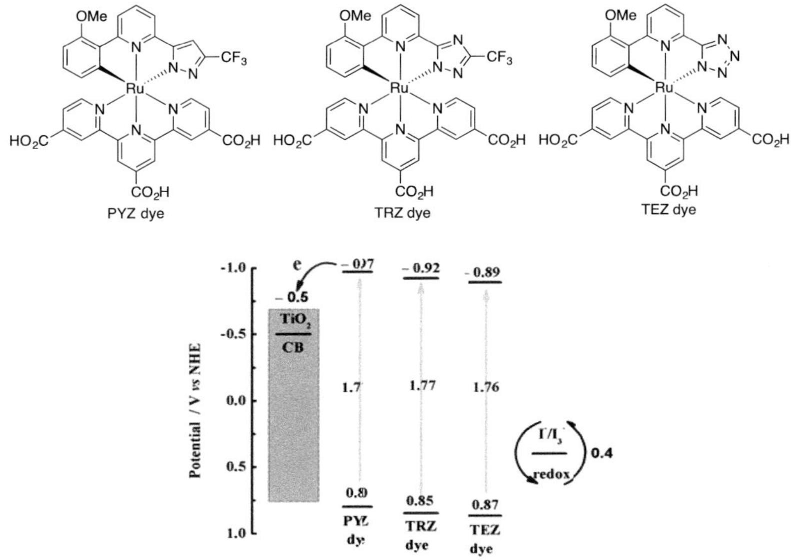

Figure 6.22 HOMO and LUMO levels of PYZ, TRZ, and TEZ dyes. Reprinted from *J. Power Sources* **2016**, 331, 100 with permission from Elsevier.

A spin forbidden singlet–triplet transition can be partially allowed by borrowing intensity from the allowed singlet–singlet transitions. The oscillator strength of a singlet–triplet transition f_T is given by

$$f_T = \left| \frac{\langle \Psi_S | H_{SO} | \Psi_T \rangle}{E_S - E_T} \right|^2 f_S,$$

where Ψ_S and Ψ_T are the singlet and triplet states, E_S and E_T are the energies of the singlet and triplet states, respectively, H_{SO} is the spin-orbit coupling matrix, and f_S is the oscillator strength of the singlet–singlet transition. The heavy ruthenium atom dominates the spin-orbit coupling and therefore the value of $\langle \Psi_S | H_{SO} | \Psi_T \rangle$ is dominated by the terms containing the contribution of the atomic orbitals of ruthenium in the expansion of Ψ_S and Ψ_T. The higher electron density on the ruthenium metal means the more contribution of the ruthenium atomic orbitals to Ψ_S and Ψ_T. As a result, significant absorption appears at the near IR range exceeding 900 nm as shown in Fig. 6.23. In the meantime, the luminescence from the black dye and DX1 overlap nearly completely, meaning that the HOMO and LUMO level are the same for these two complexes. The overall energy conversion efficiency is as high as 10.0% but the most remarkable is that the cell worked under up to 1000 nm light.

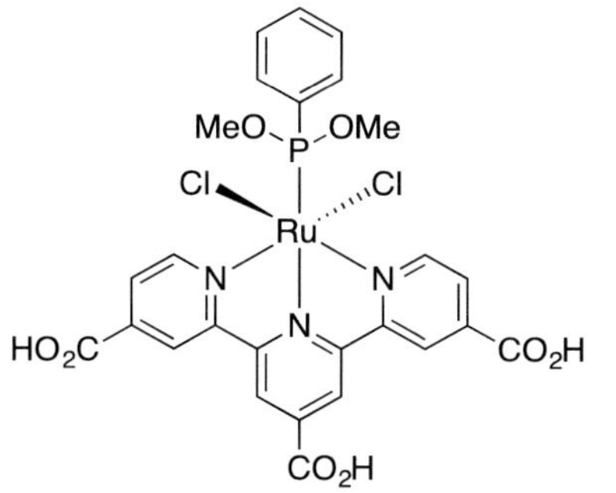

DX1

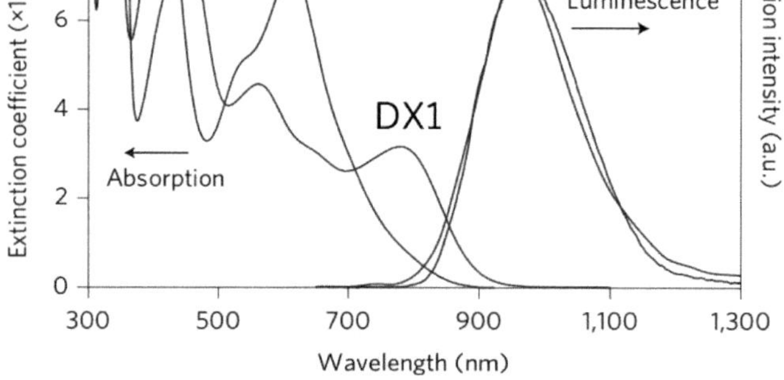

Figure 6.23 Structure of DX1 and absorption and emission spectra for DX1 and black dye (BD). Adopted from *Nat. Photonics* **2013**, 7, 535 with permission from Nature Publishing Group.

6.3.2 Organic Light-Emitting Diodes

The organic light-emitting diode (OLED) is a commercialized lighting device used for displays (Chi et al., 2007; Jou et al., 2015). In 1987, Tang and VanSlyke reported an epoch-making OLED device with an external efficiency of 1%, which became the prototype on which subsequent research and development have been built (Tang et al., 1987). Figure 6.24 shows a basic structure and a mechanism of an OLED. The molecular layers are made on a transparent conductive electrode such as ITO. The basic layer structure is constructed from a hole-transport layer, an emitting layer, and an electron-transport layer, though a number of variants are reported. The molecular layers are terminated with a cathode made of a material with a low work function such as LiF/Al. A voltage is applied between the two terminal electrodes and holes and electrons are injected into the molecular layers. A hole and an electron injected from the opposite sides encounter on a molecule in the emitting layer, producing the excited state of the emitting molecule. The excited molecules decay, a certain fraction of which produces luminescence. The excited state produced may consist of triplet and singlet states in a fraction of 3:1. Therefore, it would be favorable if we can bring out light from both the triplet and singlet states. The radiative transition from a triplet excited state to a singlet ground state is a forbidden process because the electromagnetic waves do not interact with spin. Here comes the idea of exploiting the spin-orbit coupling that mixes a spin angular momentum with an orbital angular momentum. In 1998, Forrest and co-workers reported an efficient OLED device with an external quantum efficiency of 4% utilizing platinum porphyrin as a triplet emitting molecule (Baldo et al., 1998). Heavy elements have a large spin-orbit coupling coefficients ("heavy atom effect"), which mixes the singlet and triplet spin multiplicities through orbital angular momenta, as described in the previous section.

Chi and co-workers designed and prepared ruthenium complexes with pyrazolylisoquinoline (or pyraolylpyridine) and phosphine ligands aiming at application to OLED (Lu et al., 2016; Tung et al., 2006; Tung et al., 2005). The design of these complexes have incorporated considerations on numerous factors. First of all, the complexes are neutral and have no overall charge to be used

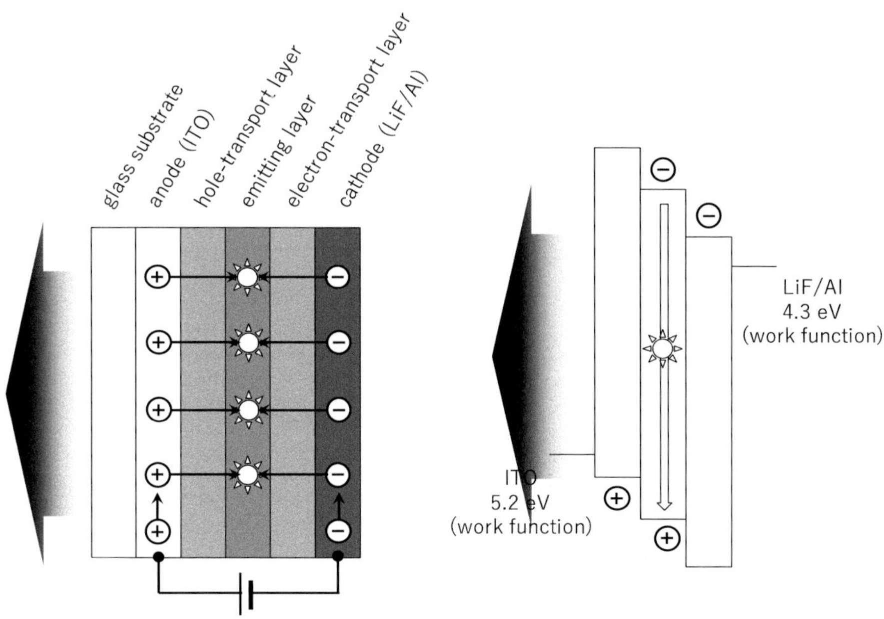

Figure 6.24 Schematic illustration of organic light-emitting diodes. (Left) Layered structure. (Right) Energy diagram.

for vacuum deposition for the fabrication of OLEDs. Two pyrazolate anions cancel out the divalent positive charge on the ruthenium ion. Strongly σ-donating phosphine ligands are used to increase the energy gap of the metal-centered dd transition, so that the nonradiative dd state is not populated in the excited state. Comparing among the *trans*-phosphine complexes with isoquinoline ligands, the metal orbitals are increasingly stabilized in the sequence of $[Ru(ibpz)_2(PPhMe_2)_2]$, $[Ru(ibpz)_2(PPh_2Me)_2]$, and $[Ru(ifpz)_2(PPh_2Me)_2]$ (1st, 2nd, and 5th entries in Fig. 6.25). The π-acidity increases from $PPhMe_2$ to PPh_2Me, resulting in lower-energy metal dπ orbitals (Tolman, 1977). Further introduction of an electron-withdrawing CF_3 group resulted in further lowering of the metal dπ orbitals. Consequently, the emission maximum blue-shifted from 718 to 700 to 636 nm in this sequence. Comparing between the isoquinoline series and pyridine series, the t_{2g} orbitals are slightly more stabilized for pyridine series (2nd and 3rd; 5th and 6th). Comparing between the *trans*-phosphine complex series and the *cis*-phosphine complex series, the t_{2g} orbitals are more stabilized for the *cis*-phosphine complexes (2nd and 4th; 5th and 7th). If we assume that the π-acidity of the bis(diphenylphosphino)ethylene is comparable to that of methyldiphenylphosphine, the *cis*-configuration induces larger stabilization of the t_{2g} orbitals. The first reduction waves for these complexes likely represent the reduction of the pyrazolylisoquinoline ligand. The reduction potentials vary in a narrower range than the oxidation potentials, except for the pyrazolylpyridine set (3rd and 6th), for which a reduction wave was not observed in the solvent window.

Chi and co-workers also looked into the mechanism behind vastly different quantum efficiency of emission for a series of cyclometalated complexes shown in Fig. 6.26 (Li et al., 2006). Among these complexes, biscarbonyl complexes $[Ru(CO)_2(BQ)_2]$ and $[Ru(CO)_2(DBQ)_2]$ are nonemissive and nearly nonemissive, respectively. The TDDFT (B3LYP/LANL2DZ, 6-31G*) calculations indicated that the lowest triplet excited state T_1 is a mixture of ^{3}MLCT and $^3\pi\pi^*$ states in the aromatic ligands. The calculation also indicated that the dd state was higher than the T_1 state as much as by 1.5 eV. On the basis of this large gap between the T_1 state and the dd state, the authors casted a doubt about commonly invoked mechanism of quenching, i.e., the nonradiative decay from the T_1 state into the ground state through the nonemissive dd state. They investigated the contribution of the ^{3}MLCT state and the $^3\pi\pi^*$ state to the T_1 state and found that the contribution of the former is significantly less (10–12%) for these complexes than those for the other complexes with a phosphine ligand displayed in Fig. 6.26 (> 40%). This originates from the fact that CO is much more π-acidic (low π^* orbital) than PR_3, leading to a stronger interaction with the dπ orbitals, producing molecular orbitals in which a larger fraction of ligand orbitals mix into the dπ orbitals. The MLCT transition is a transition in which an electron in these molecular orbitals transfers to the ligand π^* orbitals. Therefore, reduced fraction of the dπ orbital reduces the net contribution of d$\pi \rightarrow \pi^*$ transitions in the T_1 state. As the contribution of metal orbitals in the transition is reduced, the spin–orbit coupling is accordingly reduced, because the origin of the spin–orbit coupling is the heavy atom effect of ruthenium. As the spin–orbit coupling is reduced,

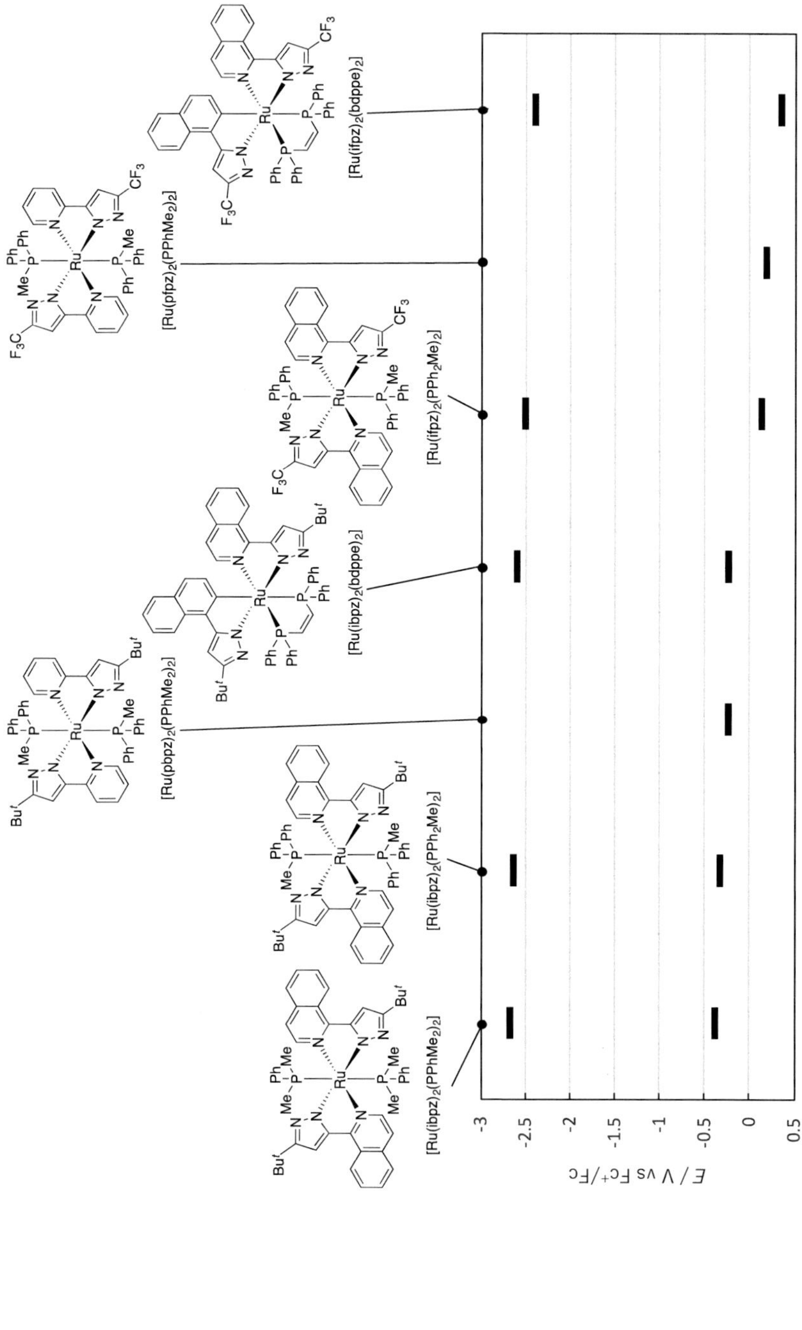

Figure 6.25 First oxidation and reduction potentials for ruthenium complexes including pyrazolate ligands.

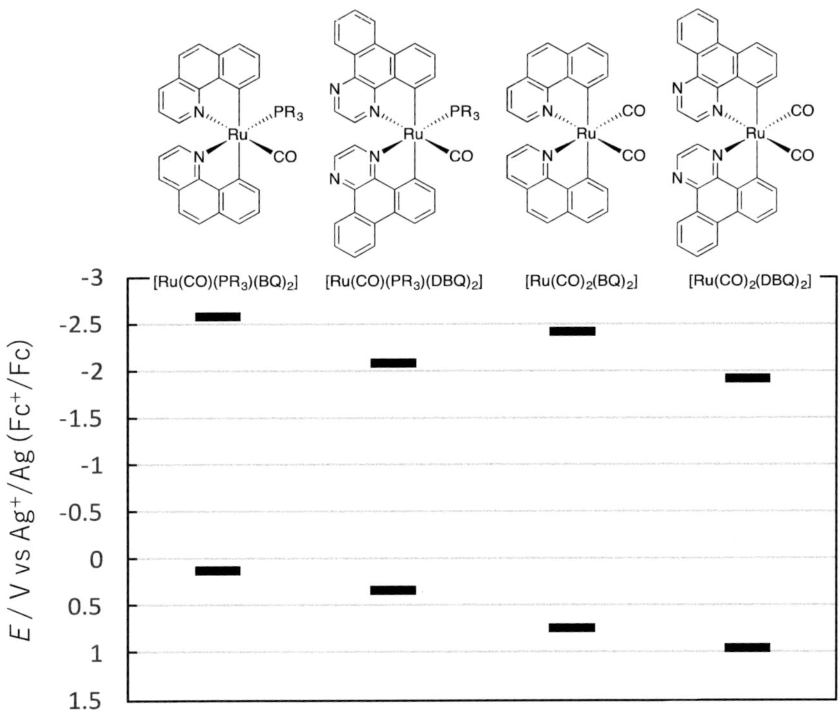

Figure 6.26 First oxidation and reduction potentials. PR_3 = PPh_2Me or $PPhMe_2$.

the mixing of the triplet and singlet states is reduced, which is required for the radiative transition from the triplet excited state to the singlet ground state. Accordingly, the radiative lifetime for $[Ru(CO)_2(DBQ)_2]$ ($\sim 1.9 \times 10^3$ s^{-1}) is an order of magnitude smaller than those for the complexes with one CO and one PR_3 ligands. As for the emissive complexes, the emission quantum yields of $[Ru(CO)(PR_3)(BQ)_2]$ (PR_3 = PPh_2Me: 0.24; $PPhMe_2$: 0.18) are much higher than those of $[Ru(CO)(PR_3)(DBQ)_2]$ (PR_3 = PPh_2Me: 0.008; $PPhMe_2$ 0.004). The difference in the emission quantum yield is due to the difference in the nonradiative decay rates because these complexes have comparable radiative lifetimes. The difference in the nonradiative decay rates is then attributed to the greater degrees of vibrational freedom for the latter larger molecules and the energy-gap law as the latter complexes emit luminescence at 655 nm, which is red-shifted from those of the former complexes by ca. 80 nm.

References

Akasaka T, Mutai T, Otsuki J, and Araki K. (2003). Photo-induced energy transfer and its switching in dyad and triad chromophore systems composed of coumarin, Ru(II) and Os(II) terpyridine-type complexes. *Dalton Trans.*, 1537–1544.

Akasaka T, Otsuki J, and Araki K. (2002). Redox-responsive molecular switches based on azoterpyridine-bridged Ru/Os complexes. *Chem. Eur. J.*, **8**: 130–136.

Albrecht M. (2010). Cyclometalation using d-block transition metals: Fundamental aspects and recent trends. *Chem. Rev.*, **110**: 576–623.

Alpeeva IS, Soukharev VS, Alexandrova L, Shilova NV, Bovin NV, Csöregi E, Ryabov AD, and Sakharov IY. (2003). Cyclometalated ruthenium(II) complexes as efficient redox mediators in peroxidase catalysis. *J. Biol. Inorg. Chem.*, **8**: 683–688.

Baldo MA, O'Brien DF, You Y, Shoustikov A, Sibley S, Thompson ME, and Forrest SR. (1998). Highly efficient phosphorescent emission from organic electroluminescent devices. *Nature*, **395**: 151–154.

Bardwell DA, Jeffery JC, Schatz E, Tilley EEM, and Ward MD. (1995). Conventional and cyclometallated complexes of ruthenium(II) with ambidentate terdentate ligands displaying N_3 or N_2C binding mode. *J. Chem. Soc. Dalton Trans.*, 825–834.

Barigelletti F, Juris A, Balzani V, Belser P, and Von Zelewsky A. (1983). Excited-state properties of complexes of the tris(diimine)ruthenium (2+) ion family. *Inorg. Chem.*, **22**: 3335–3339.

Beley M, Collin JP, and Sauvage JP. (1993). Highly coupled mixed-valence dinuclear ruthenium and osmium complexes with a bis-cyclometalating terpyridine analog as bridging ligand. *Inorg. Chem.*, **32**: 4539–4543.

Bessho T, Yoneda E, Yum J-H, Guglielmi M, Tavernelli I, Imai H, Rothlisberger U, Nazeeruddin MK, and Grätzel M. (2009). New paradigm in molecular engineering of sensitizers for solar cell applications. *J. Am. Chem. Soc.*, **131**: 5930–5934.

Bock CR, Connor JA, Gutierrez AR, Meyer TJ, Whitten DG, Sullivan BP, and Nagle JK. (1979). Estimation of excited-state redox potentials by electron-transfer quenching. Application of electron-transfer theory to excited-state redox processes. *J. Am. Chem. Soc.*, **101**: 4815–4824.

Bock CR, Meyer TJ, and Whitten DG. (1975). Photochemistry of transition metal complexes. Mechanism and efficiency of energy conversion by electron-transfer quenching. *J. Am. Chem. Soc.*, **97**: 2909–2911.

Bomben PG, Koivisto BD, and Berlinguette CP. (2010). Cyclometalated ru complexes of type $[Ru^{II}(N^\wedge N)_2(C^\wedge N)]^z$: Physicochemical response to substituents installed on the anionic ligand. *Inorg. Chem.*, **49**: 4960–4971.

Bomben PG, Robson KCD, Koivisto BD, and Berlinguette CP. (2012). Cyclometalated ruthenium chromophores for the dye-sensitized solar cell. *Coord. Chem. Rev.*, **256**: 1438–1450.

Bomben PG, Robson KCD, Sedach PA, and Berlinguette CP. (2009). On the viability of cyclometalated Ru(II) complexes for light-harvesting applications. *Inorg. Chem.*, **48**: 9631–9643.

Caspar JV, and Meyer TJ. (1983). Photochemistry of tris(2,2′-bipyridine)ruthenium(2+) ion ($Ru(bpy)_3^{2+}$). Solvent effects. *J. Am. Chem. Soc.*, **105**: 5583–5590.

Chen B-S, Chen K, Hong Y-H, Liu W-H, Li T-H, Lai C-H, Chou P-T, Chi Y, and Lee G-H. (2009). Neutral, panchromatic Ru(II) terpyridine sensitizers bearing pyridine pyrazolate chelates with superior dssc performance. *Chem. Commun.*, 5844–5846.

Chi Y, and Chou P-T. (2007). Contemporary progresses on neutral, highly emissive Os(II) and Ru(II) complexes. *Chem. Soc. Rev.*, **36**: 1421–1431.

Chi Y, Wu K-L, and Wei T-C. (2015). Ruthenium and osmium complexes that bear functional azolate chelates for dye-sensitized solar cells. *Chem. Asian J.*, **10**: 1098–1115.

Chou C-C, Wu K-L, Chi Y, Hu W-P, Yu SJ, Lee G-H, Lin C-L, and Chou P-T. (2011). Ruthenium(II) sensitizers with heteroleptic tridentate chelates for dye-sensitized solar cells. *Angew. Chem. Int. Ed.*, **50**: 2054–2058.

Collin J-P, Beley M, Sauvage J-P, and Barigelletti F. (1991). A room temperature luminescent cyclometallated ruthenium(II) complex of 6-phenyl-2,2′-bipyridine. *Inorg. Chim. Acta*, **186**: 91–93.

Constable EC, and Holmes JM. (1986). A cyclometallated analogue of tris(2,2′-bipyridine)ruthenium(II). *J. Organomet. Chem.*, **301**: 203–208.

Constable EC, and Housecroft CE. (1990). The electronic structure of some ruthenium(II) complexes related to $[Ru(bipy)_3]^{2+}$: An investigation of the stepwise replacement of *N,N*-donors by *C,N*-donors. *Polyhedron*, **9**: 1939–1947.

Creutz C, and Sutin N. (1976). Electron-transfer reactions of excited states. Reductive quenching of the tris(2,2′-bipyridine)ruthenium(II) luminescence. *Inorg. Chem.*, **15**: 496–499.

Damrauer NH, Cerullo G, Yeh A, Boussie TR, Shank CV, and McCusker JK. (1997). Femtosecond dynamics of excited-state evolution in $[Ru(bpy)_3]^{2+}$. *Science*, **275**: 54–57.

Diamantis P, Gonthier JF, Tavernelli I, and Rothlisberger U. (2014). Study of the redox properties of singlet and triplet tris(2,2′-bipyridine)ruthenium(II) ($[Ru(bpy)_3]^{2+}$) in aqueous solution by full quantum and mixed quantum/classical molecular dynamics simulations. *J. Phys. Chem. B*, **118**: 3950–3959.

Djukic J-P, Sortais J-B, Barloy L, and Pfeffer M. (2009). Cycloruthenated compounds—synthesis and applications. *Eur. J. Inorg. Chem.*, **2009**: 817–853.

Downard AJ, Steel PJ, and Steenwijk J. (1995). Syntheses of chelating tetrazole-containing ligands and studies of their palladium(II) and ruthenium(II) complexes. *Aust. J. Chem.*, **48**: 1625–1642.

Duati M, Tasca S, Lynch FC, Bohlen H, Vos JG, Stagni S, and Ward MD. (2003). Enhancement of luminescence lifetimes of mononuclear ruthenium(II)–terpyridine complexes by manipulation of the σ-donor strength of ligands. *Inorg. Chem.*, **42**: 8377–8384.

Garino C, Terenzi A, Barone G, and Salassa L. (2016). Teaching inorganic photophysics and photochemistry with three ruthenium(II) polypyridyl complexes: A computer-based excercise. *J. Chem. Educ.*, **93**: 292–298.

Grätzel M. (2005). Solar energy conversion by dye-sensitized photovoltaic cells. *Inorg. Chem.*, **44**: 6841–6851.

Hecker CR, Gushurst AKI, and McMillin DR. (1991). Phenyl substituents and excited-state lifetimes in ruthenium(II) terpyridyls. *Inorg. Chem.*, **30**: 538–541.

Holligan BM, Jeffery JC, Norgett MK, Schatz E, and Ward MD. (1992). The co-ordination chemistry of mixed pyridine–phenol ligands; spectroscopic and redox properties of mononuclear ruthenium complexes with $(pyridine)_{6-x}(phenolate)_x$ donor sets (x = 1 or 2). *J. Chem. Soc. Dalton Trans.*, 3345–3351.

Hsu C-W, Ho S-T, Wu K-L, Chi Y, Liu S-H, and Chou P-T. (2012). Ru(II) sensitizers with a tridentate heterocyclic cyclometalate for dye-sensitized solar cells. *Energy Environ. Sci.*, **5**: 7549–7554.

Jou J-H, Kumar S, Agrawal A, Li T-H, and Sahoo S. (2015). Approaches for fabricating high efficiency organic light emitting diodes. *J. Mater. Chem. C*, **3**: 2974–3002.

Juris A, Balzani V, Barigelletti F, Campagna S, Belser P, and Zelewsky AV. (1988). Ru(II) polypyridine complexes: Photophysics, photochemistry, electrochemistry, and chemiluminescence. *Coord. Chem. Rev.*, **84**: 85–277.

Kalyanasundaram K, and Nazeeruddin MK. (1992). Tuning of the CT excited state and validity of the energy gap law in mixed ligand complexes of Ru(II) containing 4,4′-dicarboxy-2,2′-bipyridine. *Chem. Phys. Lett.*, **193**: 292–297.

Kinoshita T, Dy JT, Uchida S, Kubo T, and Segawa H. (2013). Wideband dye-sensitized solar cells employing a phosphine-coordinated ruthenium sensitizer. *Nat Photon*, **7**: 535–539.

Krause RA, and Krause K. (1980). Chemistry of bipyridyl-like ligands. Isomeric complexes of ruthenium(II) with 2-(phenylazo)pyridine. *Inorg. Chem.*, **19**: 2600–2603.

Krause RA, and Krause K. (1982). Chemistry of bipyridyllike ligands. 2. Mixed complexes of ruthenium(II) with 2-(phenylazo)pyridine: A new π-bonding probe. *Inorg. Chem.*, **21**: 1714–1720.

Le Lagadec R, Alexandrova L, Estevez H, Pfeffer M, Laurinavičius V, Razumiene J, and Ryabov AD. (2006). Bis-ruthena(III)cycles $[Ru(C^\frown N)_2(N^\frown N)]PF_6$ as low-potential mediators for PQQ alcohol dehydrogenase ($C^\frown N$ = 2-phenylpyridinato or 4-(2-tolyl)pyridinato, $N^\frown N$ = bpy or phen). *Eur. J. Inorg. Chem.*, **2006**: 2735–2738.

Lever ABP. (1990). Electrochemical parametrization of metal complex redox potentials, using the ruthenium(III)/ruthenium(II) couple to generate a ligand electrochemical series. *Inorg. Chem.*, **29**: 1271–1285.

Li EY, Cheng Y-M, Hsu C-C, Chou P-T, Lee G-H, Lin IH, Chi Y, and Liu C-S. (2006). Neutral Ru^{II}-based emitting materials: A prototypical study on factors governing radiationless transition in phosphorescent metal complexes. *Inorg. Chem.*, **45**: 8041–8051.

Lu C-W, Wang Y, and Chi Y. (2016). Metal complexes with azolate-functionalized multidentate ligands: Tactical designs and optoelectronic applications. *Chem. Eur. J.*, DOI: 10.1002/chem.201601216.

Meyer TJ. (1986). Photochemistry of metal cordination complexes: Metal to ligand charge transfer excited states. *Pure Appl. Chem.*, **58**: 1193–1206.

Nazeeruddin MK, Kay A, Rodicio I, Humphry-Baker R, Müller E, Liska P, Vlachopoulos N, and Grätzel M. (1993). Conversion of light to electricity by *cis*-X_2bis(2,2′-bipyridyl-4,4′-dicarboxylate)ruthenium(II) charge-transfer sensitizers (X = $C1^-$, Br^-, I^-, CN^-, and SCN^-) on nanocrystalline TiO_2 electrodes. *J. Am. Chem. Soc.*, **115**: 6382–6390.

Nazeeruddin MK, Péchy P, Renouard T, Zakeeruddin SM, Humphry-Baker R, Comte P, Liska P, Cevey L, Costa E, Shklover V, Spiccia L, Deacon GB, Bignozzi CA, and Grätzel M. (2001). Engineering of efficient panchromatic sensitizers for nanocrystalline TiO_2-based solar cells. *J. Am. Chem. Soc.*, **123**: 1613–1624.

Nguyen PT, Lam BXT, Andersen AR, Hansen PE, and Lund T. (2011). Photovoltaic performance and characteristics of dye-sensitized solar cells prepared with the N719 thermal degradation products [Ru(LH)$_2$(NCS)(4-*tert*-butylpyridine)][N(Bu)$_4$] and [Ru(LH)$_2$(NCS)(1-methylbenzimidazole)][N(Bu)$_4$]. *Eur. J. Inorg. Chem.*, 2533–2539.

O'Regan B, and Grätzel M. (1991). A low-cost, high-efficiency solar cell based on dye-sensitized colloidal TiO_2 films. *Nature*, **353**: 737–740.

Otsuki J, Akasaka T, and Araki K. (2008a). Molecular switches for electron and energy transfer processes based on metal complexes. *Coord. Chem. Rev.*, **252**: 32–56.

Otsuki J, Imai A, Sato K, Li D-M, Hosoda M, Owa M, Akasaka T, Yoshikawa I, Araki K, Suenobu T, and Fukuzumi S. (2008b). Switchable antenna: A star-shaped ruthenium/osmium tetranuclear complex with azobis(bipyridine) bridging ligands. *Chem. Eur. J.*, **14**: 2709–2718.

Otsuki J, Li D-M, Sato K, Nakagome A, Takido T, Yoshikawa I, Akasaka T, and Araki K. (2003). Tetranuclear Ru complex linked via redox-active azobis(bipyridine) ligands as a redox-responsive photoswitch. *Bull. Chem. Soc. Jpn.*, **76**: 1185–1189.

Otsuki J, Sato K, Tsujino M, Okuda N, Araki K, and Seno M. (1996). Ruthenium complexes containing an azobipyridine ligand as redox-responsive molecular switches. *Chem. Lett.*, 847–848.

Otsuki J, Takamori Y, Sugawa K, Islam A, Ogawa K, Yamano A, Yoshikawa I, and Araki K. (2014). Heteroleptic ruthenium complexes with 6-(*ortho*-substituted phenyl)-2,2′-bipyridine derivatives. *J. Organomet. Chem.*, **749**: 312–319.

Otsuki J, Tsujino M, Iizaki T, Araki K, Seno M, Takatera K, and Watanabe T. (1997). Redox-responsive molecular switch for intramolecular energy transfer. *J. Am. Chem. Soc.*, **119**: 7895–7896.

Prier CK, Rankic DA, and MacMillan DWC. (2013). Visible light photoredox catalysis with transition metal complexes: Applications in organic synthesis. *Chem. Rev.*, **113**: 5322–5363.

Redmond C, and Fitzmaurice D. (1993). Spectroscopic determination of flatband potentials for polycrystalline TiO_2 electrodes in nonaqueous solvents. *J. Phys. Chem.*, **97**: 1426–1430.

Reveco P, Cherry WR, Medley J, Garber A, Gale RJ, and Selbin J. (1986). Cyclometalated complexes of ruthenium. 3. Spectral, electrochemical and two-dimensional proton NMR of $[Ru(bpy)_2(cyclometalating\ ligand)]^+$. *Inorg. Chem.*, **25**: 1842–1845.

Robson KCD, Bomben PG, and Berlinguette CP. (2012). Cycloruthenated sensitizers: Improving the dye-sensitized solar cell with classical inorganic chemistry principles. *Dalton Trans.*, **41**: 7814–7829.

Robson KCD, Koivisto BD, Yella A, Sporinova B, Nazeeruddin MK, Baumgartner T, Grätzel M, and Berlinguette CP. (2011). Design and development of functionalized cyclometalated ruthenium chromophores for light-harvesting applications. *Inorg. Chem.*, **50**: 5494–5508.

Saavedra-Díaz O, Cerón-Camacho R, Hernández S, Ryabov AD, and Le Lagadec R. (2008). Denial of tris(*C*,N-cyclometalated) ruthenacycle: Nine-membered η^6-*N,N*-trans or η^2-*N,N-cis* Ru^{II} chelates of 2,2′-bis(2-pyridinyl)-1,1′-biphenyl. *Eur. J. Inorg. Chem.*, 4866–4869.

Sauvage JP, Collin JP, Chambron JC, Guillerez S, Coudret C, Balzani V, Barigelletti F, Cola LD, and Flamigni L. (1994). Ruthenium(II) and osmium(II) bis(terpyridine) complexes in covalently-linked multicomponent systems: Synthesis, electrochemical behavior, absorption spectra, and photochemical and photophysical properties. *Chem. Rev.*, **94**: 993–1019.

Sun Q, Mosquera-Vazquez S, Suffren Y, Hankache J, Amstutz N, Lawson Daku LM, Vauthey E, and Hauser A. (2015). On the role of ligand-field states for the photophysical properties of ruthenium(II) polypyridyl complexes. *Coord. Chem. Rev.*, **282–283**: 87–99.

Tang CW, and VanSlyke SA. (1987). Organic electroluminescent diodes. *Appl. Phys. Lett.*, **51**: 913–915.

Thompson DW, Ito A, and Meyer TJ. (2013). $[Ru(bpy)_3]^{2+*}$ and other remarkable metal-to-ligand charge transfer (MLCT) excited states. *Pure Appl. Chem.*, **85**: 1257–1305.

Tolman CA. (1977). Steric effects of phosphorus ligands in organometallic chemistry and homogeneous catalysis. *Chem. Rev.*, **77**: 313–348.

Tung YL, Chen LS, Chi Y, Chou PT, Cheng YM, Li EY, Lee GH, Shu CF, Wu FI, and Carty AJ. (2006). Orange and red organic light-emitting devices employing neutral Ru(II) emitters: Rational design and prospects for color tuning. *Adv. Funct. Mater.*, **16**: 1615–1626.

Tung YL, Lee SW, Chi Y, Chen LS, Shu CF, Wu FI, Carty AJ, Chou PT, Peng SM, and Lee GH. (2005). Organic light-emitting diodes based on charge-neutral Ru^{II} phosphorescent emitters. *Adv. Mater.*, **17**: 1059–1064.

Tuyet Nguyen P, Degn R, Thai Nguyen H, and Lund T. (2009). Thiocyanate ligand substitution kinetics of the solar cell dye Z-907 by 3-methoxypropionitrile and 4-*tert*-butylpyridine at elevated temperatures. *Sol. Energy Mater. Sol. Cells*, **93**: 1939–1945.

Wadman SH, Kroon JM, Bakker K, Havenith RWA, van Klink GPM, and van Koten G. (2010). Cyclometalated organoruthenium complexes for application in dye-sensitized solar cells. *Organometallics*, **29**: 1569–1579.

Wadman SH, Kroon JM, Bakker K, Lutz M, Spek AL, van Klinka GPM, and van Koten G. (2007). Cyclometalated ruthenium complexes for sensitizing nanocrystalline TiO_2 solar cells. *Chem. Commun.*, 1907–1909.

Wadman SH, Lutz M, Tooke DM, Spek AL, Hartl F, Havenith RWA, Klink GPMv, and Koten Gv. (2009). Consequences of *N,C,N'*- and *C,N,N'*-coordination modes on electronic and photophysical properties of cyclometalated aryl ruthenium(II) complexes. *Inorg. Chem.*, **48**: 1887–1900.

Wang X-Y, Del Guerzo A, and Schmehl RH. (2004). Photophysical behavior of transition metal complexes having interacting ligand localized and metal-to-ligand charge transfer states. *J. Photochem. Photobiol. C*, **5**: 55–77.

Wang X, and Stanbury DM. (2006). Oxidation of iodide by a series of Fe(III) complexes in acetonitrile. *Inorg. Chem.*, **45**: 3415–3423.

Wang S-W, Wu K-L, Ghadiri E, Lobello MG, Ho S-T, Chi Y, Moser J-E, De Angelis F, Grätzel M, and Nazeeruddin MK. (2013). Engineering of thiocyanate-free Ru(II) sensitizers for high efficiency dye-sensitized solar cells. *Chem. Sci.*, **4**: 2423–2433.

Winkler JR, Netzel TL, Creutz C, and Sutin N. (1987). Direct observation of metal-to-ligand charge-transfer (MLCT) excited states of pentaammineruthenium(II) complexes. *J. Am. Chem. Soc.*, **109**: 2381–2392.

Wu K-L, Hsu H-C, Chen K, Chi Y, Chung M-W, Liu W-H, and Chou P-T. (2010). Development of thiocyanate-free, charge-neutral Ru(II) sensitizers for dye-sensitized solar cells. *Chem. Commun.*, **46**: 5124–5126.

Wu G, Kaneko R, Islam A, Zhang Y, Sugawa K, Han L, Shen Q, Bedja I, Gupta RK, and Otsuki J. (2016a). Thiocyanate-free asymmetric ruthenium(II) dye sensitizers containing azole chromophores with near-IR light-harvesting capacity. *J. Power Sources*, **331**: 100–111.

Wu G, Kaneko R, Zhang Y, Shinozaki Y, Sugawa K, Islam A, Han L, Bedja I, Gupta RK, Shen Q, and Otsuki J. (2016b). Neutral and anionic tetrazole-based ligands in designing novel ruthenium dyes for dye-sensitized solar cells. *J. Power Sources*, **307**: 416–425.

Chapter 7

The Development of Ru(II)–*p*-Cymene Complexes with Pyridine Derivatives as Anti-Cancer Agents

Nevenka Gligorijević,[a] Sandra Aranđelović,[a] Katarina K. Jovanović,[a] Sanja Grgurić-Šipka,[b] and Siniša Radulović[a]

[a]*Institute of Oncology and Radiology of Serbia, Pasterova 14, 11000 Belgrade, Serbia*
[b]*Faculty of Chemistry, University of Belgrade, Studentski trg 12-16, 11000 Belgrade, Serbia*

sinisar@ncrc.ac.rs

The synthetic versatility of the ruthenium(II)–arene complexes has led to the synthesis of a large number of structurally diverse complexes, which include different ligand systems. Some of these compounds inhibit proliferation of cancer cells and tumour growth via interaction with a variety of intracellular and extracellular targets. This chapter describes the development of Ru(II)-*p*-cymene complexes with pyridine derivatives as anti-cancer agents and achievements in investigation of their biological activity in our Laboratory for pharmacology at the Institute of Oncology and Radiology of Serbia. Our studies of

Ruthenium Chemistry
Edited by Ajay Kumar Mishra and Lallan Mishra

ISBN 978-981-4774-39-0 (Hardcover), 978-1-315-11058-5 (eBook)
www.panstanford.com

ruthenium(II) complexes started with a series of ruthenium(II)–arene compounds containing the p-cymene ligand (as arene) and a derivatives of pyridine (dicarboxylic, acetyl, amino or halogenido derivatives), coordinated in a monodentate or bidentate manner. These complexes are synthesized at the Faculty of Chemistry, University of Belgrade in the Laboratory of Professor Ž. Tešić and Professor S. Grgurić-Šipka. Among complexes of this first series, one with the coordinated picolinic acid exhibited enhanced activity on all investigated cell lines in the initial screening. Therefore, our further attention focuses on the investigation of the mechanism of action of this complex and comparison of its structure and activity to four similar complexes from the first series. These results led to the synthesis and characterization of the second series of Ru(II)-*p*-cymene complexes with derivatives of picolinic acid. Preliminary screening of anti-proliferative activity against tumour and normal cell lines identified complex with isoquinoline-3-carboxylic acid as leading ruthenium compound for further more comprehensive research. The extensive analysis of both cellular and molecular response of the HeLa cells to treatment with this complex was performed. Gene expression profiling using Whole Human Genome Agilent array technology was performed and whole transcriptome analysis was applied. The most important results regarding biological activity of these two series of ruthenium complexes, including two more series similar to aforementioned, are summarized here with focus on their anti-cancer activity.

7.1 Ruthenium Complexes as Anti-Cancer Agents

The serendipitous discovery of the anti-cancer properties of the cisplatin, *cis*-diamminedichloridoplatinum(II) (CDDP) by Professor Barnett Rosenberg in the 1960s (Rosenberg, 1978), led to a development of numerous metal-based compounds as potential anti-cancer drugs (Frezza et al., 2010; Hannon, 2007; Lippard, 1994). All drawbacks of the use of cisplatin, severe side effects and resistance, led at first to the development of many of its analogues (Farrell, 1989). However, due to relatively low success rate in discovering new platinum-based complexes, a search for alternative metals and ligands, which could potentially become part of the newly synthesized metal-based complexes, has begun.

The complexes with ligand-exchange kinetics similar to the platinum ones were examined extensively. Whereas some analogues of Pt complexes are too kinetically reactive to be of use as drugs (like direct Ni and Pd analogues), others are often too inert (Ir and Os ammine compounds) (Farrell, 2000). Although ruthenium and rhodium compounds seem like the most appropriate with regard to ligand-exchange kinetics, no direct analogues have reached to the clinic (Farrell, 2000).

It is assumed that the following properties of ruthenium as metal contribute the most to effectiveness of ruthenium complexes: slow ligand exchange kinetics, several physiologically accessible oxidation states, and the ability to mimic iron in binding and transport to key biological targets (Antonarakis et al., 2010; Jakupec et al., 2008; Süss-Fink, 2010).

Until now only three ruthenium complexes NAMI-A (imidazolium *trans*-[tetrachlorido(DMSO)(imidazole)ruthenate (III)]), KP1019 (indazolium *trans*-[tetrachloridobis(1H-indazole) ruthenate(III)]) and its analogue sodium salt NKP-1339 (sodium *trans*-[tetrachloridobis(1H-indazole)ruthenate(III)), have entered clinical trials (Antonarakis et al., 2010) (Fig. 7.1). Although structurally similar, these ruthenium complexes show extremely different anti-tumour behaviours. In preclinical studies, NAMI-A had little effect on primary tumours while inhibiting metastases formation and growth (Gianni Sava et al., 1999; Sava et al., 1992), whereas KP1019 had activity against a wide range of primary tumours (Hartinger et al., 2008; Heffeter et al., 2010).

The numerous research of the mechanism of action of ruthenium complexes showed diversity of their modes of action. The anti-cancer activity of some ruthenium compounds, like KP1019 and NAMI-A (Clarke, 2003; Clarke et al., 1999; Heffeter et al., 2010), but also of some ruthenium(II)–arene complexes (Casini et al., 2008), is not based on direct DNA damage (Jungwirth et al., 2011). Diverse classes of ruthenium complexes interact with collagens, actins, regulatory enzymes within the cell membrane and/or in the cytoplasm and DNA (Levina et al., 2009).

NAMI/N NAMI/NAMI-A 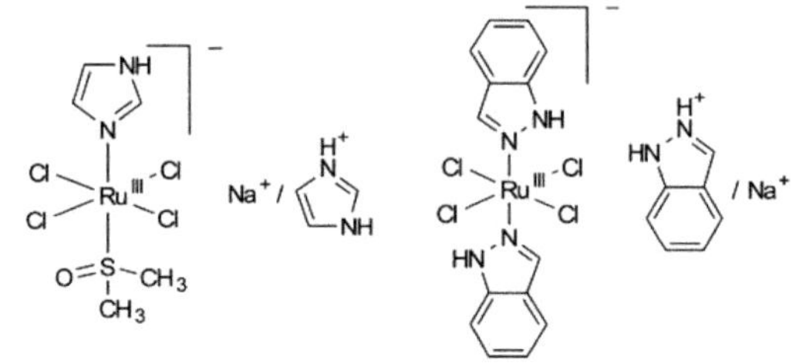 **KP1019/NKP-1339**

Figure 7.1 Chemical structures of NAMI/NAMI-A and KP1019/NKP-1339. Adapted from Trondl et al. (2014).

Previous studies of the mechanism of action of ruthenium complexes highlight very complicated and multi-target mechanism of action of these metal-based complexes (Bergamo et al., 2012). For example, the micro X-ray fluorescence studies (μ-XRF) by Aitken et al. (Aitken et al., 2012) clearly show that KP1019 and NAMI-A have different cellular fates. After the KP1019 enters the cell, ruthenium can be found in both cytosol and the nuclear region, while after the treatment with NAMI-A ruthenium could not be visualized inside the cells, indicating that this compound exerts its activity through a membrane-binding mechanism.

In addition, large-scale gene, protein and metabolite measurement ('omics') approaches are lately introduced in the drug development routine (Bergamo et al., 2015). This bottom up approach may provide 'smarter and faster' profiling (Butcher et al., 2004). To date, only few ruthenium complexes were subjected to 'omics' analyses. Bergamo et al. (Bergamo et al., 2015) determined that the changes induced in gene expression by the ruthenium-based compound NAMI-A, evaluated through whole-transcriptome analysis and RNA-sequencing in the metastatic MDA-MB-231 mammary carcinoma cells, include a set of early-response genes, involved in invasion, metastasis, cytoskeleton remodelling and cell cycle regulation. These results are in the good agreement with NAMI-A anti-metastatic potential documented by a number of literature papers (Bergamo et al., 2002; G. Sava et al., 1999; Sava et al., 2003).

In the work of Grozav and collaborators (Grozav et al., 2015) a genome profiling of the human ovarian and cisplatin-resistant human ovarian cancer cell lines (A2780 and A2780cisR), treated with hydrazinyl-thiazolo arene ruthenium complexes, showed that these compounds lead both ovarian cancer cell types through apoptotic and cell death-related processes via p53 signalling pathway. Also, one of our leading compounds (**RuT$_7$**) (Ivanović et al., 2014; Jovanović, Tanić, et al., 2016) which is discussed further in this chapter, is among a few ruthenium complexes whose mechanism of action is examined from this perspective.

Entirely new classes of ruthenium(II) compounds were developed with many different arene, phosphine, aromatic heterocycles, pyridine, pta (1,3,5-triaza-7-phosphaadamantane), pybox (bis(oxazoline) ligands with pyridine linker), pyrazolone-based β-ketoamine, nitrosyl, Schiff-base, carboxamide, carbothioamide, chalcones, thiourea, thiocarbamate thiosemicarbazone, and hydrazone ligands (Dragutan et al., 2015; Jovanović, Gligorijević, et al., 2016; Nikolić et al., 2016). The class of ruthenium complexes with outstanding anti-cancer activity is represented by ruthenium(II)–arene complexes, of which the most important are those developed by Dyson and co-workers, as well as Sadler and co-workers (Morris, Aird, Murdoch Pdel, et al., 2001; Scolaro et al., 2005). The following characteristics make these complexes particularly interesting for drug design: often-simple preparation of new derivatives, numerous structural modifications and hence, the large number of different biological effects (Biersack, 2016).

Due to the large number and great diversity of ruthenium(II)–arene complexes synthesized and tested, overview of this type of ruthenium complexes will not be the theme of this chapter. A selection of important ruthenium(II)–arene complexes with potent anti-cancer activity classified by their N-, O- and C- ligand systems is recently presented by Bernhard Biersack (Biersack, 2016).

7.1.1 Ruthenium(II)–Arene Complexes with Pyridine Derivatives as Anti-Cancer Agents

This chapter describes the development of ruthenium(II)-*p*-cymene complexes with pyridine derivatives as anti-cancer agents and achievements in investigation of their biological activity in our laboratory during the last decade. All these complexes have "piano-stool" geometry, with the *p*-cymene as arene ligand that forms the seat, and the chelating and chlorido ligand that form the legs of the piano stool (Grgurić-Šipka et al., 2010). These complexes mostly have good aqueous solubility and satisfactory lipophilicity, characteristics that allow them to enter the cell. There is no strict structure–activity relationship for these complexes, but it is known that the arene identity affect the entry into the cell and interactions with the potential targets. The monodentate and bidentate ligands can exert a strong effect on the rate of hydrolysis and activation through hydrolysis of this type of complexes and may be important for their pharmacological properties (Vock et al., 2008).

It is well known that a search for the new platinum anti-cancer compounds (Farrell, 2000) started with replacement of chloride ligands with dicarboxylato ligands, leading to a number of other active compounds. We followed the same pattern, bringing together the ruthenium–arene moiety with the potentially active dicarboxylic, acetyl or amino derivatives of pyridine. Pyridinedicarboxylic acids are known to have various biological properties: immune-suppressive and fibro-suppressive properties (2,4-pyridinedicarboxylic acid) (Dette et al., 1993), potential of protecting certain enzymes from heat inactivation (Hachisuka et al., 1967; Tochikubo et al., 1968), as well as influence on the activity of some metalloenzymes (Griggs et al., 1991; Martin, 1997). The iron(III) 2,6-pyridinedicarboxylates play a role in electron transfer in some models of biological systems (Laine et al., 1995; Mauk et al., 1979), and were recognized as specific molecular tools in DNA cleavage (Groves et al., 1993).

7.1.1.1 Ruthenium(II)–arene complexes with functionalized pyridines (I series)

The study of ruthenium(II) complexes in our laboratory has begun with the series of ruthenium(II)–arene complexes containing the *p*-cymene ligand (as arene) and a pyridine derivatives (dicarboxylic, acetyl or amino derivatives) coordinated in a monodentate or bidentate manner shown in Table 7.1 (Grgurić-Šipka et al., 2010; Ivanović et al., 2011). The aim of our work was to examine the effect of the electronic character of the substituent and its position at the pyridine ring on the biological activity of the complex (Grgurić-Šipka et al., 2010).

Table 7.1 Results of MTT assay presented as IC_{50} (μM) values obtained after 48 and 72 h treatment of HeLa cell line

	HeLa IC_{50} (μM)		
	48 h	**72 h**	**Reference**
Complexes [(η^6-*p*-cymene)Ru(L)Cl$_2$]- withligand listed below			
3-acetylpyridine (**L^1**)	/	>200	(Grgurić-Šipka et al., 2010)
4-acetylpyridine	/	>200	(Grgurić-Šipka et al., 2010)
2-amino-5-chloropyridine (**L^2**)	/	>200	(Grgurić-Šipka et al., 2010)
Complexes [(η^6-*p*-cymene)Ru(L)Cl$_2$]- with ligand listed below			
Isonicotinic acid	/	>200	(Grgurić-Šipka et al., 2010)
Nicotinic acid	/	>200	(Grgurić-Šipka et al., 2010)
Complexes [(η^6-*p*-cymene)Ru(HL)Cl]- with ligand listed below			
2,3-pyridinedicarboxylic acid (**L^3**)	/	>200	(Grgurić-Šipka et al., 2010)
2,4-pyridinedicarboxylicacid (**L^4**)	/	> 200	(Grgurić-Šipka et al., 2010)
2,5-pyridinedicarboxylic acid	/	> 200	(Grgurić-Šipka et al., 2010)
2,6-pyridinedicarboxylic acid	/	> 200	(Grgurić-Šipka et al., 2010)
2-pyridinecarboxylic acid (**L^5**)	**82.0**	**149.4**	(Ivanović et al., 2011)

Note: Values of IC_{50} (μM) represent the average of the three experiments, with each experiment performed in three replicates.

7.1.1.1.1 Analysis of cell growth inhibition

Analysis of cell growth inhibition caused by investigated ruthenium(II) complexes was determined on six tumour cell lines: human cervix cancer (HeLa), human breast cancer cell lines (MDA-MB-361, MDA-MB-453), human melanoma cells (FemX), murine melanoma cells (B16) and human colon cancer cells (LS-174) using 3-(4,5-dymethylthiazol-yl)-2,5-diphenyltetrazolium bromide (MTT) assay (Grgurić-Šipka et al., 2010). Results showed no cytotoxic activity up to 200 μM for 72 h continual agent action, except for complex with 2-pyridinecarboxylic acid (picolinic acid) as ligand (**L**^{5}) (shown as complex **5** in Fig. 7.1), which exhibited IC_{50} 82 μM for 48 h continual treatment on HeLa cells (Table 7.1) (Ivanović et al., 2011).

As mentioned in the introduction, a well-known and interesting feature of some of ruthenium complexes is activity against metastases, while having low activity on primary tumours (Antonarakis et al., 2010). Therefore, we further investigated mechanism of action of our ruthenium(II)–*p*-cymene complexes, and extended our research on examination of modulation of several steps of metastatic progression like adhesion, migration, invasion, proteolytic degradation of extracellular matrix and formation of new blood vessels.

For further examination and comparison of structure and activity we selected four complexes, of which two were with monodentate bonded pyridine ligand (complexes with 3-acetylpyridine (**1**) and 2-amino-5-chloropyridine (**2**) as ligands), and two with bidentate bonded pyridine ligand (complexes with 2,3-pyridinedicarboxylic acid (**3**) and 2,4-pyridinedicarboxylic acid (**4**) as ligands). Aforementioned compounds did not show prominent cytotoxic activity, while complex with 2-pyridinecarboxylic acid (picolinic acid) as ligand (**5**), exhibited moderate cytotoxic activity (Fig. 7.2 and Table 7.2) (Gligorijević et al., 2012).

As results of the MTT assay reveal the absence of direct cytotoxicity of novel complexes on investigated tumour cell lines, additional structure-activity comparison has been performed on endothelial cell lines (model system for in vitro angiogenesis) (Gligorijević et al., 2012; Supino, 1995). Results are shown in Table 7.2 in terms of IC_{50} values, determined from three to four independent experiments.

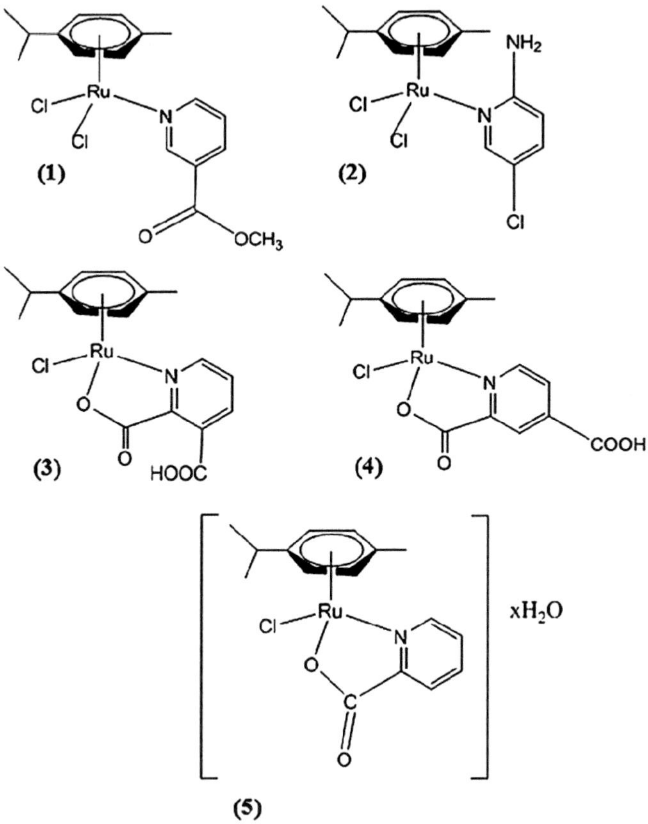

Figure 7.2 Structures of investigated ruthenium–*p*-cymene complexes (Gligorijević et al., 2012).

Table 7.2 Results of MTT assay for complexes **1–5** are presented as IC_{50} values, obtained after 48 or 72 h treatment on HeLa cells and two endothelial cell lines: EA.hy 926 (transformed human umbilical vein endothelial cells) and MS1 (murine endothelial cells)

		IC_{50} (μM)		
Complex	**Incubation time**	**EA.hy 926**	**MS1**	**HeLa**
1	48 h	> 300	> 300	275.7 ± 1.6
	72 h	> 300	270.04 ± 3.7	244.0 ± 1.9
2	48 h	> 300	220.8 ± 16.0	> 300
	72 h	258.5 ± 2.8	218.9 ± 6.2	283.2 ± 3.1

(*Continued*)

Table 7.2 (*Continued*)

Complex	Incubation time	IC$_{50}$ (μM)		
		EA.hy 926	MS1	HeLa
3	48 h	> 300	> 300	> 300
	72 h	259.4 ± 9.8	> 300	> 300
4	48 h	> 300	> 300	> 300
	72 h	> 300	> 300	> 300
5	48 h	**95.5 ± 7.4**	**78.3 ± 4.1**	**82.0 ± 14.2**
	72 h	**88.4 ± 8.1**	**108.5 ± 9.5**	**149.4 ± 12.1**

Note: Adapted from Gligorijević et al. (2012).

Anti-proliferative activity of investigated complexes on endothelial cells were low, indicated by high IC$_{50}$ values (> 200 μM), except for complex **5**. Ruthenium compound **5** exhibited cytotoxic activity on all investigated cell lines after both 48 h and 72 h of continual incubation, with IC$_{50}$ values falling in the 80–150 μM range. No significant cell type selectivity was observed.

As already mentioned the low cytotoxicity in vitro with simultaneous significant anti-metastatic properties is very common feature for some of ruthenium-based anti-cancer agents and thus encouraging for further studies of the biological activity of these complexes (Antonarakis et al., 2010).

7.1.1.1.2 Cell cycle analysis and apoptotic potential

The effect of investigated ruthenium(II)–*p*-cymene complexes (**1**–**5**) on cell cycle progression of HeLa cells was determined by flow cytometry, using staining with propidium iodide (PI) (Gligorijević et al., 2012; Ormerod, 2000). After 24 h treatment only complex **5** induced perturbations of cell cycle, with decrease of percent of cells in G1 and slight arrest in the S phase of cell cycle (Fig. 7.3). Slower progression through replication phase determined by cell cycle analysis indicates its DNA-binding potential. During this short exposure time, visible apoptotic effect was not obtained (evaluated as Sub-G1 fraction). After longer exposure (48 h) of HeLa cells to complex **5**, apoptotic fraction of cells was noted. The induction of apoptotic changes (externalization of phosphatidylserine) was detected by dual staining with fluorescein isothiocyanate labelled (FITC) Ca^{2+}-dependent phospholipid-binding protein with high affinity for phosphatidylserine Annexin-V and propidium iodide and analysis on flow cytometer (van Engeland et al., 1998) (Fig. 7.4). Complex **5** induced significant increase of percent of cells in early apoptosis up to 28% compared to control 11% (FITC+/PI-), and in late apoptosis and necrosis up to 12% compared to control 6% (FITC+/PI+) (Fig. 7.4).

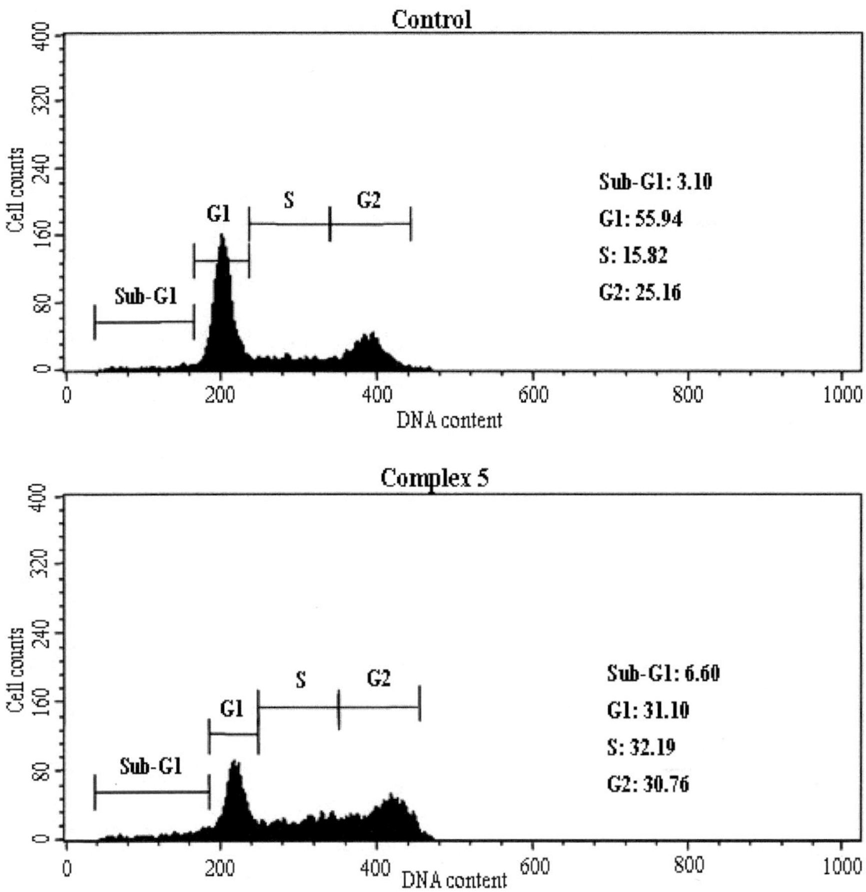

Figure 7.3 Effect of the complex **5** (IC_{50}) on cell cycle progression of HeLa cells following 24 h incubation. Representative histograms are shown. Adapted from Gligorijević et al. (2012).

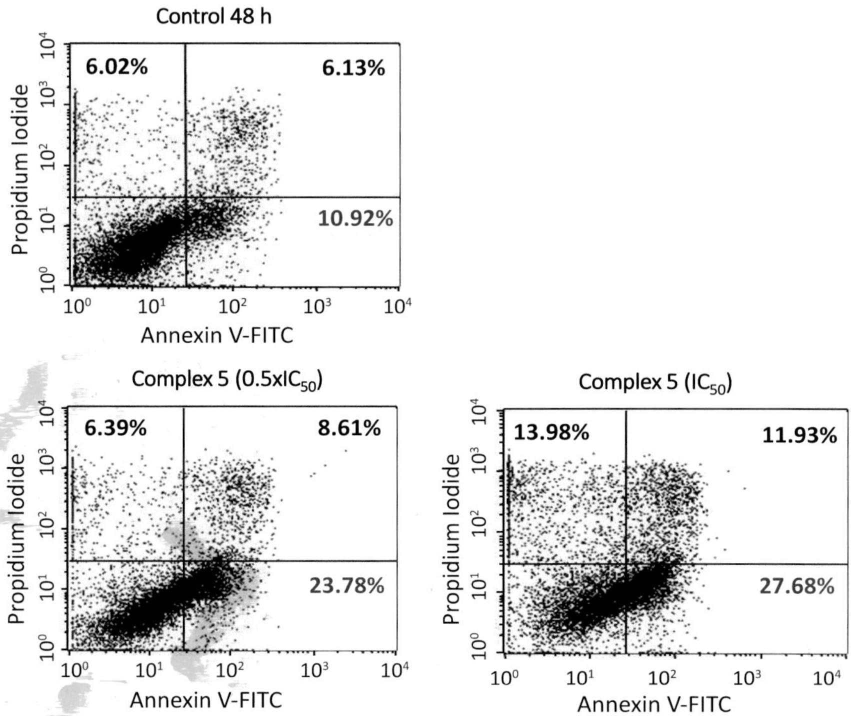

Figure 7.4 Representative dot plot diagrams obtained by flow cytometry of Annexin-V-FITC/PI double-stained HeLa cells untreated (control) or treated with complex **5**, with concentrations corresponding to 0.5 xIC_{50} and IC_{50}. FITC(-)/PI(-)(lower-left quadrant) are intact cells, FITC(+)/PI(-) (lower-right quadrant) are early apoptotic cells, FITC(+)/PI(+) (upper-right quadrant) are late apoptotic or necrotic cells and FITC(-)/PI(+) (upper-left quadrant) are dead cells.

7.1.1.1.3 Intracellular distribution of ruthenium in protein vs. DNA fraction

The ICP-OES analysis after 24 h treatment of HeLa cells with complexes **1**–**5** revealed different levels of intracellular accumulation and different affinities for protein or DNA binding (Fig. 7.5). The level of ruthenium bound to protein fraction was similar among the tested complexes. Absence of complexes **1**, **3** and **4** in DNA fraction suggests that there was no DNA binding, which indicates that differences in structure in the pyridine ligand of tested ruthenium(II) compounds significantly influence

DNA interaction potential. The structural characteristics of complexes **1** and **2**, as well as **3** and **4**, probably obstruct DNA interactions, especially drug intercalation through η^6-bonded arene. Complexes **1** and **2** with monodentately bonded pyridine ligand have potential of rotation of the ligand around the Ru-N bond which may affect the interaction with DNA. In addition, complexes **3** and **4** with N,O-bidentate bonded pyridine ligand have COOH groups on pyridine ligand. It seems that relatively planar surface of N,O-bidentate bonded pyridine ligand of complex **5**, without substituents, enables best potential for interactions with DNA, thus contributing to the observed anti-proliferative activity.

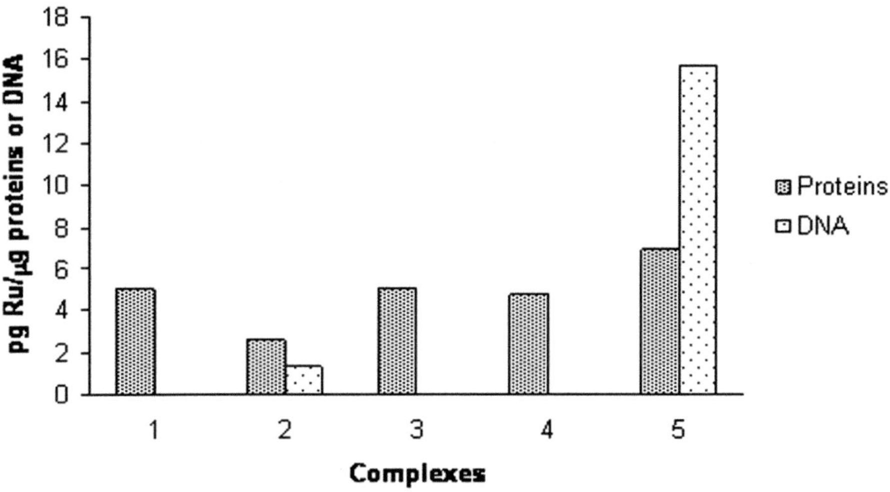

Figure 7.5 Ruthenium content in HeLa cells (DNA vs. protein fraction of cells) after 24 h treatment, measured by ICP-OES. Representative experiment is shown. Adapted from Gligorijević et al. (2012).

7.1.1.1.4 In vitro anti-metastatic potential

Some of the ruthenium-based complexes are reported to exhibit unique and impressive anti-metastatic properties regardless of their cytotoxic effect (Bergamo et al., 2012; Bergamo et al., 2015). NAMI-A shows a selective effect on lung metastases of solid metastasizing tumours, with no concomitant effect on primary tumour growth, in a number of experimental models (Bergamo et al., 2012; Bergamo et al., 2015). Numerous researches show that this anti-metastatic effect is most likely achieved by inhibition of tumour cell detachment, invasion, migration or re-adhesion (Bergamo et al., 2012; Bergamo et al., 2015). Ruthenium–arene complexes bearing the 1,3,5-triaza-7-phosphatricyclo-[3.3.1.1]decane ligand, namely, RAPTA compounds RAPTA-T, in some instances similarly to RAPTA-C, has also shown some activity in vivo in a model of solid metastasizing tumours (Bergamo et al., 2012; Bergamo et al., 2008; Scolaro et al., 2005).

So far, we have investigated potential of our ruthenium(II)-*p*-cymene complexes to inhibit several steps of metastatic progression (Gligorijević et al., 2012), such as: invasion, adhesion, inhibition of matrix metalloproteinases (MMP-2 and MMP-9) activity. Some properties of anti-metastatic potential of investigated complexes are summarized and presented in Fig. 7.6 and Table 7.3.

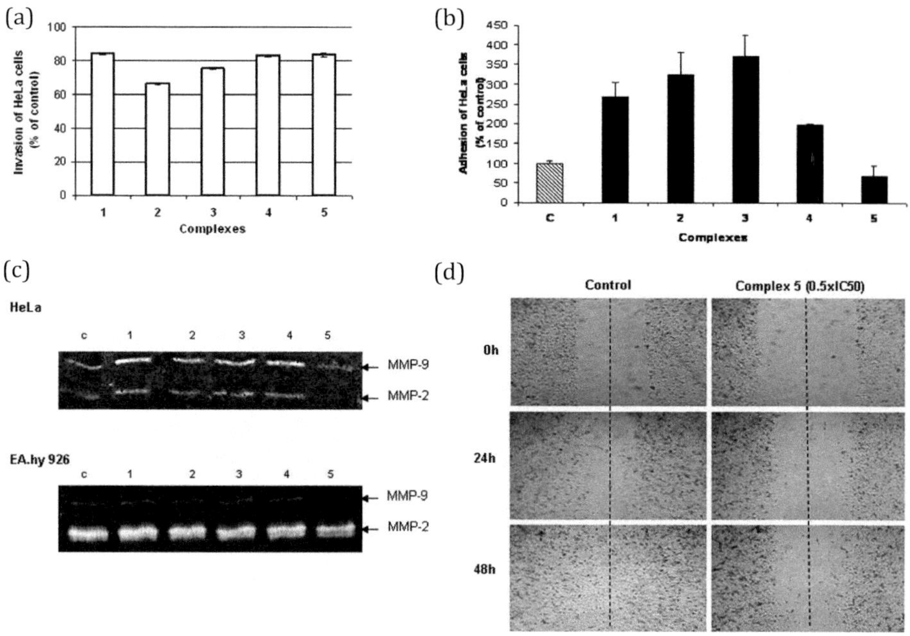

Figure 7.6 Potential of our ruthenium(II)–*p*-cymene complexes to modulate several steps of metastatic progression: (a) invasion; (b) adhesion; (c) inhibition of MMP-2 and MMP-9 activity; (d) inhibition of migration of HeLa cells following treatment with complex **5**. Adapted from Gligorijević et al. (2012).

Table 7.3 Summary of the biological activity of the four ruthenium compounds discussed in this chapter

Complexes	Cytotoxicity ($IC_{50} < 10^{-4}$ M)	Inhibition of invasion	Inhibition of adhesion	Inhibition of MMP-2/-9 activity
1	–	+	+	–
2	–	+	+	–
3	–	+	+	–
4	–	+	+	–
5	+	+	–	+

Note: Experiments performed on HeLa cells (Gligorijević et al., 2012).

Results of the gelatine zymography analysis of inhibitory activity of investigated complexes on MMP-2 and MMP-9 after 24 h treatment of HeLa and EA.hy 926 cells, showed that only complex **5** exhibited inhibitory activity on secreted MMP-2 and MMP-9 in HeLa and EA.hy 926 cells, respectively (Fig. 7.6c). In contrary, all investigated ruthenium complexes exhibited the potential of inhibiting HeLa cells invasion determined using system of BD BioCoat Matrigel invasion chambers (Fig. 7.6a). The complexes that exhibited the greatest inhibitory effect on invasion behaviour of the HeLa cells through Boyden chamber membranes, also contributed the most to the adhesion of HeLa cells (Fig. 7.6b). Additionally, only complex **5** induced inhibition of migration of HeLa cells which was prominent after 48 h incubation (Fig. 7.4d shows results of Scratch assay for complex **5** only).

7.1.1.2 Ruthenium(II)–arene complexes with substituted picolinato ligands ("II series")

Since the presence of picolinate ligand notably enhanced anti-cancer properties of investigated compound (Gligorijević et al., 2012), further efforts were made towards modification of picolinate ruthenium(II)–arene complex (complex **5** discussed above). This resulted in synthesis and characterization of seven new ruthenium(II)–arene complexes of general formula [Ru(η^6-*p*-cymene)(L)Cl], where L are fluoro, chloro, bromo or methyl derivatives of picolinic acid or isoquinoline-3-carboxylic acid (Fig. 7.7) (Ivanović et al., 2014).

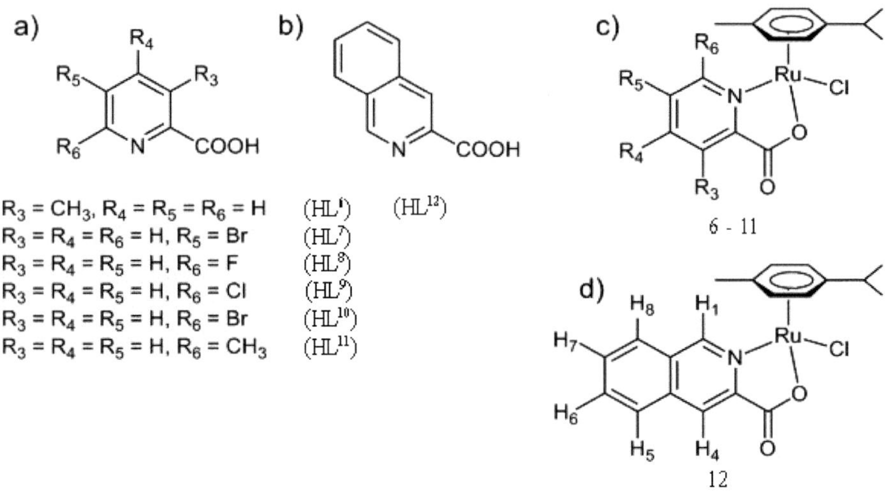

Figure 7.7 Structures of ligands (a and b) and complexes (c and d). Adapted from Ivanović et al. (2014).

7.1.1.2.1 Analysis of cell growth inhibition

The results of preliminary screening of cytotoxic potential showed that complex with isoqinoline-3-carboxylic acid as ligand (complex **12**) exhibited high cytotoxic activity in all investigated cell lines, with much lower IC_{50} values compared to those of other synthesized compounds. Moreover, introduction of isoqinoline-3-carboxylic acid resulted in significantly lower cytotoxic activity in normal human lung fibroblast cells (MRC-5) (Table 7.4). High cytotoxic activity of complex **12** against cisplatin-resistant human lung adenocarcinoma epithelial cells (A549) should be noted (Ivanović et al., 2014).

After [Ru(η^6-*p*-cymene)(isoquinoline-3-carboxylato)Cl] (complex **12**) complex has been identified as a promising drug candidate (Fig. 7.8), we have performed a comprehensive analysis of both cellular and molecular response to complex **12** treatment on HeLa cells (Jovanović, Tanić, et al., 2016).

Table 7.4 Results of MTT assay for complexes **6–12** are presented as IC_{50} values, obtained after 48 h treatment on HeLa, FemX, A549 and MRC-5 cells

	IC_{50} (μM)			
Complex	**HeLa**	**FemX**	**A549**	**MRC-5**
6	119. 9 ± 13.7	36.5 ± 1.3	128.3 ± 12.7	190.1 ± 7.7
7	114.3 ± 16.1	73.8 ± 3.7	141.3 ± 16.1	144.0 ± 17.1
8	> 300	219.5 ± 12.5	> 300	> 300
9	> 300	> 300	> 300	> 300
10	> 300	> 300	> 300	> 300
11	278.4 ± 13.5	168.7 ± 19.9	> 300	> 300
12	**45.4 ± 3.0**	**18.5 ± 1.9**	**25.8 ± 1.3**	**84.2 ± 5.7**

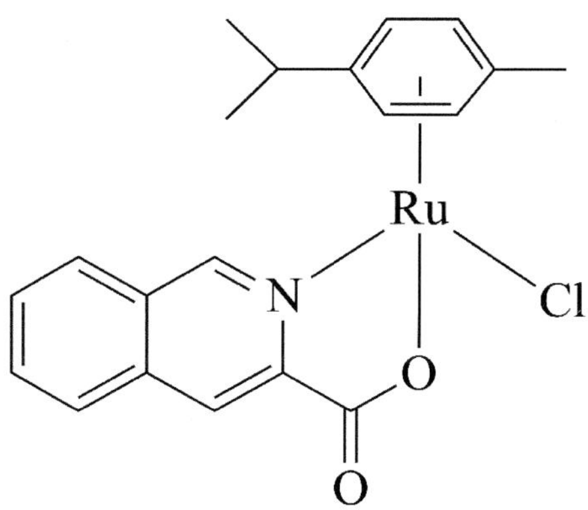

Figure 7.8 Structure of ruthenium(II)–arene complex **12** (Jovanović, Tanić, et al., 2016).

7.1.1.2.2 Cell cycle analysis and apoptotic potential

The effect of investigated complex **12** on cell cycle progression of HeLa cells was determined by flow cytometry after continual treatment for 24 and 48 h, using staining with PI (Jovanović, Tanić, et al., 2016; Ormerod, 2000), and was compared to CDDP. HeLa cells were exposed to 0.5 × IC_{50} and IC_{50} concentrations of

12 (22.7 and 45.4 μM) and CDDP (2.6 and 5.2 μM). Results are presented in Fig. 7.9.

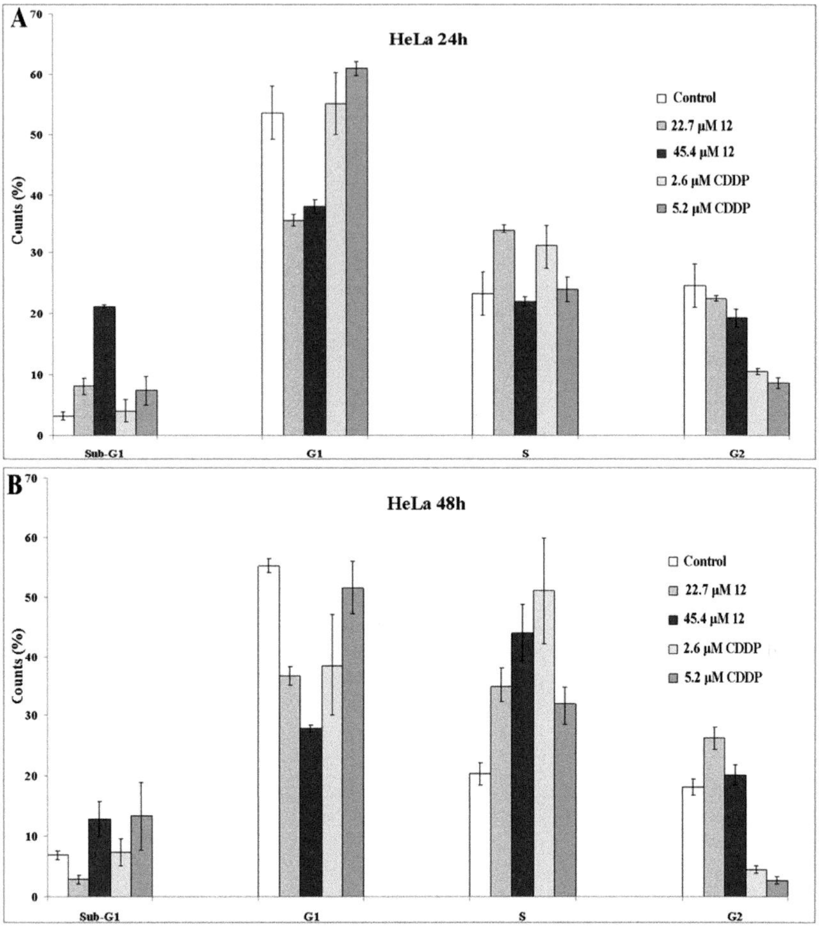

Figure 7.9 Effects of **12** and CDDP on cell cycle distribution. Untreated (control) HeLa cells or HeLa cells treated for 24 h (A) and 48 h (B) with 0.5 × IC_{50} and IC_{50} concentrations of **12** (22.7 and 45.4 μM) and CDDP (2.6 and 5.2 μM). The results are expressed as mean ± standard deviations of three independent experiments (Jovanović, Tanić, et al., 2016).

After CDDP treatment, accumulation of cells in S phase occurs, which is a well-known mechanism of action of cisplatin (Chu, 1994). Similarly, arrest in the S phase of cell cycle induced by complex **12** indicates its possible binding to and/or damage of DNA, leading to consequent replication inhibition of cancer cells. These results are in accordance with our previous study of the biological activity of similar ruthenium complexes of the I series, where the complex with picolinic acid as the ligand (complex **5**) showed its cytotoxic effect through cell cycle arrest in the S phase and significant intracellular distribution in DNA rather than protein fraction of treated cells, indicating DNA as target of its action (Gligorijević et al., 2012). In addition, after 24 h of continual treatment of HeLa cells with **12**, we observed an increase in Sub-G1 phase, which denotes apoptotic and/or necrotic population of cells. From these results, we can conclude that growth inhibition of HeLa cells by **12** involves a profound cell cycle arrest in S phase, indicating its potential of interaction with DNA. These results comply with the results of our previous DNA binding study by UV spectroscopic titration, which demonstrated DNA binding potential of complex **12** (Ivanović et al., 2014).

The potential of **12** and CDDP to induce apoptotic cell death was determined by staining treated HeLa cells with two dyes, Annexin V–FITC and propidium iodide, and analysis on a flow cytometer.

After 24 h treatment with **12**, HeLa cells exposing PS on its surface were monitored. Compared to control, an increase in early apoptotic cell fraction (FITC(+)/PI(−)) was not obtained, neither for **12** nor CDDP. However, there is a prominent increase in late apoptotic and/or already dead cells (FITC(+)/PI(+)), as well as increase in dead cells number (FITC(−)/PI(+)) after **12** treatment (Fig. 7.10). Absence of early apoptotic fraction of cells can be explained by possible fast and immediate action of **12**, where after 24 h of incubation period majority of early apoptotic cells have already proceeded to late apoptotic stage. These results are in correlation with the appearance of sub-G1 fraction of cells in cell cycle phase distribution analyses (Fig. 7.9).

In order to confirm the apoptotic potential of the investigated complex, a morphological analysis by fluorescence microscopy of acridine orange/ethidium bromide-stained HeLa cells exposed for 3–24 h to **12** (0.5 × IC_{50} concentration) was performed.

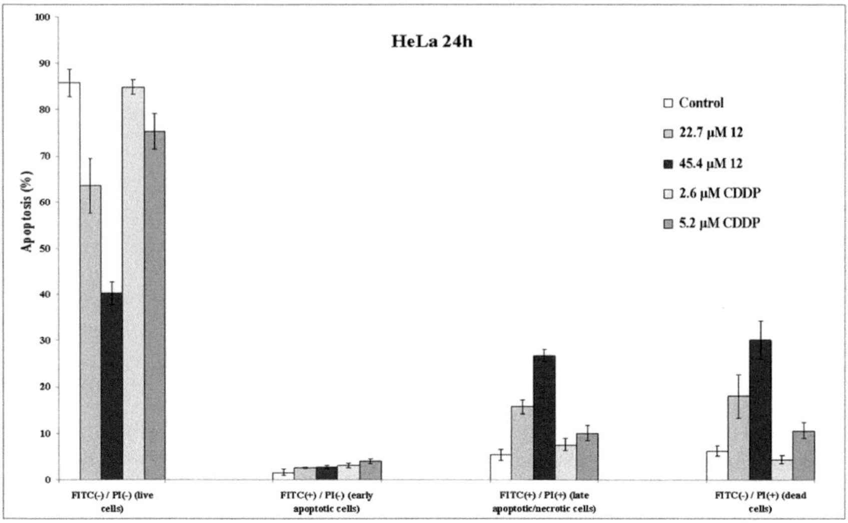

Figure 7.10 Apoptosis induction potential of **12** and CDDP. Cells were treated for 24 h with 0.5 × IC_{50} and IC_{50} concentrations of **12** (22.7 and 45.4 μM) and CDDP (2.6 and 5.2 μM); FITC(−)/PI(−) are live cells, FITC(+)/PI(−) are early apoptotic cells, FITC(+)/PI(+) are late apoptotic or necrotic cells and FITC(−)/PI(+) are dead cells. The results are expressed as mean ± standard deviations of three independent experiments (Jovanović, Tanić, et al., 2016).

Typical morphological features of the apoptotic cells are highly condensed nuclei, cell shrinkage and formation of apoptotic bodies. In the early apoptosis, only acridine orange enters the cell with ethidium bromide excluded and the nucleus stains green. This also applies to living cells. In the late apoptosis along with the loss of membrane integrity, both dyes enter the cell and the nucleus becomes orange-red (Kasibhatla et al., 2006). After 3 h of treatment with **12**, changes in morphological features of HeLa cells are obvious, when compared to control population (Fig. 7.11). Decreased density/cell number, round shape and beginning of nuclear condensation are the most prominent features after 3 h of treatment. Already after 5 h of incubation with **12**, highly condensed nuclei and membrane blebbing are predominant hallmarks. These findings, together with photomicrographs obtained after 24 h of continual treatment are in concordance with the results of Annexin V-FITC assay, and represent confirmation of hypothesized fast **12** action. The continuous presence of cells with apoptotic morphology can be explained with persistent intracellular accumulation of **12** (Fig. 7.12).

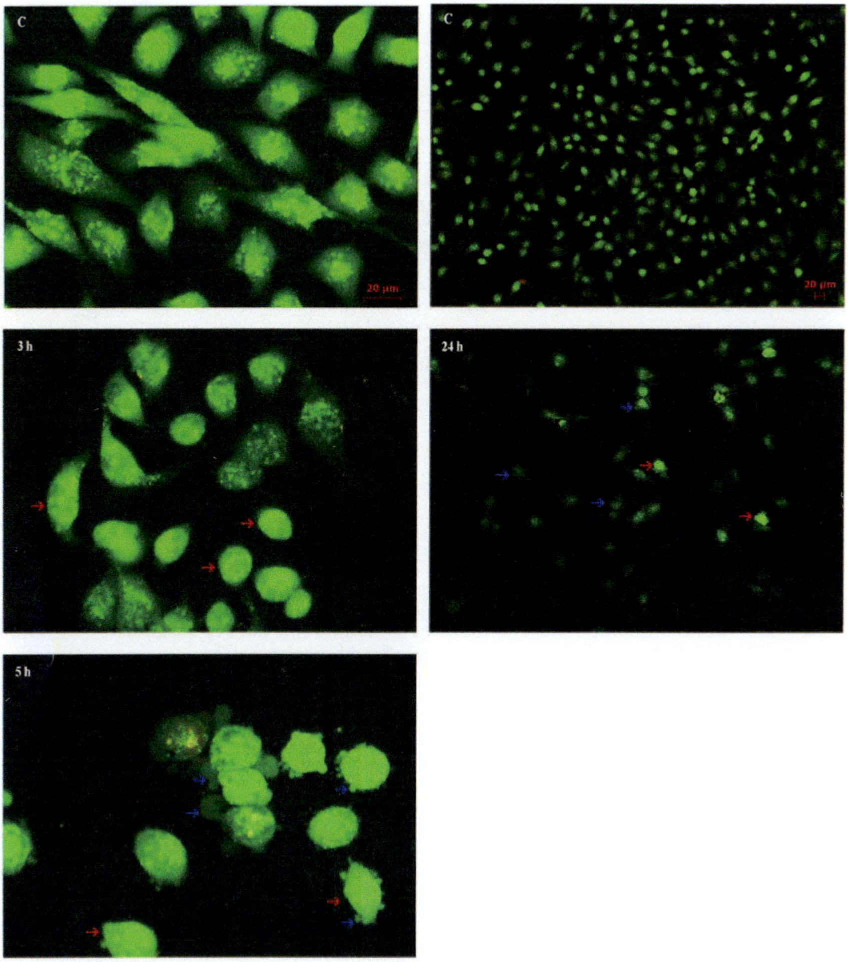

Figure 7.11 Photomicrographs of acridine orange/ethidium bromide-stained control (C) HeLa cells and HeLa cells exposed for 3 and 5 h to complex **12** (left), and photomicrographs of acridine orange/ethidium bromide-stained control (C) HeLa cells and HeLa cells exposed for 24 h to complex **12** (right). Red arrows point to cytoplasmic shrinkage and nuclear condensation; blue arrows denote membrane blebbing structures. Scale bars in the upper images correspond to magnification of photomicrographs below. Applied concentrations of **12** corresponded to 22.7 μM (0.5 × IC_{50}) values determined for 48 h incubation period (Jovanović, Tanić, et al., 2016).

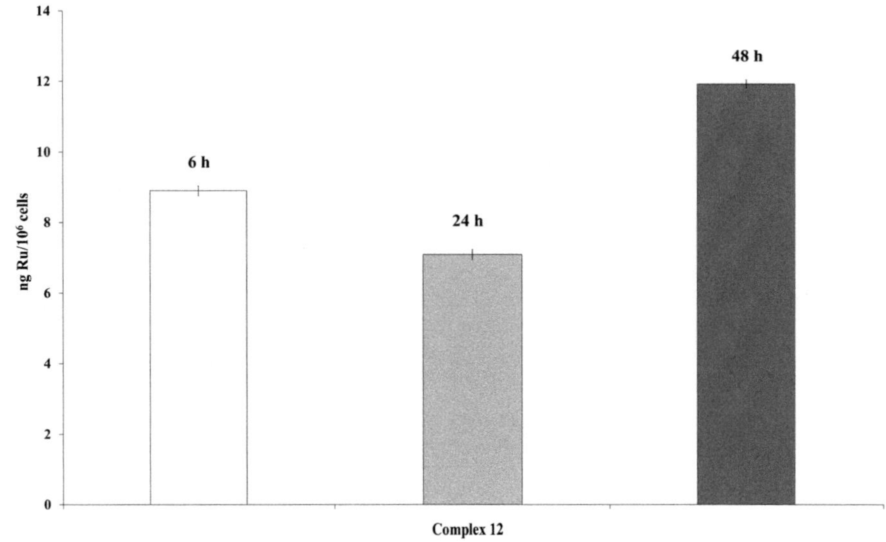

Figure 7.12 Total intracellular accumulation of ruthenium in HeLa cells after 6, 24 and 48 h, measured by ICP-MS. All experiments were performed in triplicate and presented with corresponding standard deviations (SD) (Jovanović, Tanić, et al., 2016).

7.1.1.2.3 Intracellular accumulation of ruthenium

The total intracellular accumulation of ruthenium after 6, 24 and 48 h of treatment with IC_{50} concentration of **12** was analyzed with ICP-MS (Heitland et al., 2006).

ICP-MS analysis of total intracellular accumulation of ruthenium has shown time-dependent increase of intracellular ruthenium content. High intracellular ruthenium concentration already after 6 h of incubation (8.83 ± 0.09 ng Ru/10^6 cells) is in agreement with previous conclusions about very fast action of **12**. Our assumption is that **12** enters cell very efficiently, which implicates possible passive transmembrane transport of this compound. Due to the presence of arene ligand (Giannini et al., 2015; Morris, Aird, Murdoch, et al., 2001; Su et al., 2015), as well as isoquinoline-3-carboxylic acid (derivative of picolinic acid with fused phenyl group), which contribute together to lipophilicity of the complex (Chakov et al., 1999; Ding et al., 1996; Hepburn et al., 2003; Liang et al., 2000; Morris et al., 1995; Song et al., 1999; Stearns et al., 1992; Stearns et al., 2002), passive transport through hydrophobic lipid bilayer is possible. Through the course of time, ruthenium holds and even increases its concentration inside cells (24 and 48 h incubation, respectively, Fig. 7.12). This implicates binding of **12** to intracellular structures and its continuous accumulation in intracellular environment (Kuhn et al., 2015).

7.1.1.2.4 Molecular response of HeLa cells to complex 12 treatment

We tested the effect of the complex **12** on the gene expression profile of HeLa cells at two time points, 12 and 24 h, using Agilent SurePrint G3 Hmn GE 8 × 60 K microarray slides, covering 27958 Entrez genes and 7419 lncRNAs (Fig. 7.13A). The changes in gene expression induced by the **12** were compared to HeLa cells treated with cisplatin (CDDP) at both time points. Treatment with **12** resulted in subtler gene expression perturbation compared to the changes induced by CDDP treatment at both time points (Fig. 7.13B). Treatment with **12** induced significant expression changes at 12 h compared to control cells in 2415 unique annotated genes, with 1184 upregulated and 1243 downregulated genes. This effect was potentiated at 24 h, with 5988 differentially expressed genes (DEG), of which 5475 were upregulated and only 545 downregulated.

Conversely, cisplatin treatment induced a strikingly greater level of gene expression perturbations at both 12 and 24 h, with 8103 (96 upregulated and 8025 downregulated) and 11263 differentially expressed genes (998 upregulated and 10416 downregulated), respectively. At 12 h, cisplatin induced extensive gene expression perturbation with 7347 CDDP-specific DEG, in contrast to moderate changes in gene expression in **12** treated cells (1659 **12**-specific DEG), with only 756 genes commonly deregulated genes between **12** and CDDP treated cells (Fig. 7.13C). After 24 h treatment, more prominent changes in gene expression were observed in **12** treated cells, with ~60% of **12**-dependent genes being also deregulated in cisplatin treated cells.

It is clear from the results above that, on cellular level, **12** exhibited cytotoxic effect comparable to CDDP. However, this effect is reflected on molecular level as subtler perturbation of HeLa cell transcriptome. From schematic representation of specific and commonly deregulated genes in HeLa cells at 12 and 24 h upon treatment with **12** and CDDP (Venn diagrams, Fig. 7.13C) we can speculate about different mechanism of action of **12**. After 12 h of treatment, 1659 **12**-specific DEG has been found, which were not deregulated in CDDP-treated population. These results provide evidence that, during this 'early response' to treatment (12 h), different molecular mechanisms are active after **12** and CDDP treatment.

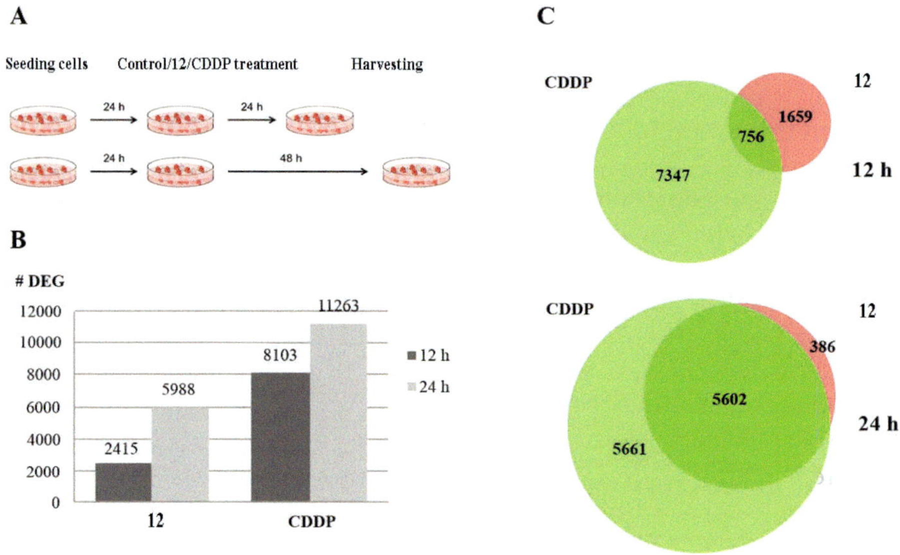

Figure 7.13 **A**. Experimental procedure outline **B**. Number of differentially expressed genes (DEG) at 12 and 24 h in HeLa cells treated with **12** or CDDP compared to control. **C**. Venn diagrams representing specific and commonly deregulated genes (FDR < 0.05 and FC > 2) in HeLa cells at 12 and 24 h upon treatment with **12** (in red) and CDDP (in green) (Jovanović, Tanić, et al., 2016).

The analysis of gene clusters with similar expression patterns gave us valuable information concerning the mechanism of **12** action. Analysis of involvement of DEGs in specific biological functions and canonical pathways has revealed that, besides direct binding to DNA, the investigated complex induces lesions in DNA and other biomolecules indirectly, with generation of free radical species inside the cells (Jovanović, Tanić, et al., 2016). This analysis has also showed that, after treatment with **12** complex, intrinsic (mitochondrial) apoptotic pathway is initiated in HeLa cells. The assessment of potential systemic toxicity of the investigated compound, performed using IPA Tox Comparison Analysis on differentially expressed genes at 12 and 24 h, has revealed absence of systemic toxicity for genes involved in cardiac arrhythmia, cardiac infarction and kidney failure, for which CDDP-treated cells showed strong enrichment for 24 h treatment.

7.1.1.3 Ruthenium–*p*-cymene complexes carrying pyridine or 7-chloroquinolin derivatives

With a wide range of potential ligands and diverse chemistry, ruthenium metal complexes represent adjustable scaffolds that can be tailored to overcome the specific limitations of a broad spectrum of drugs. In this context, nitrogen-containing aromatic heterocycles such as pyridines, imidazoles and quinolines, represent interesting class of ubiquitous functional groups, found in prevalence in many bioactive compounds and natural products, which makes them attractive tools, for rational drug design in organometallic chemistry (Altaf et al., 2015). Introducing the bioactive compounds as ligands, into the structure of organometallics, was successful strategy for enhancing the biological activity of an already active molecule, such as tamoxifen and staurosporine (Pizarro et al., 2010; Renfrew, 2014). In order to target more specifically cancer cells in vivo, an alternative approach is to deliver the active biomolecule, with a chaperone that encapsulates or complexes it. Subsequently, activation of organometallic complexes towards substitution reactions in physiological conditions may be mediated through different pathways (extensively reviewed by Pizarro et al., 2010). Activation of ruthenium organometallic "prodrugs" inside the cells may be triggered by oxidation of a less labile ligand, by chelate ring opening, or by an external source, such as light of a certain wavelength. For example, some "piano-stool" ruthenium(II)–arene complexes of the type [Ru(η-*p*-cym)(bpm)(py)]$(PF_6)_2$, (where bpm = 2,2-bipyrimidine, and py = pyridine), or ruthenium(II) polypyridine complexes of structure $[Ru(tpy)(LL)(L)]^{2+}$, carrying L = monodentate ligand, or LL = crowded bidentate ligand (where tpy = 2,2′:2′,6″-terpyridine), are capable of undergoing photoinduced reactions with DNA bases (Li et al., 2016; Sainuddin et al., 2016).

Visible light selectively induce photo-dissociation of a monodentate ligand (py), conventionally non-labile, generating a reactive aqua species, able to bind to DNA bases. Overall, possibility to control the hydrolysis reaction of the ruthenium drugs (on a timescale and under conditions relevant in vivo), along with possibility to consequently control kinetics of drug binding to biomolecules, present strong rational for further developments in the expanding field of ruthenium-organometallics (Chen et al., 2013; Dragutan et al., 2015; Mishra et al., 2000; Pantić et al., 2016).

In this context, of particular interest are the ruthenium(II)-arene piano stool complexes of the type $[Ru(\eta^6\text{-arene})(XY)(Z)]^{n+}$, that carry a labile group, prone to substitution, such as XY-chelating ligand, or Z-labile halide. Their activation occurs primarily in the cell, through hydrolysis of the Ru–Z bond, in a manner similar to cisplatin (Nazarov et al., 2014; Pizarro et al., 2010). The hydrolysis of metal–chloride bonds as the chloride concentration diminishes from extracellular to intracellular (cytoplasmic and nuclear) compartments represents the key element in kinetics of ligand substitution events that can occur in vivo. The aqua adduct, further interacts with the biological target, such as DNA, and the drug exerts its biological effect.

We have undertaken synthesis of ruthenium(II)–*p*-cymene complex (**14**), carrying 3-(4,5-dihydro-1H-imidazol-2-yl)-pyridine hydrochloride ($\mathbf{L_{14}}$), and two chlorides as leaving ligands (Fig. 7.14), designed to achieve coordination to Ru(II) centre directly, via pyridine moiety (Nikolić et al., 2015). In addition, we prepared different complex (**13**), with a 1-(7-chloroquinolin-4-yl) thiourea ligand ($\mathbf{L_{13}}$) (Fig. 7.14), and one chloride as leaving ligands. Ligands $\mathbf{L_{13}}$ and $\mathbf{L_{14}}$ significantly differ in structure and mode of coordination. Ligand $\mathbf{L_{13}}$ was designed to be coordinated to a central ion via the thioamide group, and to leave the quinoline core available for interactions with the target.

From the literature, it is well established that the biologically active compounds derived from quinoline, are prone to π-stacking interactions with biomolecules, such protoporphyrine IX (malaria), or DNA/ RNA (cancer). Quinoline derivatives have been found to exert an anti-malarial and anti-microbial effect, as well as anti-proliferative activity on human tumour cell lines derived from ovarian and colorectal cancer (Kumar et al., 2014). Recently, one 7-chloroquinoline based drug (4-[3-(2-Nitro-1-imidazolyl)propylamino]-7-chloroquinoline hydrochloride), (NLCQ-1), demonstrated interesting DNA intercalating/hypoxia selective cytotoxicity, and showed promising profile as an adjuvant to (radio/chemo)therapy in the clinics, presenting a candidate to enter phase I clinical trials (Papadopoulou et al., 2008; Reid et al., 2003).

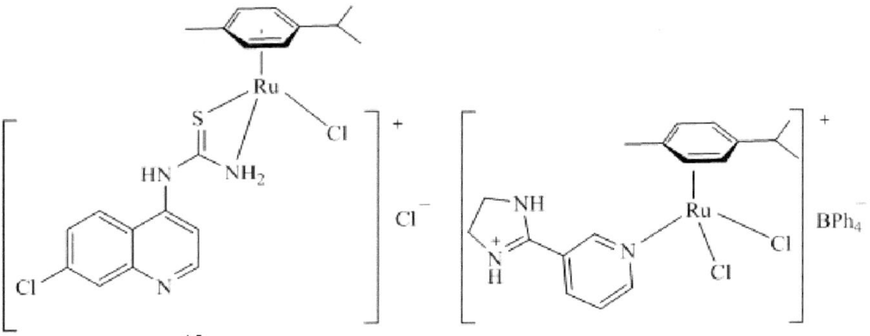

Figure 7.14 Proposed structures of complexes **13** and **14**. Adapted from Nikolić et al. (2015).

Pyridine derivatives as ligands coordinated to ruthenium-arenes typically slow down the hydrolysis, and cause unique effect in biologically relevant time scales (Li et al., 2016). For metal based complexes that carry bulky ligands, in general, it was shown (Buss et al., 2012; Savic et al., 2014), that both lipophilic characteristics and reactivity of complexes should be balanced in order to obtain drug with better cellular uptake properties, and efficiency to interact with cellular targets, such as DNA. Thus, it was interesting to investigate the potential of novel (**13**, **14**) complexes for in vitro cytotoxicity, intracellular accumulation and DNA binding, in comparison to cisplatin, used in clinics. Results of biological studies will be briefly presented.

The structures of complexes **13** and **14** are presented in Fig. 7.14. Prior to biological testing, the complexes were examined with respect to stability in DMSO. ^{1}HNMR showed that complexes remained intact in DMSO solution for 24 h.

7.1.1.3.1 Analysis of cell growth inhibition

Cytotoxicity study performed on several human neoplastic cell lines (HeLa, K562, A549, MDA-MB-231, EA.hy 926) and in one non-tumour human foetal lung fibroblast cell line (MRC-5), revealed that complexes exhibited cytotoxic activity, with IC_{50} values in the range 11.0–56.5 μM (Table 7.5). Interestingly, complex **13** bearing 7-chloroquinolin derivative exerted higher cytotoxicity when compared to **14**, particularly in human erythromyeloblastoid leukaemia cells (K562), and human transformed endothelial cells (EA.hy 926), with the IC_{50} values (μM) being 11.0 ± 1.4 and 13.8 ± 2.5, respectively. Observed activity (IC_{50} values) for **13** in K562 and EA.hy 926 cells was comparable to CDDP (10.9 ± 0.6 and 7.8 ± 1.2, respectively). In addition, **13** showed approximately three times less cytotoxicity in MRC-5 cells (Fig. 7.15b), than in cancer K562 and EA.hy 926 cells, which suggested its cytoselective toxicity towards cancer cells. Ligands itself were devoid of cytotoxic activity (IC_{50}) up to 100 μM and could be considered as inactive. This clearly demonstrated substantial contribution of coordination of ligands to ruthenium(II) centre, for complex bioactivation.

Table 7.5 In vitro anti-proliferative activities of compounds **13**, **14**, **L^{13}**, **L^{14} × HCl** and CDDP against human cell lines after 72 h of continuous drug action

Cell line[a]	**Compound**				
IC_{50} (μM)	**13**	**14**	**L^{13}**	**L^{14} × HCl**	**CDDP**
HeLa	56.5 ± 1.6	28.6 ± 1.7	> 100[b]	> 100	7.6 ± 0.0
K562	**11.0 ± 1.4**	54.9 ± 6.6	> 100	> 100	10.9 ± 0.6
MDA-MB-231	26.6 ± 7.2	47.7 ± 1.0	> 100	> 100	13.2 ± 0.4
EA.hy 926	13.8 ± 2.5	25.3 ± 3.6	> 100	> 100	7.8 ± 1.2
A549	31.5 ± 5.6	39.8 ± 0.8	> 100	> 100	17.2 ± 0.7
MRC-5	33.7 ± 5.5	42.5 ± 6.7	> 100	> 100	11.5 ± 0.5

[a]IC_{50} values (μM) are represented as mean values with corresponding standard deviations (SD) from three or more independent experiments (Nikolić et al., 2015).
[b]> 100 denotes that IC_{50} was not reached in the range of concentrations tested up to 100 μM.

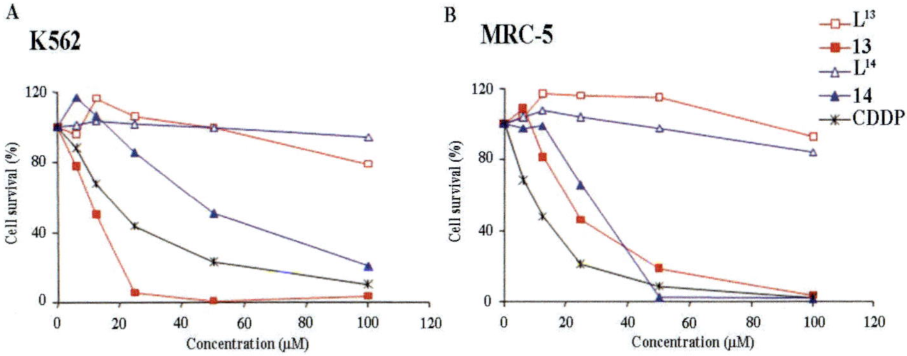

Figure 7.15 Cell survival after 72 h treatment with ligands **L**^{13}, **L**^{14}, complexes **13**, **14** and CDDP of: (A) K562 cell line; (B) MRC-5 cell line.

7.1.1.3.2 Cell cycle analysis and apoptotic potential

Investigation of the effect of **13** and **14** on cell cycle in K562 cells was performed following 24 h drug action, using flow cytometry and staining with propidium iodide (PI). It was noticed that **13** and **14** did not cause any significant alterations of the cell cycle. However, both **13** and **14** induced concentration dependent accumulation of cells in sub-G1 fraction (Fig. 7.16, Table 7.6), which is considered as a hallmark of internucleosomal DNA cleavage, in the course of apoptotic cell death (Huang et al., 2005; Vock et al., 1998). This result may not be surprising, since the ability of some ruthenium(II)–arene complexes to generate cytotoxic effect with just moderate or no effect on cell cycle, was

Table 7.6 Percentage of cells in sub-G1 fraction of cell cycle, after 24 h treatment with **L**^{13}, **L**^{14}, CDDP, **13** and **14** at concentration corresponding to IC_{50} and 2 × IC_{50}

	Sub G1 fraction (%)	
	IC_{50}	2 × IC_{50}
Control	8.1 ± 0.0	8.1 ± 0.0
L^{13}	6.6 ± 0.0	6.5 ± 0.6
L^{14}	10.4 ± 0.4	6.9 ± 0.1
CDDP	18.5 ± 1.4	24.4 ± 1.4
13	10.8 ± 0.7	18.4 ± 0.3
14	11.8 ± 0.0	17.9 ± 1.5

reported previously (Ginzinger et al., 2012; Gligorijević et al., 2012). Free ligand did not induce changes in the sub-G1 fraction relative to control, which is consistent with its lack of cytotoxicity.

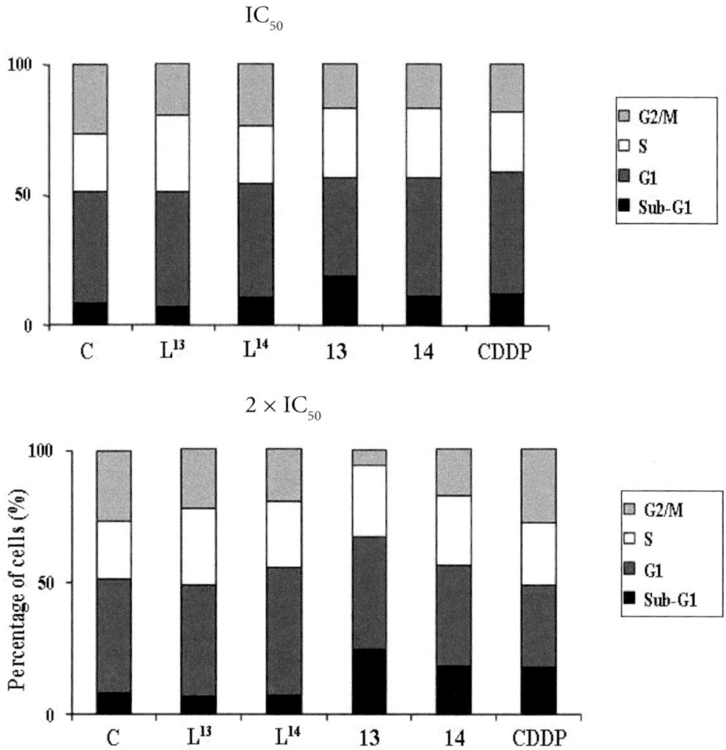

Figure 7.16 Diagrams of cell cycle phase distribution of treated K562 cells after 24 h treatment with $\mathbf{L^{13}}$, $\mathbf{L^{14}}$, CDDP, **13** and **14** at concentration corresponding to (A) IC_{50} and (B) 2 × IC_{50}. Adapted from Nikolić et al. (2015).

7.1.1.3.3 Intracellular accumulation and DNA binding study

The most intriguing results were obtained by ICP-MS study of intracellular accumulation and DNA binding, in K562 cells. Results obtained following 24 h treatment with equal concentrations (50 μM) of compounds demonstrated significantly higher intracellular accumulation of **13** versus **14** and CDDP, and clearly indicated the positive effect of the quinoline moiety on cellular uptake of ruthenium (η^6-*p*-cymene) complexes (Fig. 7.17A). Intracellular ruthenium concentration (**13**) 1.4 ± 0.2

(pg Ru/10^6 cells) exceeded that of CDDP 0.3 ± 0.0 (pg Pt/10^6 cells), by approximately 4.6 times.

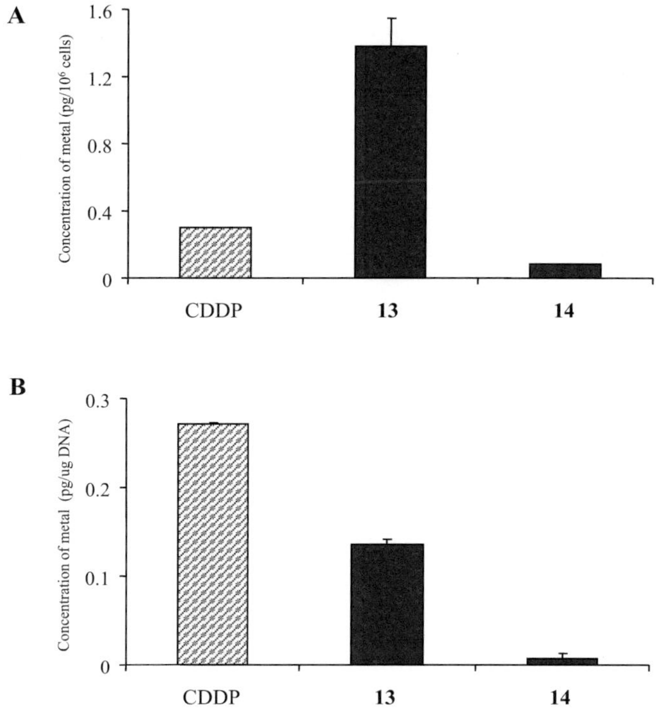

Figure 7.17 Concentrations of platinum(II)/ruthenium(II) after 24 h treatment of K562 cells with equimolar concentrations (50 μM) of **13, 14** and CDDP: A) intracellular concentrations (pg/10^6 cells); B) concentrations (pg/μg) bounded to cellular DNA. Bar graphs represent mean values of three independent measurements. Obtained results are presented as plot of metal concentration Ru(II) or Pt(II), recorded inside the cells, normalized upon the cell number (pg/10^6 cells).

However, ICP-MS study of the DNA binding (pg Ru(II) or Pt(II)/μg DNA) in K562 cells, under the same treatment conditions, revealed that both ruthenium complexes bind to cellular DNA less efficiently when compared to CDDP (Fig. 7.17B). The ratio of DNA binding (relative to the amount of complex present in the cell) decreased in order: CDDP > **13** > **14**. The level of DNA platination, 0.271 ± 0.001 pg Pt/μg DNA, exceeded twice the level of DNA-ruthenium binding caused by **13**: 0.135 ± 0.005 pg Ru/μg DNA. Poor DNA binding of **14**: 0.007 ± 0.006 pg Ru/μg DNA, was in accordance to its lower intracellular accumulation (0.081 ± 0.004 (pg/10^6 cells), and to its generally lower cytotoxicity. Although **13** exhibited comparable in vitro activity to CDDP in K562, our conclusion is that its mechanism of cytotoxic action could not be explained only by the DNA binding capability, or at least that DNA might not be the only intracellular target, for this type of complexes. Bulky lipophilic 7-chloroquinolin ligand may be able to hamper formation of ruthenium DNA adducts, and favour complex interactions to different cellular targets (Buss et al., 2012). This particular study, however, showed clear correlation between drug intracellular accumulation and cytotoxicity in K562 cell lines, and pointed out the contribution of quinoline substituent to enhancement of drug influx and cytotoxicity.

Ruthenium-arene complexes of similar structures (half-sandwich metallocenes) containing quinoline derivatives have been initially synthesized as successful anti-bacterial and anti-malarial agents, and were characterized by balanced structural and lipophilic properties (Forsgren et al., 1987; Singh et al., 2014). Ruthenium complexes with *p*-cymene as arene have been shown to exert cytotoxic activity in a number of cell lines, in level comparable to that of cisplatin (Dougan et al., 2008; Savić et al., 2011; van Zutphen et al., 2005). It is also relevant to mention that nuclear DNA might not represent the single target in the mechanism of cytotoxicity of these compounds.

7.1.1.4 Ruthenium(II)–arene complexes bearing hydrazides and the corresponding (thio)semicarbazones of 3- and 4-acetylpyridine

Among the different series of ruthenium(II)–*p*-cymene complexes with pyridine derivatives investigated in our laboratory as potential anti-cancer agents, we investigated group of ruthenium(II)–arene complexes bearing hydrazides and the corresponding (thio)semicarbazones of 3- and 4-acetylpyridine (Ivanović et al., 2013). The ruthenium(II)–arene complexes containing different types of ligands, such as caprylic hydrazide (a hydrazide with a long hydrocarbon chain), isonicotinic acid hydrazide (a hydrazide with an aromatic pyridine ring), thiosemicarbazones and semicarbazones (derived from the reaction of 3- and 4-acetylpyridine with thiosemicarbazide or caprylic hydrazide), were investigated in order to study the influence of structural modifications of the ligands on the cytotoxic activity (Fig. 7.18).

The caprylic hydrazide in complex **15** coordinates to ruthenium(II) in a bidentate fashion through the N-atom of the primary amino group and the O-atom of the carbonyl group in the neutral form. The complexes **16–20** have ligands coordinated in a monodentate mode via the pyridine nitrogen. All these complexes have a typical "piano-stool" geometry, the same as the previous series.

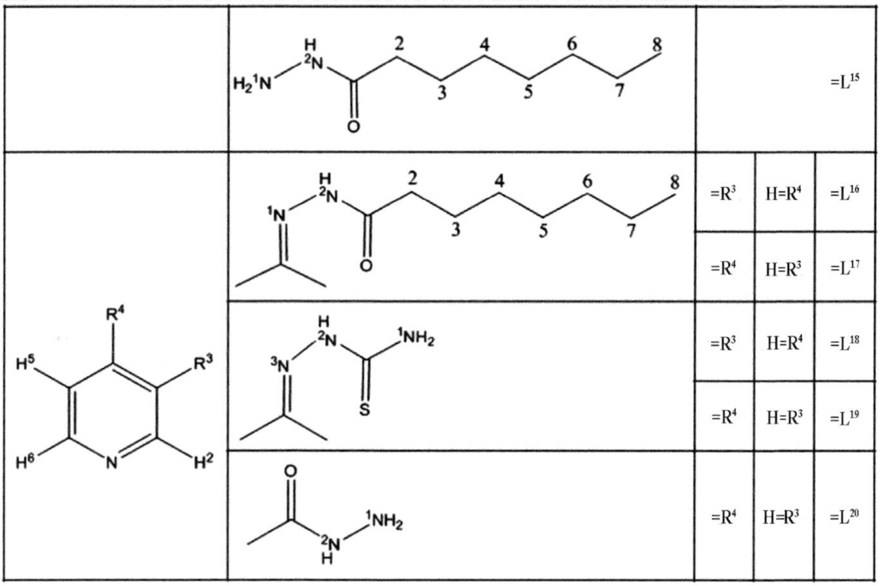

Figure 7.18 Structures of the ligands used for synthesis of complexes **15–20**. Adapted from Ivanović et al. (2013).

7.1.1.4.1 Analysis of cell growth inhibition

The anti-proliferative activity of the investigated complexes **15–20** on tumour cells was low after 48 h incubation (IC_{50} > 100 μM) (Table 7.7). Prolongation of the incubation time to 72 h contributed to reaching mild cytotoxic activity against HeLa and A549 cells, with IC_{50} values ranging from 85 to 91 μM for complexes

16 and **17**. The change of position of the acetyl group on the pyridine part of the ligand, from position 3 to position 4 did not changed the cytotoxicity of the complexes **16** and **17**. The introduction of thiosemicarbazone and semicarbazone parts in ligand significantly lowered the anti-proliferative potential of the complexes.

Table 7.7 Results of MTT assay for complexes **15**–**20** are presented as IC_{50} values, obtained after 72 h treatment on HeLa, A549 and LS-174 cells

	IC_{50} (μM)		
Complex	**HeLa**	**A549**	**LS-174**
15	> 100	> 100	> 100
16	85.4 ± 4.3	90.7 ± 8.1	> 100
17	87.4 ± 8.6	87.8 ± 5.1	> 100
18	> 100	n.d.	> 100
19	241.3 ± 12.1	243.1 ± 18.2	> 300
20	> 300	> 300	> 300

Source: Adapted from Ivanovic et al. (2013).

7.1.1.4.2 In vitro anti-metastatic potential

The potential of complexes **15–17** to interfere with the MMP-2 and MMP-9 was analyzed by gelatine zymography. The results revealed no significant influence on the MMP-2/MMP-9 activity for investigated complexes.

7.2 Conclusions

The chemical and biological results collected over the years of investigation of anti-cancer properties of ruthenium Ru(II)-*p*-cymene complexes with pyridine derivatives in our laboratory, led us to two leading compounds for further research and important data with regard to structure activity relationship. Structure–activity comparison of ruthenium(II) complexes of general formula [(η^6-*p*-cymene)Ru(L)Cl_2] and [(η^6-*p*-cymene)Ru(HL)Cl] revealed heterogenic mode of activity with regard to cytotoxic and anti-metastatic potential. Variations of the nature of monodentate or bidentate ligand influenced the pharmacological properties of the complexes. Investigation of in vitro anti-cancer properties of first series of complexes of general formula [(η^6-*p*-cymene)Ru(L)Cl_2] and [(η^6-*p*-cymene)Ru(HL)Cl], L-pyridine derivatives (dicarboxylic, acetyl or amino derivatives), show that additional ionizing groups within pyridine ligand had negative impact on activity of complexes (IC_{50} > 200 μM). Only complex with picolinic acid ([(η^6-*p*-cymene)Ru(L)Cl], H_2L'2-pyridinecarboxylicacid (**5**) exhibited IC_{50} 82 μM for 48 h continual agent action. The complex **5** distinguished among the tested ruthenium complexes, according to its highest cytotoxicity exerted through preferential DNA binding inducing S phase cell cycle arrest and apoptosis in HeLa cells. Complex **5** also demonstrated important anti-metastatic potential in vitro, through inhibition of MMP-s activity, inhibition of migration and invasion of HeLa cells and anti-angiogenic effect. In order to investigate multi-target potential of ruthenium complexes of this type our further investigation included coordinatively diversificated derivatives of pyridine. Activity comparison suggested, that introduction of 7-chloroquionoline-4-thiourea moiety to ruthenium(II)–*p*-cymene, markedly enhanced intracellular uptake of complex, and contributed to the cell-specific cytotoxicity. Complexes **13**, carrying 1-(7-chloroquinolin-4-yl) thiourea ligand and **14**, carrying 3-(4,5-dihydro-1H-imidazol-2-yl)-pyridine hydrochloride, exerted cytotoxic activity in micromolar range (up to 100 μM), in panel of tumour cell lines. Interesting results were obtained in K562 cells (erythromyeloblastoid leukaemia), where **13**, exerted significantly higher cytotoxicity, comparing to both **14** and cisplatin. Biological studies revealed that the highest cytotoxic potential of **13** was correlated to its highest intracellular accumulation. In addition, its mechanism of biological action might not be the only DNA-damage related. Certainly, factors that favour efficient cellular uptake of metal complex, are pre-request to its biological activity in both sensitive and resistant tumour cell lines. Our studies suggested at least, that ruthenium arenes of general structure [Ru(η^6-arene)(L)Cl_2], where L is 7-chloroquinolin derivative, may provide a way to the design of complexes with an encouraging cellular-uptake properties. Particularly important results are obtained with the complexes of second series, [Ru(η^6-*p*-cymene)(L)Cl], with non-ionizable functional groups within picolinic acid ligand (fluoro, chloro, bromo or methyl or isoquinoline-3-carboxylic acid). The complex with isoquinoline-3-carboxylic acid as ligand (**12**) has shown selective and high cytotoxic activity in the panel of investigated tumour cell lines, even cisplatin-resistant variant (A549), compared to the non-tumour cell line. According to the number of interesting biological properties, it has become a leading compound for further research. Our further more comprehensive studies of anti-cancer activity of this complex by using 'omics' analysis confirmed DNA damage and apoptotic potential and showed that it leads the cells through the intrinsic (mitochondrial) apoptotic pathway, via indirect DNA damage due to the action of reactive oxygen species, and through direct DNA binding. This confirmed that future studies of the mechanism of action of novel anti-cancer agents will require the use of a wide variety of techniques in pre-clinical research. This approach gave us time and money saving guidelines for further anti-tumour studies of our leading compound in an effort to determine its exact mechanism of action.

Conflict of Interest

The authors declare that there is no conflict of interest.

Acknowledgement

This work was supported by the Ministry of Science, Republic of Serbia, Grant, No. III 41026 and Grant, No 172035.

References

Aitken JB, Antony S, Weekley CM, Lai B, Spiccia L, Harris HH. (2012). Distinct cellular fates for KP1019 and NAMI-A determined by X-ray fluorescence imaging of single cells. *Metallomics,* **4**(10): 1051–1056, 1007.

Altaf AA, Shahzad A, Gul Z, Rasool N, Badshah A, Lal B, Khan E. (2015). A review on the medicinal importance of pyridine derivatives. *J Drug Des Med Chem,* **1**(1): 1–11.

Antonarakis ES, Emadi A. (2010). Ruthenium-based chemotherapeutics: Are they ready for prime time? *Cancer Chemother Pharmacol,* **66**(1): 1–9.

Bergamo A, Gaiddon C, Schellens JH, Beijnen JH, Sava G. (2012). Approaching tumour therapy beyond platinum drugs: Status of the art and perspectives of ruthenium drug candidates. *J Inorg Biochem,* **106**(1): 90–99.

Bergamo A, Gava B, Alessio E, Mestroni G, Serli B, Cocchietto M, Zorzet S, Sava G. (2002). Ruthenium-based NAMI-A type complexes with in vivo selective metastasis reduction and in vitro invasion inhibition unrelated to cell cytotoxicity. *Int J Oncol,* **21**(6): 1331–1338.

Bergamo A, Gerdol M, Lucafo M, Pelillo C, Battaglia M, Pallavicini A, Sava G. (2015). RNA-seq analysis of the whole transcriptome of MDA-MB-231 mammary carcinoma cells exposed to the antimetastatic drug NAMI-A. *Metallomics,* **7**(10): 1439–1450.

Bergamo A, Masi A, Dyson PJ, Sava G. (2008). Modulation of the metastatic progression of breast cancer with an organometallic ruthenium compound. *Int J Oncol,* **33**(6): 1281–1289.

Biersack B. (2016). Anticancer activity and modes of action of (arene) ruthenium(ii) complexes coordinated to C-, N-, and O-ligands. *Mini Rev Med Chem,* **16**(10): 804–814.

Buss I, Kalayda GV, Lindauer A, Reithofer MR, Galanski M, Keppler BK, Jaehde U. (2012). Effect of reactivity on cellular accumulation and cytotoxicity of oxaliplatin analogues. *J Biol Inorg Chem,* **17**(5): 699–708.

Butcher EC, Berg EL, Kunkel EJ. (2004). Systems biology in drug discovery. *Nat Biotechnol,* **22**(10): 1253–1259.

Casciola-Rosen L, Rosen A, Petri M, Schlissel M. (1996). Surface blebs on apoptotic cells are sites of enhanced procoagulant activity: Implications for coagulation events and antigenic spread in systemic lupus erythematosus. *Proc Natl Acad Sci U S A,* **93**(4): 1624–1629.

Casini A, Gabbiani C, Sorrentino F, Rigobello MP, Bindoli A, Geldbach TJ, Marrone A, Re N, Hartinger CG, Dyson PJ, Messori L. (2008). Emerging protein targets for anticancer metallodrugs: Inhibition of thioredoxin reductase and cathepsin B by antitumor ruthenium(II)–arene compounds. *J Med Chem,* **51**(21): 6773–6781.

Chakov NE, Collins RA, Vincent JB. (1999). A re-investigation the electronic spectra of chromium(III) picolinate complexes and high yield synthesis and characterization of $Cr_{-2}(\mu\text{-}OH)_{(2)}(pic)_{(4)}$ center dot $5H_{(2)}O$ (Hpic=picolinic acid). *Polyhedron,* **18**(22): 2891–2897.

Chen YJ, Lei WH, Jiang GY, Zhou QX, Hou YJ, Li C, Zhang BW, Wang XS. (2013). A ruthenium(II) arene complex showing emission enhancement and photocleavage activity towards DNA from singlet and triplet excited states respectively. *Dalton Trans,* **42**(16): 5924–5931.

Chu G. (1994). Cellular responses to cisplatin. The roles of DNA-binding proteins and DNA repair. *J Biol Chem,* **269**(2): 787–790.

Clarke MJ. (2003). Ruthenium metallopharmaceuticals. *Coord Chem Rev,* **236**(1): 209–233.

Clarke MJ, Zhu F, Frasca DR. (1999). Non-platinum chemotherapeutic metallopharmaceuticals. *Chem Rev,* **99**(9): 2511–2534.

Dette C, Wätzig H, Uhl H. (1993). Characterisation of the important pharmaceutical intermediate pyridine-2,4-dicarboxylic acid using capillary electrophoresis (CE). *Pharmazie,* **48**(4): 276–280.

Ding H, Olson LK, Caruso JA. (1996). Elemental speciation for chromium in chromium picolinate products. *Spectrochim Acta B,* **51**(14): 1801–1812.

Dougan SJ, Habtemariam A, McHale SE, Parsons S, Sadler PJ. (2008). Catalytic organometallic anticancer complexes. *Proc Natl Acad Sci U S A,* **105**(33): 11628–11633.

Dragutan I, Dragutan V, Demonceau A. (2015). Editorial of special issue ruthenium complex: The expanding chemistry of the ruthenium complexes. *Molecules,* **20**(9): 17244–17274.

Farrell N. (1989). Platinum—amine complexes as anticancer agents. *Transition Metal Complexes as Drugs and Chemotherapeutic Agents.* Dordrecht, The Netherlands, Kluwer Academic Publishers, p. 46.

Farrell N. (2000). Polynuclear charged platinum compounds as a new class of anticancer agents. *Platinum-Based Drugs in Cancer Therapy.* New York, Humana Press, p. 321–338.

Forsgren A, Schlossman SF, Tedder TF. (1987). 4-Quinolone drugs affect cell cycle progression and function of human lymphocytes in vitro. *Antimicrob Agents Chemother,* **31**(5): 768–773.

Frezza M, Hindo S, Chen D, Davenport A, Schmitt S, Tomco D, Dou QP. (2010). Novel metals and metal complexes as platforms for cancer therapy. *Curr Pharm Des,* **16**(16): 1813–1825.

Giannini F, Geiser L, Paul LEH, Roder T, Therrien B, Suss-Fink G, Furrer J. (2015). Tuning the in vitro cell cytotoxicity of dinuclear arene ruthenium trithiolato complexes: Influence of the arene ligand. *J Organomet Chem,* **783**: 40–45.

Ginzinger W, Muhlgassner G, Arion VB, Jakupec MA, Roller A, Galanski M, Reithofer M, Berger W, Keppler BK. (2012). A SAR study of novel antiproliferative ruthenium and osmium complexes with quinoxalinone ligands in human cancer cell lines. *J Med Chem,* **55**(7): 3398–3413.

Gligorijević N, Aranđelović S, Filipović L, Jakovljević K, Janković R, Grgurić-Sipka S, Ivanović I, Radulović S, Tesić Z. (2012). Picolinate ruthenium(II)–arene complex with in vitro antiproliferative and antimetastatic properties: Comparison to a series of ruthenium(II)–arene complexes with similar structure. *J Inorg Biochem,* **108**: 53–61.

Grguric-Sipka S, Ivanovic I, Rakic G, Todorovic N, Gligorijevic N, Radulovic S, Arion VB, Keppler BK, Tesic Z. (2010). Ruthenium(II)–arene complexes with functionalized pyridines: Synthesis, characterization and cytotoxic activity. *Eur J Med Chem,* **45**(3): 1051–1058.

Griggs DL, Hedden P, Temple-Smith KE, Rademacher W. (1991). Inhibition of gibberellin 2β-hydroxylases by acylcyclohexanedione derivatives. *Phytochemistry,* **30**(8): 2513–2517.

Groves JT, Kady IO. (1993). Sequence-specific cleavage of DNA by oligonucleotide-bound metal complexes. *Inorg Chem,* **32**(18): 3868–3872.

Grozav A, Balacescu O, Balacescu L, Cheminel T, Berindan-Neagoe I, Therrien B. (2015). Synthesis, anticancer activity, and genome profiling of thiazolo arene ruthenium complexes. *J Med Chem,* **58**(21): 8475–8490.

Hachisuka Y, Tochikubo K, Yokoi Y, Murachi T. (1967). The action of dipicolinic acid and its chemical analogues on the heat stability of glucose dehydrogenase of Bacillus subtilis spores. *J Biochem,* **61**(5): 659–661.

Hannon MJ. (2007). Metal-based anticancer drugs: From a past anchored in platinum chemistry to a post-genomic future of diverse chemistry and biology. *Pure Appl Chem,* **79**(12): 2243–2261.

Hartinger CG, Jakupec MA, Zorbas-Seifried S, Groessl M, Egger A, Berger W, Zorbas H, Dyson PJ, Keppler BK. (2008). KP1019, a new redox-active anticancer agent--preclinical development and results of a clinical phase I study in tumor patients. *Chem Biodivers,* **5**(10): 2140–2155.

Heffeter P, Bock K, Atil B, Reza Hoda MA, Korner W, Bartel C, Jungwirth U, Keppler BK, Micksche M, Berger W, Koellensperger G. (2010). Intracellular protein binding patterns of the anticancer ruthenium drugs KP1019 and KP1339. *J Biol Inorg Chem,* **15**(5): 737–748.

Heitland P, Koster HD. (2006). Biomonitoring of 30 trace elements in urine of children and adults by ICP-MS. *Clin Chim Acta,* **365**(1–2): 310–318.

Hepburn DDD, Xiao JR, Bindom S, Vincent JB, O'Donnell J. (2003). Nutritional supplement chromium picolinate causes sterility and lethal mutations in Drosophila melanogaster. *Proc Natl Acad Sci U S A,* **100**(7): 3766–3771.

Huang X, Halicka HD, Traganos F, Tanaka T, Kurose A, Darzynkiewicz Z. (2005). Cytometric assessment of DNA damage in relation to cell cycle phase and apoptosis. *Cell Prolif,* **38**(4): 223–243.

Ivanovic I, Gligorijevic N, Arandelovic S, Radulovic S, Roller A, Keppler BK, Tesic ZL, Grguric-Sipka S. (2013). New ruthenium(II)–arene complexes bearing hydrazides and the corresponding (thio)semicarbazones of 3-and 4-acetylpyridine: Synthesis, characterization, crystal structure determination and antiproliferative activity. *Polyhedron,* **61**: 112–118.

Ivanovic I, Grguric-Sipka S, Gligorijevic N, Radulovic S, Roller A, Tesic ZL, Keppler BK. (2011). X-Ray structure and cytotoxic activity of a picolinate ruthenium(II)–arene complex. *J Serb Chem Soc,* **76**(1): 53–61.

Ivanovic I, Jovanovic KK, Gligorijevic N, Radulovic S, Arion VB, Sheweshein KSAM, Tesic ZL, Grguric-Sipka S. (2014). Ruthenium(II)–arene complexes with substituted picolinato ligands: Synthesis, structure, spectroscopic properties and antiproliferative activity. *J Organomet Chem,* **749**: 343–349.

Jakupec MA, Galanski M, Arion VB, Hartinger CG, Keppler BK. (2008). Antitumour metal compounds: More than theme and variations. *Dalton Trans,* **2**: 183–194.

Jovanovic KK, Gligorijevic N, Gaur R, Mishra L, Radulovic S. (2016). Anticancer activity of two ruthenium(II)-DMSO-chalcone complexes: Comparison of cytotoxic, pro-apoptotic and antimetastatic potential. *J BUON,* **21**(2): 482–490.

Jovanovic KK, Tanic M, Ivanovic I, Gligorijevic N, Dojcinovic BP, Radulovic S. (2016). Cell cycle, apoptosis, cellular uptake and whole-transcriptome microarray gene expression analysis of HeLa cells treated with a ruthenium(II)–arene complex with an isoquinoline-3-carboxylic acid ligand. *J Inorg Biochem,* **163**: 362–373.

Jungwirth U, Kowol CR, Keppler BK, Hartinger CG, Berger W, Heffeter P. (2011). Anticancer activity of metal complexes: Involvement of redox processes. *Antioxid Redox Signal,* **15**(4): 1085–1127.

Kasibhatla S, Amarante-Mendes GP, Finucane D, Brunner T, Bossy-Wetzel E, Green DR. (2006). Acridine orange/ethidium bromide (AO/EB) staining to detect apoptosis. *CSH Protoc,* **2006**(3), doi: 10.1101/pdb.prot4493.

Kuhn PS, Cremer L, Gavriluta A, Jovanovic KK, Filipovic L, Hummer AA, Buchel GE, Dojcinovic BP, Meier SM, Rompel A, Radulovic S, Tommasino JB, Luneau D, Arion VB. (2015). Heteropentanuclear oxalato-bridged *n*d-4f (*n* = 4, 5) metal complexes with NO ligand: Synthesis, crystal structures, aqueous stability and antiproliferative activity. *Chem Eur J,* **21**(39): 13703–13713.

Kumar K, Schniper S, Gonzalez-Sarrias A, Holder AA, Sanders N, Sullivan D, Jarrett WL, Davis K, Bai FW, Seeram NP, Kumar V. (2014). Highly potent anti-proliferative effects of a gallium(III) complex with 7-chloroquinoline thiosemicarbazone as a ligand: Synthesis, cytotoxic and antimalarial evaluation. *Eur J Med Chem,* **86**: 81–86.

Laine P, Gourdon A, Launay J-P. (1995). Chemistry of iron with dipicolinic acid. 1. Mononuclear complexes of iron (II) or iron (III). *Inorg Chem,* **34**(21): 5129–5137.

Li A, White JK, Arora K, Herroon MK, Martin PD, Schlegel HB, Podgorski I, Turro C, Kodanko JJ. (2016). Selective release of aromatic heterocycles from ruthenium tris(2-pyridylmethyl)amine with visible light. *Inorg Chem,* **55**(1): 10–12.

Liang Y, Noda LK, Sala O. (2000). Polarizability and concentration effects on the Raman spectra of picolinic acid species in aqueous solution. *J Mol. Struct,* **554**(2–3): 271–277.

Lippard SJ. (1994). Metals in medicine. *Bioinorganic Chemistry*. Mill Valley, California, University Science Books, pp. 505–583.

Martin BL. (1997). Selective activation of calcineurin by dipicolinic acid. *Arch Biochem Biophys,* **345**(2): 332–338.

Mauk AG, Coyle CL, Bordignon E, Gray HB. (1979). Bis (dipicolinate) complexes of cobalt (III) and iron (II) as new probes of metalloprotein electron-transfer reactivity. Analysis of reactions involving cytochrome c and cytochrome c551. *J Am Chem Soc,* **101**(17): 5054–5056.

Mishra L, Sinha R. (2000). Mononuclear and binuclear ruthenium(III) polypyridyl complexes containing 2,6-bis(2′,-benzimidazyl)-pyridine as co-ligand: Synthesis, spectroscopic properties and redox activity. *Indian J Chem A,* **39**(11): 1131–1139.

Morris RE, Aird RE, Murdoch Pdel S, Chen H, Cummings J, Hughes ND, Parsons S, Parkin A, Boyd G, Jodrell DI, Sadler PJ. (2001). Inhibition of cancer cell growth by ruthenium(II) arene complexes. *J Med Chem,* **44**(22): 3616–3621.

Morris GS, Guidry KA, Hegsted M, Hasten DL. (1995). Effects of dietary chromium supplementation on cardiac mass, metabolic enzymes, and contractile proteins. *Nutr Res,* **15**(7): 1045–1052.

Nazarov AA, Hartinger CG, Dyson PJ. (2014). Opening the lid on piano-stool complexes: An account of ruthenium(II)–arene complexes with medicinal applications. *J Organomet Chem,* **751**: 251–260.

Nikolić S, Opsenica DM, Filipović V, Dojčinović B, Arandelović S, Radulović S, Grgurić-Sipket S. (2015). Strong in vitro cytotoxic potential of new ruthenium-cymene complexes. *Organometallics,* **34**(14): 3464–3473.

Nikolić S, Rangasamy L, Gligorijević N, Arandelović S, Radulović S, Gasser G, Grgurić-Sipka S. (2016). Synthesis, characterization and biological evaluation of novel Ru(II)-arene complexes containing intercalating ligands. *J Inorg Biochem,* **160**: 156–165.

Ormerod M. (2000). Analysis of DNA–general methods. *Flow Cytometry: A Practical Approach.* Oxford, England, Oxford University Press, pp. 83–97.

Pantić DN, Arandelović S, Radulović S, Roller A, Arion VB, Grgurić-Šipka S. (2016). Synthesis, characterisation and cytotoxic activity of organoruthenium (II)-halido complexes with 1H-benzimidazole-2-carboxylic acid. *J Organomet Chem,* **819**: 61–68.

Papadopoulou MV, Bloomer WD. (2008). NLNQ-1, a 2-[3-(2-nitro-1-imidazolyl)propylamino]-3-chloro-1,4-naphthoquinone, as a hypoxiaselective cytotoxin and radiosensitizer. *In Vivo,* **22**(3): 285–288.

Pizarro AM, Habtemariam A, Sadler PJ. (2010). Activation mechanisms for organometallic anticancer complexes. *Top Organometal Chem,* **32**: 21–56.

Reid JM, Squillace DP, Ames MM. (2003). Single-dose pharmacokinetics of the DNA-binding bioreductive agent NLCQ-1 (NSC 709257) in CD2F(1) mice. *Cancer Chemother Pharmacol,* **51**(6): 483–487.

Renfrew AK. (2014). Transition metal complexes with bioactive ligands: Mechanisms for selective ligand release and applications for drug delivery. *Metallomics,* **6**(8): 1324–1335.

Rosenberg B. (1978). Platinum complexes for the treatment of cancer. *Interdiscip Sci Rev,* **3**(2): 134–147.

Sainuddin T, Pinto M, Yin HM, Hetu M, Colpitts J, McFarland SA. (2016). Strained ruthenium metal-organic dyads as photocisplatin agents with dual action. *J Inorg Biochem,* **158**: 45–54.

Sava G, Alessio E, Bergamo A, Mestroni G. (1999). Sulfoxide ruthenium complexes: Non-toxic tools for the selective treatment of solid tumour metastases. *Metallopharmaceuticals I*. Berlin, Heidelberg, Springer-Verlag.

Sava G, Gagliardi R, Bergamo A, Alessio E, Mestroni G. (1999). Treatment of metastases of solid mouse tumours by NAMI-A: Comparison with cisplatin, cyclophosphamide and dacarbazine. *Anticancer Res*, **19**(2A): 969–972.

Sava G, Pacor S, Mestroni G, Alessio E. (1992). Na[trans-$RuCl_4$(DMSO)Im], a metal complex of ruthenium with antimetastatic properties. *Clin Exp Metastasis*, **10**(4): 273–280.

Sava G, Zorzet S, Turrin C, Vita F, Soranzo M, Zabucchi G, Cocchietto M, Bergamo A, DiGiovine S, Pezzoni G, Sartor L, Garbisa S. (2003). Dual Action of NAMI-A in inhibition of solid tumor metastasis: Selective targeting of metastatic cells and binding to collagen. *Clin Cancer Res*, **9**(5): 1898–1905.

Savić A, Dulović M, Poljarevic JM, Misirlic-Dencic S, Jovanovic M, Bogdanovic A, Trajkovic V, Sabo TJ, Grguric-Sipka S, Markovic I. (2011). Synthesis and in vitro anticancer activity of ruthenium-cymene complexes with cyclohexyl-functionalized ethylenediamine-*N,N'*-diacetate-type ligands. *Chemmedchem*, **6**(10): 1884–1891.

Savić A, Filipović L, Aranđelović S, Dojčinović B, Radulović S, Sabo TJ, Grgurić-Sipka S. (2014). Synthesis, characterization and cytotoxic activity of novel platinum(II) iodido complexes. *Eur J Med Chem*, **82**: 372–384.

Scolaro C, Bergamo A, Brescacin L, Delfino R, Cocchietto M, Laurenczy G, Geldbach TJ, Sava G, Dyson PJ. (2005). In vitro and in vivo evaluation of ruthenium(II)–arene PTA complexes. *J Med Chem*, **48**(12): 4161–4171.

Singh AK, Pandey DS, Xu Q, Braunstein P. (2014). Recent advances in supramolecular and biological aspects of arene ruthenium(II) complexes. *Coord Chem Rev*, **270**: 31–56.

Song R, Kim KM, Sohn YS. (1999). Synthesis and properties of (diamine)platinum(II) complexes of pyridine carboxylate isomers and their antitumor activity. *Inorg Chim Acta*, **292**(2): 238–243.

Stearns DM, Armstrong WH. (1992). Mononuclear and binuclear chromium(III) picolinate complexes. *Inorg Chem*, **31**(25): 5178–5184.

Stearns DM, Silveira SM, Wolf KK, Luke AM. (2002). Chromium(III) tris(picolinate) is mutagenic at the hypoxanthine (guanine) phosphoribosyltransferase locus in Chinese hamster ovary cells. *Mutat Res Genet Toxicol Environ Mutag*, **513**(1–2): 135–142.

Su W, Tang ZF, Xiao Q, Li PY, Qian QQ, Lei XL, Huang S, Peng BH, Cui JG, Huang CS. (2015). Synthesis, structures, antiproliferative activity of a series of ruthenium(II) arene derivatives of thiosemicarbazones ligands. *J Organomet Chem*, **783**: 10–16.

Süss-Fink G. (2010). Arene ruthenium complexes as anticancer agents. *Dalton Trans*, **39**(7): 1673–1688.

Supino R. (1995). MTT Assays. In O'Hare S, Atterwill CK, eds., *In Vitro Toxicity Testing Protocols*. Totowa, NJ, Humana Press, pp. 137–149.

Tochikubo K, Hachisuka Y, Murachi T. (1968). Properties of glucose dehydrogenase from vegetative cells of Bacillus subtilis and effect of dipicolinic acid and its chemical analogues on the enzyme. *Jpn J Microbiol*, **12**(4): 435–440.

Trondl R, Heffeter P, Kowol CR, Jakupec MA, Berger W, Keppler BK. (2014). NKP-1339, the first ruthenium-based anticancer drug on the edge to clinical application. *Chem Sci*, **5**(8): 2925–2932.

van Engeland M, Nieland LJ, Ramaekers FC, Schutte B, Reutelingsperger CP. (1998). Annexin V-affinity assay: A review on an apoptosis detection system based on phosphatidylserine exposure. *Cytometry*, **31**(1): 1–9.

van Zutphen S, Reedijk J. (2005). Targeting platinum anti-tumour drugs: Overview of strategies employed to reduce systemic toxicity. *Coord Chem Rev*, **249**(24): 2845–2853.

Vermes I, Haanen C, Steffens-Nakken H, Reutelingsperger C. (1995). A novel assay for apoptosis. Flow cytometric detection of phosphatidylserine expression on early apoptotic cells using fluorescein labelled Annexin V. *J Immunol Methods*, **184**(1): 39–51.

Vock EH, Lutz WK, Hormes P, Hoffmann HD, Vamvakas S. (1998). Discrimination between genotoxicity and cytotoxicity in the induction of DNA double-strand breaks in cells treated with etoposide, melphalan, cisplatin, potassium cyanide, Triton X-100, and gamma-irradiation. *Mutat Res Genet Toxicol Environ Mutag*, **413**(1): 83–94.

Vock CA, Renfrew AK, Scopelliti R, Juillerat-Jeanneret L, Dyson PJ. (2008). Influence of the diketonato ligand on the cytotoxicities of [Ru(eta(6)-*p*-cymene)-(R(2)acac)(PTA)](+) complexes (PTA = 1,3,5-triaza-7-phosphaadamantane). *Eur J Inorg Chem*, **10**: 1661–1671.

Chapter 8

Approaching Cancer Therapy with Ruthenium Complexes by Their Interaction with DNA

Niraj Kumari* **and Lallan Mishra**[a]

Department of Chemistry, Institute of Science, Banaras Hindu University, Varanasi 221005, India

lmishrabhu@yahoo.co.in

Metal ions can be introduced into a biological system either for therapeutic effect or as diagnostic aids. In this context, ruthenium coordination compounds are currently the most promising metal-based diagnostic and chemotherapeutic agents and some of them are being extensively studied and either are in clinical trials or have passed the clinical stage in the treatment of metastases and colon cancers. The novel Ru(II) complexes have been directed as targets for selectively expressed tumor metastases, lower side effects, easy absorption, and excretion by body with their cytotoxicity. They form stable complexes with predictable structure and shape selectivity of complexes and can be improved by the incorporation of functionalized ligands

*Present address: Kalinga Institute of Industrial Technology, Bhubaneswar, Orissa 751024, India

Ruthenium Chemistry
Edited by Ajay Kumar Mishra and Lallan Mishra

ISBN 978-981-4774-39-0 (Hardcover), 978-1-315-11058-5 (eBook)
www.panstanford.com

around metal ion. However, this area of research is still demanding. This chapter embodies the several facets of ruthenium complexes with greater emphasis on the cytotoxicity correlated with their binding with DNA and subsequently DNA cleavage.

8.1 Introduction

The historic discovery of cisplatin by a physicist, Dr. Barnett Rosenberg, and his associate, a biologist, Loretta Van Camp, had a significant impact on cancer sufferers worldwide. This finding opened a door to the development of several new metal-based drugs. The search for "non-classical" metal-based antitumor drugs has since long stimulated investigations into the field of non-platinum metal drugs. Non-platinum active compounds are likely to have mechanism of action, biodistribution and toxicity which are different from those of platinum drugs. They might show reduced host toxicity. Among metal ions, several non-platinum complexes occupied a significant position and ruthenium complex stood on a firm footings owing to being congener of iron. These complexes are recognized to have great prospect in the treatment of several types of cancer cell lines. Among them, NAMI-A {[ImH] [*trans*-Ru(dmso-S)(Im)Cl_4], Im = imidazole} and KP1019 {[*trans*–Ru$(Ind)_2Cl_4$], Ind = Indazole} have entered clinical trials. These complexes showed different antitumor properties (Sava et al., 2003). Owing to potential emission, absorption, and redox activity, ruthenium complexes were found easier to be explored. Among several mechanisms of actions, these complexes normally prefer to bind with DNA of cell and create distortion in DNA helix and hinder the process like transcription and thereby create cell death following a variety of mechanisms. However, killing of cancer cell because of DNA binding cannot be only accountable. The identification of cellular/molecular targets is of great importance while formulating a novel anticancer agent.

This chapter embodies the introduction of ruthenium complexes, techniques employed to detect binding of newly developed ruthenium complexes with DNA, DNA cleavage studies using spectroscopic techniques and gel electrophoretic mobility assay, brief description of DNA structure with its constitutional components, mode of binding of complexes of with specific component of DNA and cytotoxicity of the complexes.

The advantage of using ruthenium in the development of metal-based antitumor drugs has been considered in a number of excellent reviews (Clarke, 2007; Allardyce and Dyson, 2001; Reisner et al., 2007). A steady growth of interest in Ru anticancer drugs over the last 20 years is reflected in the accelerating growth of publications in this area (Fig. 8.1, based on a search in the Chemical Abstracts database).

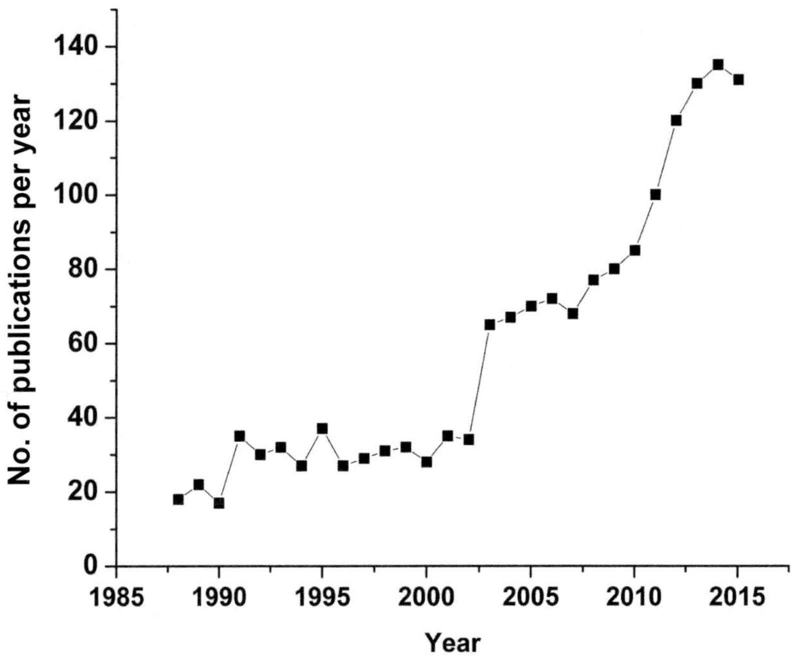

Figure 8.1 Growth in publications on ruthenium complexes as anticancer drugs (searched in the Chemical Abstracts database).

Briefly, the benefits of exploiting ruthenium include (i) a well-developed preparative coordination chemistry of this transition metal, providing reliable routes to novel compounds; (ii) a rate of ligand exchange often comparable to that of platinum or which can be tuned by coordination of appropriate ancillary ligands; (iii) octahedral coordination geometry in contrast to the square-planar geometry of platinum(II) complexes, implying a reactivity and mode of function different from cisplatin; (iv) accessibility of oxidation states 2^+, 3^+ and 4^+ under physiological conditions and the ability to tune the electron transfer rates and redox potentials; (v) the ability of ruthenium to mimic iron in binding to biomolecules such as human serum transferrin and other proteins, which makes ruthenium-based agents markedly less toxic than platinum drugs; and (vi) increasing knowledge about the biological effects of ruthenium complexes. Ruthenium-based drugs are much less toxic than the worldwide approved platinum-based drugs. This can be explained by the ability of ruthenium to mimic iron in binding to biological molecules, such as human serum albumin and transferrin (Pongratz et al., 2004). These are present in human serum at concentrations of 35–50 mg ml^{-1} and 2.5–3.5 mg ml^{-1}, correspondingly (Kratz and Beyer, 1998; Sadler et al., 1999). Platinum(II)-based antitumor agents are also capable of binding to these proteins; their coordination geometry (square-planar) is, however, distinct from that of ruthenium (III) or iron(III) (octahedral). This difference between platinum(II) and ruthenium(III) makes the delivery of platinum(II)-based drugs into cells via transferrin receptor-mediated endocytosis at least less likely. The "activation-by-reduction" mechanism could also be responsible for the lower toxicity of some ruthenium-containing agents (Keppler et al., 2005; Clarke, 1980). This mechanism, proposed about three decades ago (Clarke, 2001) is supposed to be operative in solid tumor with low oxygen level as compared to the normal tissue, enabling the reduction of Ru(III) to the kinetically more reactive ruthenium(II) species. The reductive microenvironment arises in rapidly growing tumors because of insufficient formation of new blood vessels and poor blood supply. Tumor hypoxia is a main factor contributing to failure of radiotherapy or chemotherapy (Brown, 1999, Wardman, 2001). The low oxygen content, together with the lower extracellular pH and the presence of appreciable amounts of cellular reducing agents such as glutathione, provides favorable conditions for a selective reduction of drugs with physiologically accessible RuIII/RuII redox potential. The reducing capability of ruthenium (III)-based drugs depends on their ligand environment. The knowledge of the net electron donation from ligand to metal enables the prediction of metal-centered redox potentials and assists the creation of drugs with desired redox parameters (Reisner et al., 2007; Reisner et al., 2004; Reisner et al., 2005). However, Ru(III) pro drug can undergo hydrolysis or binding with protein prior to reduction, changing the redox properties of the resulting metabolite significantly. In particular, $[Ru(III)Cl_3(H_2O)(1H\text{-indazole})_2]$ ($E_{1/2}$ = −0.16 V vs. NHE) is significantly easier to reduce than $[Ru(III)Cl_4(1H\text{-indazole})_2]^-$ (Reisner et al., 2007). Since all major genes involved in iron metabolism respond to oxygen depletion, it is tempting to assume that hypoxia generally sensitizes tumor cells to compounds interfering with iron-dependent processes. In particular, hypoxia has been shown to induce elevated transferrin receptor expression in tumor cells (Tacchini et al., 1999). It suggests that hypoxia-activated pro drugs capable of transferrin-mediated cellular uptake attain their tumor selectivity by a cooperative mechanism. The "activation-by reduction" hypothesis implies that the compound should be much more readily reducible under the hypoxic conditions of solid tumors than under the normoxic conditions of normal tissues, in order to obtain a tumor-selective activity and high therapeutic index. For Ru(III) compounds, an appropriate redox potential is therefore probably a more important parameter than cytotoxic potency under normal conditions. All the clinically established ruthenium-based drug candidates KP1019-indazolium [*trans*-tetrachlorobis(1H-indazole) ruthenate(III)] are administered intravenously and, therefore, proteins are among the first available binding partners in the blood stream. The binding of platinum complexes to serum proteins is thought to contribute to the side effects, but the binding of KP1019 to transferrin seems to be an important step in the mode of action (Timerbaev et al., 2006). Considering the enhanced permeability and retention (EPR) effect and the higher expression of transferrin receptors in tumor cells, binding of drugs to serum proteins appears promising target. Human serum albumin and transferrin as the most important transport proteins in the blood were extensively exploited for studies as potential target molecules for a variety of metal complexes (Timerbaev et al., 2006). In general, the metal complexes were found to bind with lower binding constants to these proteins than, for example, organic drugs do. For the drug candidate KP1019, the attachment to transferrin and its effect on the mode of action is well characterized (Hartinger et al., 2006). The protein was shown to preferentially bind two ruthenium moieties to the iron binding sites which most probably influence the protein's structure (Pongratz et al., 2004, Smith et al., 1996). The reaction of KP1019 with transferrin is slightly faster than with human serum albumin. The structural change imposed by attaching two ruthenium moieties to transferrin probably prevents the protein from binding to its receptor. It is reflected by a lower accumulation of ruthenium in the cell. However, loading transferrin with physiologically normal amounts of iron prior to binding of one ruthenium unit, led to a markedly increased cellular uptake (Pongratz et al., 2004). The release from the protein is thought to take place in the endosomes at a lower pH in the presence of biological chelators (Kratz et al., 1994). In an initial, simplified assumption of analogy with platinum drugs, DNA has been considered a critical target of KP1019, and binding to nucleotides and DNA has therefore been investigated in various studies (Schulga et al., 2006; Hartinger et al., 2006; Kung et al., 2001; Malina et al., 2001; Egger et al., 2005). The sensitization of cells to the sodium salt analogue KP1339 by inhibitors of DNA repair seems to argue for this assumption, but since the pattern of sensitization differs from that of cisplatin, cytotoxic DNA lesions are probably processed differently by the cell (Hartinger et al., 2006). The induction of apoptosis in cancer cells by the intrinsic mitochondrial pathway (Kapitza et al., 2005) does not exclude the possibility that DNA binding triggers the apoptotic process, but the comparatively rapid onset of membrane depolarization in mitochondria suggests that a direct interaction with mitochondria might be involved in the mechanism of action of KP1019. Taking into account the similarities between ruthenium and iron, it is tempting to assume an interference with iron-dependent metabolic processes. In this context, the capacity to bind to cytochrome c is noteworthy. Although a direct interaction with the heme group has not been reported, binding to this protein results in marked conformational changes in the heme environment which might affect biological functionality (Trynda-Lemiesz, 2004). Despite its moderate cytotoxicity, KP1019 exerts impressive effects in non-toxic doses in a variety of tumor models, probably reflecting a high degree of tumor selectivity mediated by the mechanisms mentioned earlier. The predictive power of the chosen models, in particular chemically induced autochthonous colorectal carcinoma in the rat, responded with a complete remission of one third of tumors (Berger et al., 1989). It also responded primary cultures of human tumor cells, a high proportion of which (including clinically chemo resistant specimens and cells from metastatic lesions) proved sensitive to this compound (Depenbrock et al., 1997) and excites confidence that a therapeutic benefit will be confirmed by further clinical studies. The phase I dose escalation study conducted in patients already with advanced solid tumors yielded encouraging results with disease stabilizations in five of six evaluable patients despite the fact that the majority of patients are not treated with the therapeutically optimal dose in this type of study. Another ruthenium complex under clinical investigation, NAMI-A, imidazolium [*trans*-tetrachloro (dimethylsulfoxide)-imidazoleruthenate(III)], (Rademaker-Lakhai et al., 2004), exhibits a quite different biological activity despite its structural relationship to KP1019. This compound reduces the formation of metastases and appears to inhibit their growth as a result of a delayed process of metastasis but has little impact on primary tumors in animal models (Sava et al., 2003). NAMI-A interferes with the extracellular matrix with the interactions of tumor cells, including an increase of actin-dependent cell adhesion, (Sava et al., 2004; Frausin et al., 2005). The inhibition of matrix degradation by matrix metalloproteinases (Pacor et al., 2004) and reduction of cell invasiveness and migration, resulting in a less malignant cell phenotype. A contribution of antiangiogenic effects to the antimetastatic properties has also been suggested (Vacca et al., 2002). However, the low capacity of DNA binding (Pluim et al., 2004) is unlikely to account for the antimetastatic activity. The unique properties of NAMI-A imply that its effects are mainly directed against the process of metastasis, while its inhibitory effects on established tumor lesions are much less pronounced, corresponding to its negligible cytotoxicity. Clinical experience suggests that established metastases can only be treated effectively with compounds that have the capacity to exert their effects on the primary tumor and are comprehensible from the common tissue characteristics shared by primary and secondary tumors. Beside Ru(III) compounds in clinical trials, organometallic Ru(II) complexes have attracted interest in recent years, some of which show similar antimetastatic activity to NAMI-A and RAPTA (Ang and Dyson, 2006). Others were proven to exert their activity by a DNA binding mechanism (Yan et al., 2005). Furthermore, the disadvantage of the small therapeutic window of multinuclear platinum complexes might be overcome by using multinuclear ruthenium (or other non-platinum) complexes, since ruthenium complexes are usually less toxic than platinum compounds.

8.2 Brief Introduction to DNA

8.2.1 General Structure

DNA is comprised of deoxyribose sugars linked together in a chain through phosphate groups, and with a nucleotide-base attached to each sugar (Fig. 8.2) (Berg et al., 2002; Branden et al., 2002). Four different bases are present (guanine, G; adenine, A; cytosine, C; thymine, T), two of which are purines (G, A) and two pyrimidines (C, T).

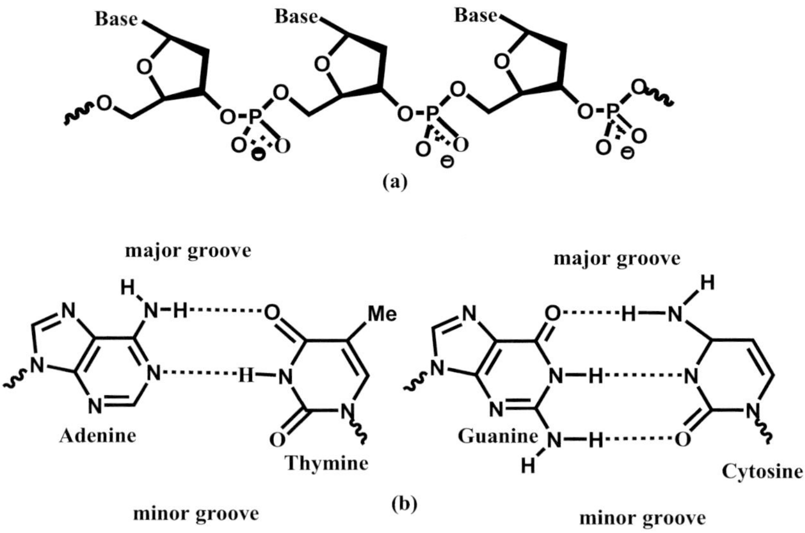

Figure 8.2 The components of DNA: (a) deoxyribose phosphate backbone. (b) DNA bases and their Watson–Crick hydrogen-bonded pattern.

The double-stranded double-helical structure of DNA, elucidated by Watson and Crick, is one of the scientific icons of the 20th century (Watson and Crick, 1953). The DNA structure termed B-DNA has two anionic sugar phosphate backbones enfolded around each other in a right-handed double-helix, with the bases hydrogen-bonded together in pairs (A with T and G with C) in the heart of the helix. The hydrophilic sugar-phosphate units point out into solution while the more hydrophobic bases are in the core. The bases are perpendicular to the helical axis and are stacked in a parallel fashion upon each other (face–face, π–π interactions) with a regular inter-planar separation of 3.5A° as shown in Fig. 8.3.

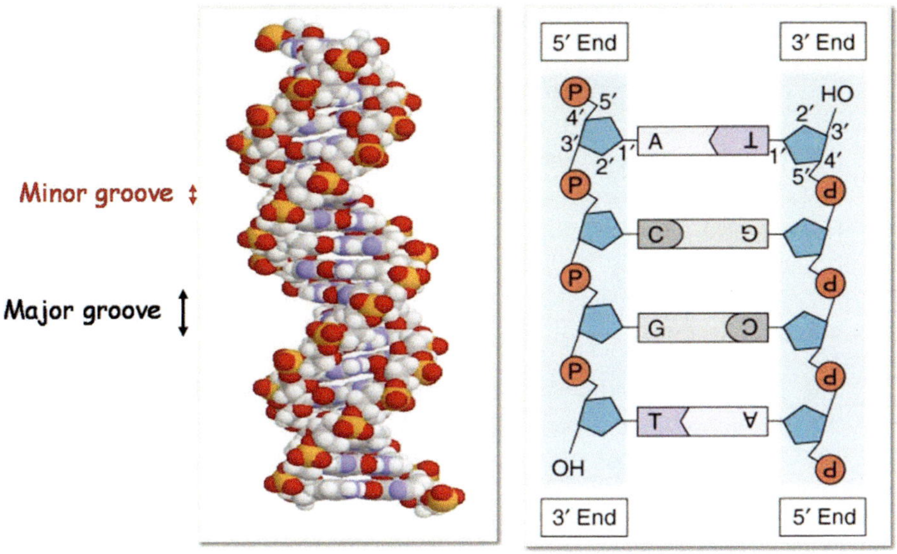

Figure 8.3 View of DNA double helix. Adapted from Pray (2008).

The attachment points of the bases to the sugar-phosphate backbone are offset with respect to the hydrogen bonds between the bases. The pairing of the non-symmetric bases and their stacking, at approximately 36° torsion angle gives rise in B-DNA to a small (minor) groove and a large (major) groove as shown earlier in Fig. 8.2. The sequence of the DNA bases along the polymeric sugar-phosphate chain encodes the genetic information and the Watson–Crick hydrogen bond recognition of G with C and A with T is key to the accurate reading and replication of the genetic information. B-DNA is believed to be the most prevalent form of DNA in biological systems although other double-helical forms such as the left handed Z-DNA and right-handed A-DNA (shorter and flatter than B-DNA) have been observed by crystallography and may have a more limited biological relevance.

8.2.2 Probable Interaction Sites

In practice, DNA is enfolded tightly like a ball of anionic (negatively charged) knitting wool. It needs some cationic species to enable it to hold together. Much of this role in cell nuclei is provided by histone proteins; however, metal salts and other small molecules such as polyamines contribute significantly to this structure. There are several mechanisms based on the structure and chemical composition of DNA, by which molecular substrates can bind to DNA and these interactions can be reversible or irreversible. Three different modes of DNA binding by metallomolecules are identified: non-specific external association, (major and minor) groove binding, and intercalation between bases of DNA.

8.2.2.1 Irreversible binding to DNA

The compound under investigation can form covalent bonds to the phosphodiester backbone, sugar residues, or bases of the DNA. The damage of DNA by the molecule can be taken as a basis of its chemotherapeutic investigation. This binding event can cause genetic instability due to which cancer cells are unable to effect the correct cell cycle checkpoint responses to induced damage and consequently undergo cell cycle arrest which may ultimately lead to apoptosis, or programmed cell death. Cisplatin and its derivatives interact with DNA through coordination bonds formed between the Pt(II) centers and available nitrogen atoms on nucleotides, commonly N7 atom on adjacent purine base. Intrastrand cross-link has the effect of kinking the structure of the DNA molecule and it is thought that this is the key event in the therapeutic action of these drugs (Gelasco et al., 1998).

8.2.2.2 Reversible binding to DNA

Coordination complexes that reversibly bind to DNA are becoming increasingly interesting in biological relevance. In such systems, the three-dimensional arrangement of ligands provides a means by which binding to DNA can be achieved. Since ligands can be easily interchanged or modified, this provides a mechanism to control hydrophobicity, binding affinity, and selectivity in addition to cellular uptake.

8.2.2.2.1 External association

Cationic molecules intend to associate with negatively charged DNA. Naturally occurring polyamines such as spermine bind to DNA through electrostatic interactions. In 1968, Eichhorn and Shin described the ability of metal ions to interact with DNA and suggested two probable binding modes: external association with the negatively charged phosphate backbone and interactions with the electron donor groups of the bases (Eichhorn and Shin, 1968). Electrostatic interactions, ligand hydrophobicity and the total size of the metal ions predominately influence the backbone association of DNA. Metals can also bind covalently to purine group. Simple metal complexes, such as $[Ru(bpy)_3]^{2+}$ (bpy = 2,2′bipyridine), that interact with DNA solely through electrostatic interactions usually possess very low binding affinities.

8.2.2.2.2 Major and minor groove binding

Another mode of reversible binding of a molecule to DNA could be associated with its minor or major groove of double helix. Molecules that bind to DNA grooves can span many base pairs and hence they can exhibit very high levels of DNA sequence-specific recognition. Molecules may approach within van der Waals contact distances of the groove walls and then occupy the DNA grooves. Recognition and sequence-selective binding is dependent on a combination of hydrogen bonding, van der Waals forces, hydrophobic contacts and electrostatic interactions. Classical groove binding agents are usually cationic and composed of aromatic rings connected by bonds with torsional freedom so that they are able to twist and become isohelical with the DNA groove. Such systems tend to reside in minor groove of B-DNA and selectively bind to narrower and more electronegative A-T rich sequences where electrostatic and van der Waals contacts are maximized. Any additional hydrogen bonding sites in the groove binder enhance binding affinities and largely drive more specific sequence preferences. In 1981, Gale et al. (1981) noted that most small molecules that bind to B-DNA grooves do so either in or via the minor groove of the double helix, while DNA binding proteins or gene-targeted oligonucleotides interact with the major groove. Owing to different dimensions of the two grooves, targeting them requires dissimilar and different shaped molecules. The major groove is much wider than the minor groove. Due to this dimensional difference, the major grooves are mainly the site of binding of many DNA binding proteins (Schleif, 1988). They recognize and bind in or around the major groove of DNA and form specific hydrogen-bonded contacts to the edges of the base pairs. This is because this groove not only shows the greater variation in size and shape with base sequence but also has a greater number and variation in the pattern of hydrogen-bond donor and acceptor units to which the protein can bind. Major groove recognition usually involves cylindrical binding units based on alpha helices, and the units are just the right size and shape to fit into the major groove as they are found too large for the minor groove. Synthetic or natural oligonucleotides are also found to recognize the major groove of DNA (Da Ros et al., 2005). They bind by forming hydrogen-bonds to the major groove edges of the purine nucleobases. Methyl green is a well-known example of major groove binder and is depicted in Fig. 8.4a. The minor groove of DNA can present as a selective binding sites for small and flat cationic molecules (Neidle, 1994). Lateron, Goodsell and Dickerson found that minor-groove binding drugs have a characteristic structure, usually contain two to five aromatic heterocycles linked by amide or vinyl groups with cationic groups at either ends (Goodsell and Dicherson, 1986). Some metal complexes do not fit neatly into the minor groove, but a part of such a molecule may slot into minor groove (Coggan et al., 1999). The minor groove binders stabilize DNA by van der Waals interactions with the walls and floor of the groove, as well as hydrogen bonds with the concave curvature of the inner surface of the molecule complementing the convex surface of the floor of the DNA minor groove. Such molecules have a preference for A–T rich sequences for which the groove has a deeper electrostatic potential and sterically less hindered. Among well-known minor groove binder, DAPI (4′,6-diamidino-2-phenylindole) is considered most simple and well-exploited example as shown in Fig. 8.4b.

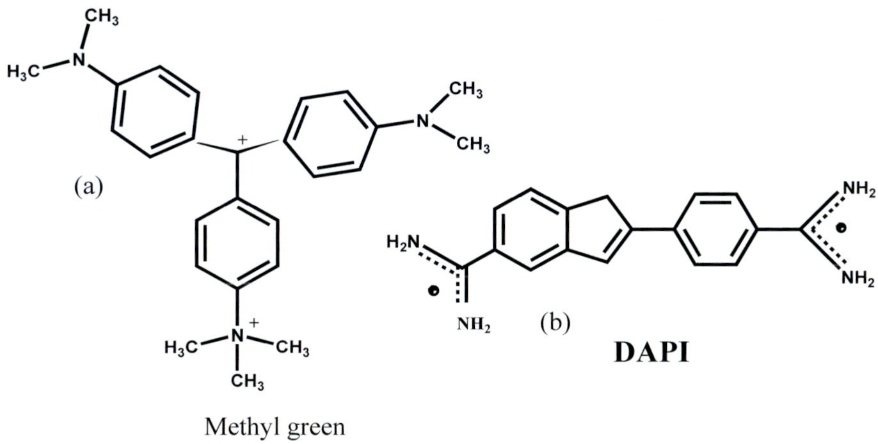

Figure 8.4 Major and minor groove binding agents.

DAPI was used as an antiparasitic agent although its use is limited by side effects. Its more common use is as a blue-fluorescent stain for DNA. It is readily transported across membranes into cells and has acquired significant popularity in microscopic studies. It binds to AT-rich regions of double stranded DNA in the minor groove and inhibits DNA and RNA polymerase. The A-T region of B-DNA has a narrower minor groove than G-C region and hence allows fitting of a drug against the walls of the groove (Fig. 8.5).

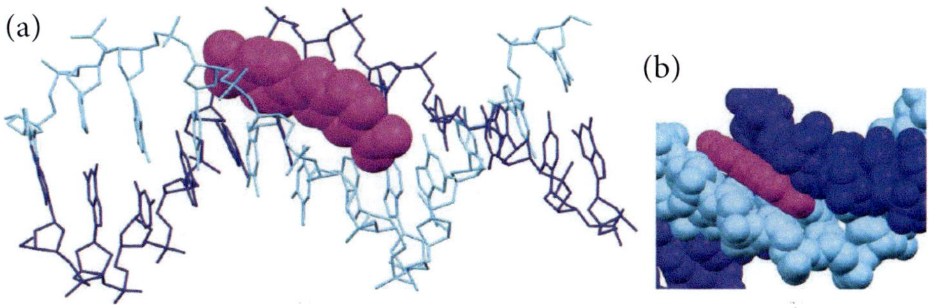

Figure 8.5 (a) View of DAPI (pink) binding in the minor groove of DNA. (b) Close-up space-filling view showing the snug fit in the narrow and deep AT-rich minor groove. Adapted from Larsen et al. (1989).

The complex $[Ru(phen)_3]^{2+}$ (Phen = 1,10 phenanthroline) was shown as a groove binder by Barton et al. (Barton et al., 1984).

It was initially, thought that this complex intercalate into B-DNA, with the D-enantiomer displaying the highest affinity. However, later work revealed that the complex is not an intercalator, but a groove binder (Satyanarayana et al., 1993). Furthermore, it was found that the L-enantiomer is a major groove binder at all loading ratios—with one phenanthroline ligand approximately parallel to the base pair planes (partial insertion). In contrast, the D-enantiomer exclusively bind to a minor groove with high loading but some partial insertion into major groove may also occurs (Coggan et al., 1999).

8.2.2.2.3 Intercalation between DNA bases

Mainwaring et al. defined the intercalation as the sandwich of a molecule between two adjacent pairs of bases in the DNA double helix (Mainwaring et al., 1982). Intercalators push the DNA-base apart, and intercalation could be attributed to the extended electron-deficient aromatic planar ring system. They extend and unwind the deoxyribose–phosphate backbone and are stabilized by π–π stacking interactions with the planar aromatic bases (Lerman, 1961). As the bases are pushed away by intercalation, they can lead to hydrodynamic changes in the DNA and lead lengthening of DNA beside decrease in twisting between the base pair layers, the stiffening of the helix, and the decrease in mass per unit length. These effects are fully reversible upon removal of the intercalator as long as the DNA duplex structure is not destroyed by the process of removal. Lerman showed that a bound intercalator lies in a plane perpendicular to the helix axis, and perpendicularity of the base pairs to the helix is not significantly altered (Lerman, 1963). Thus, for a metal complex to be an intercalator, it must either be planar or have an extended planar component which can slot between base pairs. The intercalation in ruthenium-based metallo-system is illustrated by dipyridophenanzine-bi(phenanthroline) ruthenium(II), $[Ru(phen)_2dppz]^{2+}$ (Haq et al., 1995) where a phenanthroline of $[Ru(phen)_3]^{2+}$ is extended by a planar aromatic group and prefers an intercalation. Its binding to DNA causes an elongation of the rod-like DNA molecule in consistence with classical intercalation (Haq et al., 1995).

8.2.3 Mode of Binding and Their Detection Using Spectroscopic and Biochemical Techniques

Metal-containing complexes can have dramatic effects on DNA structure. There are various ways of measuring the deviation of the helix backbone from a straight line and various measures of DNA base displacement from the canonical B-DNA positions. Such measures really rely on atomic level structural information from nuclear magnetic resonance spectroscopy (NMR), X-ray crystallography or molecular modeling and are valuable when such data are available. A description of the main techniques used to detect and analyze DNA structural changes in this context is given below.

8.2.3.1 UV-visible absorption spectroscopy

UV–Visible absorption spectroscopy is the simplest and most commonly employed instrumental technique for studying both the stability of DNA and their interactions with small molecules. The study of complex–DNA interactions could be carried out by UV–visible absorption spectroscopy by monitoring the changes in the absorption properties of the complex or the DNA molecules. Usually, complexes show an absorption band that can clearly be distinguished in the visible region. The intercalation of compound and DNA can be monitored by the changes observed in the position of absorption in free state and on binding with DNA. It has been assumed that the magnitude of this shifting could be correlated with the strength of the interaction between DNA and molecule under consideration (Sun et al., 2011; Jaumot et al., 2012; Wei et al., 2010; Bhadra et al., 2011). The UV–visible absorption spectrum of DNA shows a broad band at λ_{max} 200–350 nm in the UV region with a maximum absorption at λ_{max} 260 nm. This maximum observed is considered a consequence of the chromophoric groups present in purine and pyrimidine bases of DNA. Complex–DNA interactions can be studied by a comparison of UV–visible absorption spectra of a free complex and that of complex–DNA adduct. Compounds binding with DNA through intercalation usually results hypochromic and bathochromic shift. The extent of hypochromism is usually found consistent with the strength of intercalation (Liu et al., 2002). The strength of this electronic interaction is expected to decrease as the cube of the distance between the chromophore and the DNA bases decreases. By decreasing the distance between intercalated compound (drug) and DNA bases, hypochromism take place apparently. Thus, this is consistent with the combination of compound π-electrons and π-electrons of DNA bases. Consequently, the energy level of the π–π electron transition decreases, which causes a red shift in absorption maxima. This contributes to the hypochromic effect (Sirajuddin et al., 2012; Sirajuddin et al., 2012). In case of electrostatic attraction between the compound and DNA, hyperchromic effect is observed that reflects the corresponding changes of DNA in its conformation and structure after the complex–DNA interaction has occurred. The hyperchromic effect is the outstanding increase in absorbance of DNA upon denaturation. The two strands of DNA are held together mainly by the stacking interactions, hydrogen bonds and hydrophobic effect between the complementary bases. When DNA double helix is treated with denaturing agents, the interaction force holding the double helical structure is disrupted. The double helix then separates into two single strands which are in the random coiled conformation. At this time, the inter base interaction will be reduced, increasing the UV absorbance of DNA solution because many bases are in free form and do not form hydrogen bonds with complementary bases. As a result, the absorbance for single-stranded DNA is found 40% higher than that for double stranded DNA at the same concentration. Furthermore, the hyperchromic effect arises mainly due to the presence of charged cations which bind to DNA via electrostatic attraction to the phosphate group of DNA backbone and thereby causing a contraction and overall damage to the secondary structure of DNA (Arjmand et al., 2011). The hyperchromic effect may also be attributed to external contact (electrostatic binding) (Pratviel et al., 1998) or to partial uncoiling of the helix structure of DNA, exposing more bases of the DNA (Shahabadi et al., 2010). If there is a weaker interaction then only hypochromic or hyperchromic effects are observed without significant changes of shifts in the spectral profiles (Gonzalez-Ruiz et al., 2011; Kumar et al., 2009). The absorption spectral titration are used to determine the binding constants (K_b) of the complexes using the expression $[DNA]/(\varepsilon_a - \varepsilon_f) = [DNA]/(\varepsilon_b - \varepsilon_f) + 1/K_b(\varepsilon_b - \varepsilon_f)$, where ε_a, ε_f, and ε_b are the apparent absorption coefficient, ε of the complex in free form, and ε of the complex in fully bound form, respectively. The K_b value was obtained from the $[DNA]/(\varepsilon_a - \varepsilon_f)$ vs. [DNA] plot. Spectral changes of two Ru(II) complexes $[RuH(HL)(PPh_3)_2(CO)]$ and $[RuH(HL)(AsPh_3)_2(CO)]$ (HL = 2,2′-bipyridine-5,5′-dicarboxylic acid) (Kamatchi et al., 2013) is shown in Fig. 8.6 after addition of CT DNA.

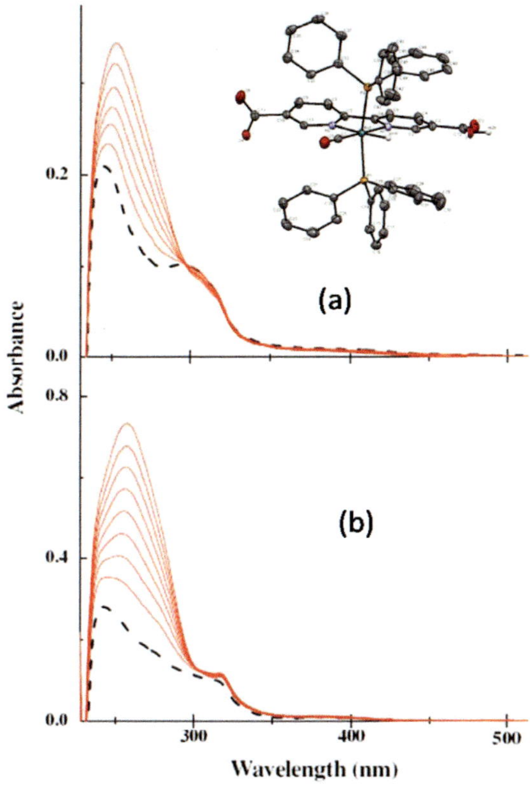

Figure 8.6 UV-visible spectra of complexes (a) $[RuH(HL)(PPh_3)_2(CO)]$ and (b) $[RuH(HL)(AsPh_3)_2(CO)]$ in the absence (- - -) and in the presence of CT-DNA in increasing amounts, [Complex] = 10 μM, [DNA] = 0–100 μM. Adapted from Kamatchi et al. (2013).

The absorption spectrum of complex $[RuH(HL)(PPh_3)_2(CO)]$ on the addition of DNA showed hypochromism with a slight red shift in π–π ligand-centered band at λ_{max} 310 nm. For complex

$[RuH(HL)(AsPh_3)_2(CO)]$, the addition of increasing amount of DNA resulted in hyperchromic effect in the absorption. These results suggested a relatively weak association of $[RuH(HL)(PPh_3)_2(CO)]$ and $[RuH(HL)(AsPh_3)_2(CO)]$ to the helix. It is obvious from Fig. 8.6 that as DNA concentration is increased, main absorption shows hypochromic effect without red-shift in the band position of $[RuH(HL)(PPh_3)_2(CO)]$. Hence, it clearly binds to DNA by non-intercalative mode.

8.2.3.2 Fluorescence spectroscopy

Fluorescence spectroscopy is one of the most common, sensitive, and selective techniques used to study interactions between molecules with DNA. The intense and most useful fluorescence is observed from compounds containing aromatic functional groups with low-energy π–π^* transition levels. Compounds containing aliphatic and alicyclic carbonyl structures or highly conjugated double-bond structures may also show fluorescence, but the number of these transitions is small compared to those observed from aromatic systems (Jaumot et al., 2012; Lakowicz, 2006). The orientation of fluorophoric ligands and their closeness to DNA base pairs can be studied by fluorescence anisotropy or fluorescence resonance energy transfer. Fluorescence quenching experiments give additional information concerning the localization of the drugs and their mode of interaction with DNA (Gonzalez-Ruiz et al., 2011). Fluorescence emission is very sensitive to the environment, and hence the fluorophore transfer from high to low polarity environments commonly causes spectral shifts 10–20 nm in the excitation and emission spectra of drugs (Suh et al., 1995). Moreover, the effective interaction with DNA commonly causes a significant enhancement of the fluorescence intensity as a consequence of different factors. Thus, in the case of intercalating drugs, the molecules are inserted into the base stack of the helix. The rotation of the free molecules favors the nonradiative deactivation of the excited states, but if the drugs are bound to DNA the deactivation through fluorescence emission is favored, and a significant increase in the fluorescence emission is normally observed. In case of groove binding agents, electrostatic, hydrogen bonding or hydrophobic interactions are involved and the molecules are close to the sugar-phosphate backbone. It is possible to observe a decrease in fluorescence intensity in the presence of DNA. The use of well-established quenchers, such as halide ions, provides further information about the binding of drugs to DNA. The groove binders are more sensitive to the quenching effect by halides than the intercalating agents because the pairs of bases hinder the accessibility of the drug by the quenchers. Besides, the electrostatic repulsion among phosphate groups on DNA and anionic quenchers protect the drug from the quencher effects. Thus, in case of intercalation, reduction in K_{SV} occurs. K_{SV} is the Stern–Volmer quenching constant. It can be obtained from the slope of the plot of F_o/F vs. [DNA] using the equation, F_o/F (or I_0/I) = 1 + K_{SV} [Q] where F_o or I_0 is the fluorescence intensity in the absence of quencher, while F or I is the fluorescence intensity in the presence of quencher (Gonzalez-Ruiz et al., 2011; Li et al., 2005).

The emission intensity of ethidium bromide (EB) is used as a spectral probe, as it shows enhanced emission intensity when it is bound to the hydrophobic part of DNA. The binding of the complexes to DNA could result in the displacement of the bound EB and could cause a decrease in emission intensity due to quenching by the paramagnetic complexes. The DNA binding propensity of complexes is measured from the reduction of the emission intensity of EB at different complex concentrations. The fluorescence spectral method using EB as a reference was used to determine the relative DNA binding properties of the complexes to calf thymus (CT) DNA in Tris-HCl/NaCl buffer. DNA concentrations, expressed with respect to mononucleotides, were determined spectrophotometrically using the reported data for a molar absorption coefficient of 6600 M^{-1} cm^{-1} at 260 nm. Fluorescence intensities of EB at λ_{max} 601 nm with an excitation wavelength of λ_{max} 510 nm were measured at different complex concentrations. Reduction in emission intensity was observed with the addition of complexes. The relative binding tendency of the complexes to CT DNA was determined from a comparison of the slopes of the lines in the fluorescence intensity versus complex concentration plot. The apparent binding constant (K_{app}) was calculated using the equation K_{EB} [EB] = K_{app} [complex], where the complex concentration was the value at a 50% reduction of the fluorescence intensity of EB and K_{EB} = 1.0 × 10^7 M^{-1}. Binding of Ruthenium(III) complexes of formula mer-[$RuCl_3$(dmso)(1,10-phenanthroline)] and mer-[$RuCl_3$(dmso) (dipyrido[3,2-a:2′,3′-c] phenazine)] (dmso = dimethyl sulfoxide) with DNA has been investigated using competitive binding with EB (Tan et al., 2008). The addition of complexes to DNA, pretreated with EB, causes appreciable reduction in emission intensity relative to that observed in the absence of the complex as shown in Fig. 8.7, and shows intercalative mode of binding.

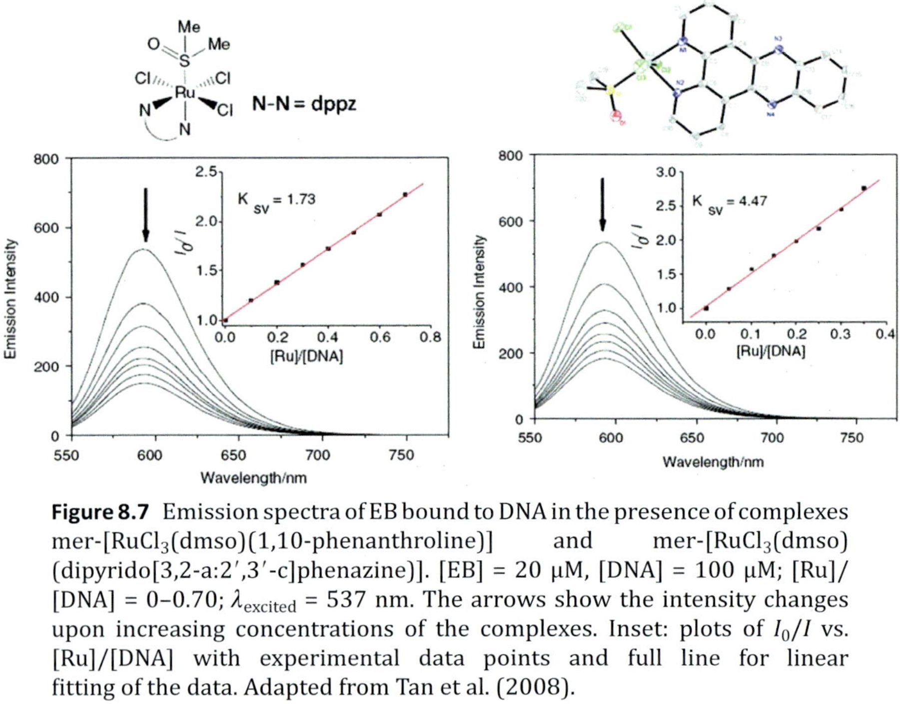

Figure 8.7 Emission spectra of EB bound to DNA in the presence of complexes mer-[$RuCl_3$(dmso)(1,10-phenanthroline)] and mer-[$RuCl_3$(dmso) (dipyrido[3,2-a:2′,3′-c]phenazine)]. [EB] = 20 μM, [DNA] = 100 μM; [Ru]/[DNA] = 0–0.70; $\lambda_{excited}$ = 537 nm. The arrows show the intensity changes upon increasing concentrations of the complexes. Inset: plots of I_0/I vs. [Ru]/[DNA] with experimental data points and full line for linear fitting of the data. Adapted from Tan et al. (2008).

8.2.3.3 Viscosity measurements

Hydrodynamic measurements that are sensitive to length change (i.e., viscosity and sedimentation) are regarded as the least ambiguous and the most critical tests of the binding mode of a small molecule to DNA. A classical intercalative mode causes significant increase in viscosity of the DNA solution due to an increase in the separation of base pairs at the intercalation sites, hence an increase in overall DNA length occurs. In contrast, complex that binds with DNA grooves by partial and/or non-classical intercalation causes less pronounced changes in the viscosity (Satyanaryana et al., 1993). Under appropriate conditions, EB usually increases the relative specific viscosity due the lengthening of the DNA double helix which results from intercalation. On increasing the concentration of $[RuH(HL)(PPh_3)_2(CO)]$ and $[RuH(HL)(AsPh_3)_2(CO)]$, the relative viscosity of DNA did not alter significantly and found consistence with the well-known $[Ru(bpy)_3]^{3+}$ as shown in Fig. 8.8. Such a trend is typical of electrostatic mode (Lerman, 1961)).

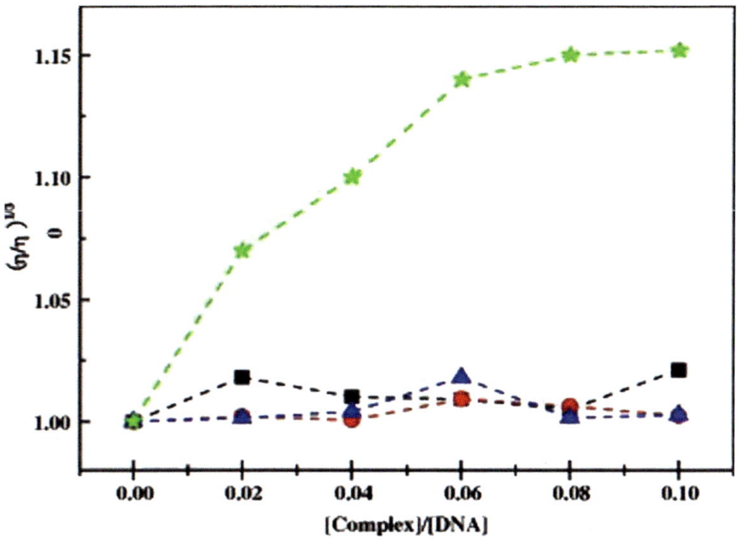

Figure 8.8 Effect of increasing amounts of complexes and EB on the relative viscosities of CT-DNA in 5 mM Tris–HCl buffer (pH 7.0). $[Ru(bpy)^3]^{2+}$ (■), $[RuH(HL)(PPh_3)_2(CO)]$ (●), $[RuH(HL)(AsPh_3)_2(CO)]$ (▲), EB(★). (Kamatchi et al., 2013).

8.2.3.4 Circular dichroism (CD)

Circular dichroism (CD) is a useful technique and allows to assess whether nucleic acids undergo conformational changes as a result of complex formation or changes occurred owing to change in its environment. Circular dichroism, the difference in absorption of left and right circularly polarized light, is uniquely sensitive to chirality or helicity of molecules. Thus, CD depends on the arrangement of the DNA bases and if this is changed, CD spectrum changes. The source of chirality in DNA is the ribose sugar backbone of DNA. Between 190 and 300 nm, there are DNA signals observed from coupling of the stacked planar bases, which adopt chirality (helical arrangement in space) structures. CD can be used to probe DNA structures. Since right-handed B-DNA and the left-handed Z-DNA, for example, have quite different CD spectra (not mirror images of one another since the structures are not enantiomeric), the effect of DNA on the CD of bound ligands can also be a useful probing technique. This technique is useful in diagnosis of changes in DNA morphology during drug–DNA interactions. The band due to base stacking at λ_{max} (~275 nm) and that due to right-handed helicity λ_{max} (~245 nm) are quite sensitive to the mode of DNA interactions with small molecules. Therefore, changes in CD spectral pattern of DNA upon interaction with complex are often assigned to corresponding changes in DNA structure. The simple groove binding and electrostatic interactions of small molecules show less or no perturbation on base-stacking and helicity bands.

8.2.3.5 Nuclear magnetic resonance spectroscopy

NMR spectroscopy is a powerful technique that enables geometrical details of the structure of DNA on an atomic level in solution to be elucidated. In addition, changes in chemical shifts induced by temperature or concentration variation can provide information about DNA–complex association. When a nucleus is placed in an external magnetic field, its spin magnetic moment moves around the direction of the field at a frequency termed the Larmor frequency, which is directly proportional to the strength of the magnetic field experienced by the particle. With metallo-systems binding to DNA, the extent of changes in chemical shifts for DNA protons upon complex binding have often been used as an indication of binding site. A better approach is to use a range of alternative NMR techniques which give through-bond or through-space distances between atoms. Nuclear Overhauser effect spectroscopy (NOESY) provides the ability to plot through space connectivity between atoms with the strength of the signal observed depending on their distance. NOESY couplings arise because the use of a small radio frequency field at Larmor frequency of one nuclear spin results in an enhancement of the magnetization of some of non-radiated nuclear spins (Levitt, 2001). Thus, one can identify whether given atoms are near each other in three-dimensional structures. The NOESY data gives certain inter atomic distances, which can be used as restraints in molecular modeling and one can therefore sometimes determine 3D structures. NOESY data cannot be used to analyze the structure of a long DNA molecule as the peak-assignments on the resulting spectra become complicated and the chemical shifts for various atoms get overlapped. Caspar et al. observed the binding of optically pure ruthenium complexes Δ- or Λ-$[Ru(bpy)_2(L\text{–}L)][PF_6]_2$ [where, L–L Hcmbpy = 4-carboxy-4′-methyl-2,2′-bipyridine] to DNA. The binding of the two enantiomeric Δ and Λ complexes to DNA were estimated from ^1H NMR spectroscopic titrations. 2D transferred NOESY experiments support the conclusion that Δ and Λ bind to DNA and that an intermediate-to-fast exchange occurs between bound and free Ru(II) complex (Caspar et al., 2004).

8.2.3.6 X-Ray crystallography

X-Ray crystallography relies on Bragg's law, which states that X-ray beam reflecting from a surface layer of a crystalline material travels a shorter distance than those reflected by the inner layers. The beams are in-phase if the difference in these distances is an integer value of wavelengths of the incident radiation and hence produces an enhanced signal compared to that out-of-phase. Bragg suggested that the differences in distances relies upon the angle of incidence of the beam so that by changing this angle, a diffraction pattern can be built up which can then be Fourier-transformed and interpreted to give atomic-level structural information. It is extremely attractive to have such data for metal complex–DNA systems, however, it is challenging to crystallize the sample. This is often not a trivial exercise, and always involves carefully chosen short DNA sequences. The biological significance of such structures is not clear. There is also the fact that solid-state crystalline structures do not necessarily bear any relationship to those adopted in biological systems.

8.2.3.7 Gel electrophoresis

Gel electrophoresis involves application of an electric field across the length of the gel causing the migration of charged molecules toward the oppositely charged electrode. At neutral pH, DNA is negatively charged so moves through the gel from cathode towards the anode. The electrophoretic mobility of a piece of DNA on a gel is dependent on its flexibility, size and charge. If, a molecule binds to DNA and affects the size or shape of DNA, it then will affect its electrophoretic mobility. For example, an intercalator is expected to unwind the DNA, which lengthens and stiffens it. Using circular DNA of known supercoiling, the degree of unwinding per ligand bound can be determined. Ethidium bromide, an archetypical intercalator, unwinds DNA by an average of 26° per ligand (which is 10° less than the base–base twist in canonical B-DNA). The degree of unwinding induced by a given metal complex has been shown to indicate binding mode it adopts: mono-adduct platinum(II) complexes generally afford little unwinding, whereas bifunctional adducts such as cisplatin (unwinding angle 13°) have more effect, though significantly less than that of an intercalator. An alternative application of gel electrophoresis has been described by Carle and Olson. They used it as a method of fractionating DNA into size-based fractions by applying a series of orthogonal electric fields on a gel. By altering the duration of the electric field pulses in each direction, a greater accuracy in the DNA separation for a given size range can be achieved. The technique can also be used to check for DNA charge reversal which can occur when DNA is re solubilized in the presence of multivalent cations. Charge reversal is said to have occurred if the DNA migrates towards the cathode terminal. Ruthenium(II) arene anticancer complex $[(\eta^6\text{-p-terp})\text{RuII(en)Cl}]^+$ (*p*-terp = paraterphenyl, en = 1,2-diaminoethane) has been reported and shown that this complex exhibits promising toxic effects in several human tumor cell lines and concomitantly its DNA binding mode involves combined intercalative and monofunctional (coordination) binding modes (Bugarcic et al., 2008). The unwinding of supercoiled plasmid pSP73KB DNA induced on binding complex was determined by incubating the plasmid with complex for 24 h at 37°C at various r_b (different lanes in the gel). The native agarose gels resulting from DNA modified by the complex, is shown in Fig. 8.9. A decrease in the rate of migration is the result of unwinding the DNA as this reduces the number of supercoils.

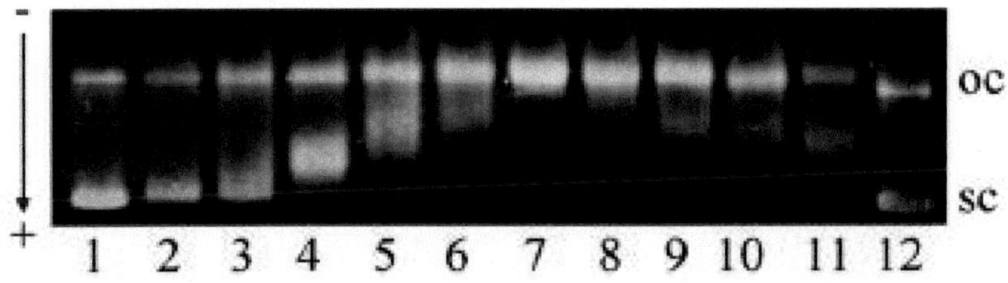

Figure 8.9 The unwinding of supercoiled pSP73KB plasmid DNA by complex. The plasmid was incubated with $[(\eta^6\text{-p-terp})\text{RuII(en)Cl}]^+$ complex in 10 mM $NaClO_4$, at pH 6 for 24 h at 37°C. Lanes in the top panel: 1 and 12, control, unmodified DNA; 2, r_b = 0.008; 3, r_b = 0.016; 4, r_b = 0.024; 5, r_b = 0.031; 6, r_b = 0.039; 7, r_b = 0.047; 8, r_b = 0.055; 9, r_b = 0.063; 10, r_b = 0.071; 11, r_b = 0.079. The top bands in each panel correspond to the form of nicked plasmid and the bottom bands to the closed, negatively supercoiled plasmid. (r_b values are defined as the number of atoms of the metal bound per nucleotide residue). Adapted from Bugarcic et al. (2008).

8.3 Cytotoxicity of Ruthenium Complexes

Studies of medicinal applications of ruthenium complexes have been facilitated by the wide diversity of the coordination and organometallic chemistry of ruthenium. Briefly, earliest types of anticancer ruthenium complexes, proposed by Clarke and co-workers in the 1980s, were chlorido-ammine Ru(II)/Ru(III) complexes which were thought to act primarily by binding to DNA (Clarke, 2003). In this chapter, the cytotoxicity of a few ruthenium complexes based on dimethyl sulfoxide, heterocycles, polypyridyl, and arene derivatives has been described as follows.

8.3.1 Dimethyl Sulfoxide Complexes

Dimethyl sulfoxide (DMSO) complexes of both ruthenium(II) and ruthenium(III) exhibit antitumor activity comparable to cisplatin at equitoxic dosage in animal models of metastasizing tumors, but with less severe side effects and prolonged host survival times

(Sava et al., 1999). A small series of complexes whose parent compounds are *cis*- and *trans*-[Ru(II)$(DMSO)_4Cl_2$] (compounds **1** and **2**) constitute one class of dimethyl sulfoxide ruthenium compounds (Clarke et al., 1999; Sava et al., 1999). Examination of their effects on primary tumor and on metastasis has revealed antimetastatic activities superior to effects on primary tumor growth. The initial studies were performed with *cis*-[Ru(II)$(DMSO)_4Cl_2$] because of its similarity to cisplatin. However, the comparison of the antitumor effects of *cis* and *trans*-[Ru(II)$(DMSO)_4Cl_2$] has revealed superiority of the latter.

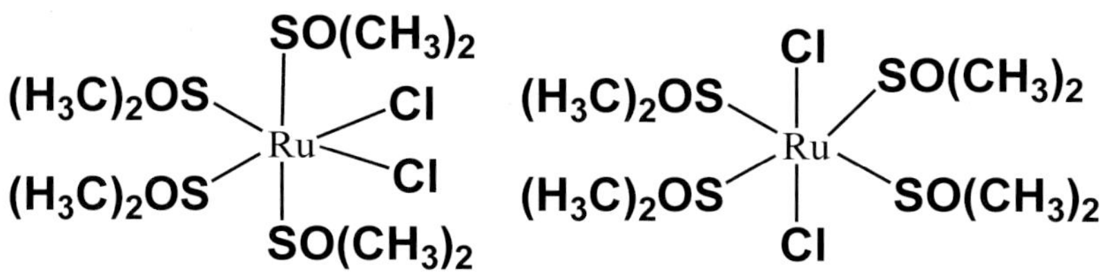

1 **2**

Cis- and *trans*-[Ru(II)$(DMSO)_4Cl_2$] contain two chlorides in the octahedral structure (Mestroni et al., 1989). In *cis*-[Ru(II)$(DMSO)_4Cl_2$] the three DMSO molecules are S-bound in a facial configuration and the fourth is O bonded. In *trans*-[Ru(II)$(DMSO)_4Cl_2$] all the DMSO are S-bound. When dissolved in water, the *cis* isomer immediately undergoes loss of the O-bonded DMSO ligand whereas the *trans* compound rapidly loses two S-bonded DMSO ligands yielding *cis*-di aqua species. Both hydrolyzed isomers then undergo slow reversible chloride dissociation forming cationic compounds. After this step, the *trans* compound contains three reactive groups while the *cis* isomer only two (Mestroni et al., 1989). In addition, the three remaining DMSO ligands in the *cis* isomer represent a considerable steric hindrance, which makes the *cis* aqua species inert relative to the *trans* isomer. This difference correlates with a higher potency of the *trans* isomer as an antitumor agent (Loseto et al., 1991). Both *cis*- and *trans*-[Ru(II)$(DMSO)_4Cl_2$] bind to DNA in cell-free media (Loseto et al., 1991; Novakova et al., 2000). Some early studies based on the analysis of circular dichroism (CD) spectra of DNA have suggested that the coordination of the *cis* isomer to

DNA does not significantly alter the conformation of B-DNA (Mestroni et al., 1989). The *trans* isomer binds to DNA more rapidly with some changes in the CD spectra indicating conformational alterations (Mestroni et al., 1989). Both isomers have a limited preference for bifunctional binding to neighboring guanine residues at their N7 atoms with the *trans* isomer being more effective (Novakova et al., 2000). The DNA binding mode of *trans*-[Ru(II)$(DMSO)_4Cl_2$] includes formation of bifunctional adducts such as intrastrand cross-links between neighboring purine residues and a small amount (~1%) of interstrand cross-links. *Cis*-[Ru(II)$(DMSO)_4Cl_2$] forms on natural DNA mainly monofunctional lesions. These findings are consistent with the results of the binding of *cis*- and *trans*-[Ru(II)$(DMSO)_4Cl_2$] to the dinucleotide demonstrating a considerably slower binding of the *cis* isomer (Anagnostopoulou et al., 1999). Both ruthenium isomers induce conformational alterations in DNA, the *trans* compound being more effective. In addition, DNA adducts of *trans*- [Ru(II)$(DMSO)_4Cl_2$] are capable of inhibiting RNA synthesis by DNA-dependent RNA polymerases, while the adducts of the *cis* isomer are not. Thus, several features of the DNA binding mode of *trans*-[Ru(II)$(DMSO)_4Cl_2$] are similar to those of antitumor cisplatin (Novakova et al., 2000), which may be relevant to the biological effects of this antitumor ruthenium drug. On the other hand, the different DNA binding mode of *cis*-[Ru(II)$(DMSO)_4Cl_2$] is consistent with its less pronounced biological effects. The cytotoxicity and photocytotoxicity of *cis*- and *trans*-[Ru(II)$(DMSO)_4Cl_2$] complexes was tested in two melanoma cell lines, human (SK-MEL 188) and mouse (S91) (Brindell et al., 2005). The *trans* isomer was found to be more effective for cell growth inhibition than its *cis* analogue both in the presence and in the absence of light. However, the antiproliferative activity of both isomers was significantly enhanced after irradiation with UVA light in comparison with their activity in the dark. Interesting results have been also obtained in studies of the mechanism of antitumor activity of Ru(II)(C_6H_6)(DMSO)Cl_2 complex (compound 3) (Gopal et al., 1999). This compound exhibits a strong DNA-binding affinity, but binding does not affect substantially DNA conformation. On the other hand, it could completely inhibit DNA relaxation activity of topoisomerase II by trapping it into a ternary complex with DNA. A model has been proposed for this ternary complex, in which the ruthenium atom is coordinately bound to DNA and its ligands are cross-linked to topoisomerase II.

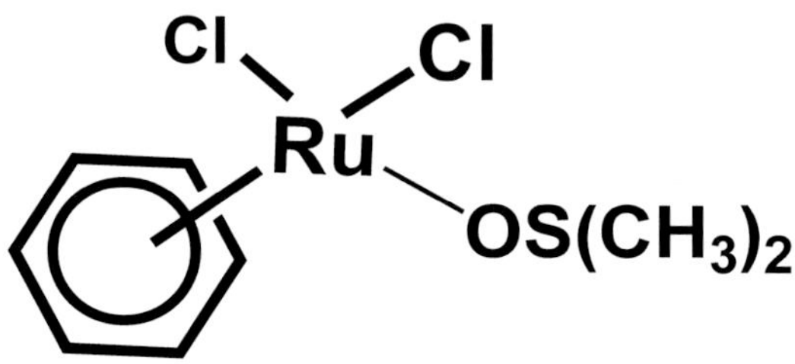

3

8.3.2 Heterocyclic Complexes

Heterocyclic complexes of Ru(III) constitute a relatively new group of potential anticancer compounds (Keppler et al., 1993; Clarke et al., 1999). The general formula of this structural class is (HB)[Ru(III)B_2Cl_4], where B stands for a heterocyclic base, such as imidazole (Im) or indazole (compounds **4** and **5**, respectively). These complexes exhibit activity in various tumor models and are particularly effective against cisplatin-resistant colorectal tumors (Keppler et al., 1993). The complex (HInd)[Ru(III)Cl_4 $(Ind)_2$] (KP1019) (compound **5**) is highly active against a colorectal tumor cells both in vivo and in vitro (Berger et al., 1989; Galeano et al., 1992; Seelig et al., 1992; Sava and Bergamo, 2000; Galanski et al., 2003), is completely devoid of side effects and drug induced lethality at therapeutically relevant doses (Keppler et al., 1993). Its therapeutic index is better than that of (HIm)[Ru(III)$Cl_4(Im)_2$]. The complex (HInd)[Ru(III)$Cl_4(Ind)_2$] has been shown to be efficiently taken up into the cells probably via interaction with transferrin (Kratz et al., 1994; Frasca et al., 2001; Pongratz et al., 2004). It induces apoptosis (Kapitza et al., 2005), but the cellular mechanisms of the apoptosis induction are still largely unknown.

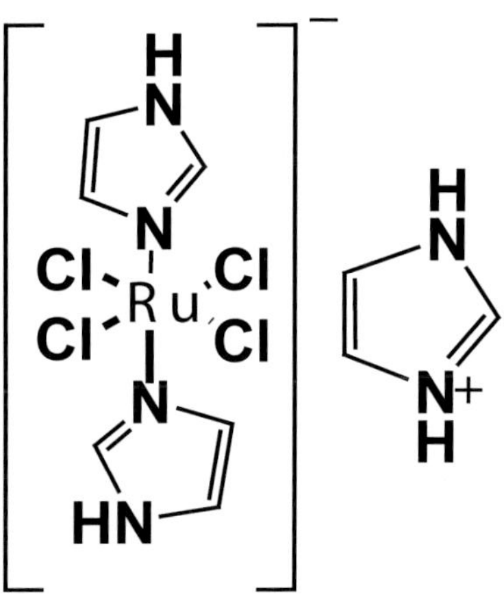

4

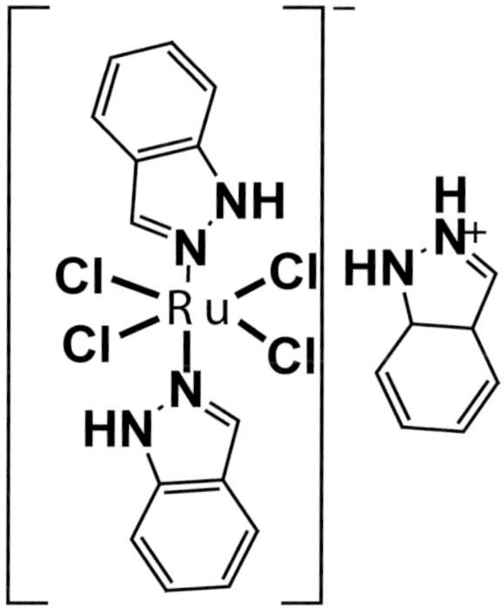

5

Ruthenium as a congener of iron, shows strong affinity for transferrin and displays reductive activation (Frasca et al., 1996; Clarke, 2003) in cells. It can be assumed that Ru(III) can substitute Fe(III) and can induce Fenton type redox processes and intracellular radical formation. This may well result in cellular damage that induces apoptosis. Hence, induction of oxidative stress seems to be an essential component of the cytotoxic effect of Ru(III) complexes. This cellular mechanism may provide rapid onset of Ru(III) induced apoptosis. It also allows the protection of normal tissues by the combination with moderate amounts of antioxidants. The complex (HInd)[Ru(III)$Cl_4$$(Ind)_2$] interacts with DNA and forms cross-links or induces strand breaks. This complex induces formation of H_2O_2 and DNA-strand breaks in colorectal tumor cells in a dose-dependent way (Kapitza et al., 2005). Both effects are inhibited by *N*-acetylcysteine and concomitantly cytotoxicity is reduced. Induction of apoptosis has been shown by loss of mitochondrial membrane potential and by caspase-dependent cleavage of poly-(ADP-ribose)-polymerase (PARP). Both effects were inhibited by *N*-acetylcysteine, which reduced the population with depolarized mitochondrial membranes and prevented cleavage by poly-(ADP-ribose)-polymerase. It indicates an important role of oxidative stress in apoptosis induced by

(HInd)[Ru(III)$Cl_4$$(Ind)_2$]. Later on, analogues of these complexes were developed, namely Na[*trans*-Ru(III)((DMSO)Cl_4(Im)] (NAMI) and (H_2Im)[*trans*-Ru(DMSO)Cl_4(Im)] (NAMI-A) (compound **6**) was synthesized with the aim of improving the solid state stability of the complex. Both NAMI and NAMI-A exhibit encouraging antitumor and antimetastatic properties; NAMI-A exhibits high efficiency in vivo against lung metastasis and is currently on clinical trial as an antimetastatic drug (Sava et al., 1999; Sava and Bergamo, 2000; Rademaker-Lakhai et al., 2004). Thus, the spectrum of the antitumor effects of these ruthenium compounds differs significantly from that of cisplatin while showing lower systemic toxicity than platinum(II) compounds (Keppler, 1993).

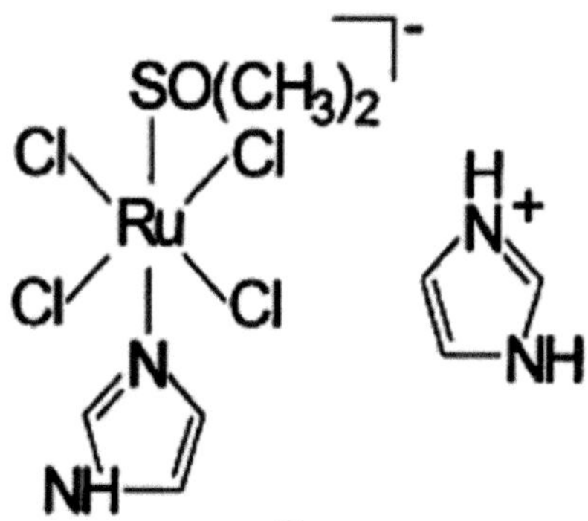

6

8.3.3 Polypyridyl Compounds

A third group of antitumor ruthenium compounds discussed in this chapter are the ruthenium complexes of polypyridyl ligands. The DNA binding and cleavage properties of various ruthenium polypyridyl complexes have been investigated, because they have been proposed as useful probes of tDNA conformation (Barton, 1986) or DNA cleavage agents (Grover et al., 1994; Neyhart et al., 1995). Analogues of these ruthenium complexes containing aqua or chloro groups have also been synthesized and it was found that they bind DNA covalently in cell-free media (Barton and Lolis, 1985; Grover et al., 1994). The aqua or chloro-ligands in these complexes represent leaving-ligands in contrast to the kinetically more stable pyridyl groups. The cytotoxicity of chloropolypyridyl ruthenium complexes of structural formulas [Ru(II)Cl(bpy)(terpy)]

Cl, *cis*-[Ru(II)(bpy)$_2$Cl$_2$] and *mer*-[Ru(II)Cl$_3$(terpy)] (**7**) (bpy = 2,2′-bipyridyl, terpy = 2,2′:6′,2″-terpyridine) has been demonstrated in murine and human tumor cell lines (Novakova et al., 1995). *mer*-[Ru(II)Cl$_3$(terpy)] exhibits a remarkably higher cytotoxicity than the other complexes. Moreover, investigations of antitumor activity (studied on murine lymphosarcoma LS/BL ascitic tumor) have revealed the highest efficiency for *mer*-[Ru(II)Cl$_3$(terpy)]. In a cell free medium, the ruthenium complexes coordinate to DNA preferentially at guanine residues (Novakova et al., 1995; van Vliet et al., 1995). The resulting adducts terminate DNA synthesis in vitro. The reactivity of the complexes to DNA, their efficiency to unwind supercoiled DNA and a sequence preference of their DNA adducts do not show a correlation with biological activity. On the other hand, the cytotoxic *mer*-[Ru(II)Cl$_3$(terpy)] exhibits a significant DNA interstrand cross-linking, in contrast to the less active complexes. Thus, this potential new class of metal-based antitumor compounds may act by a mechanism involving DNA interstrand cross-linking.

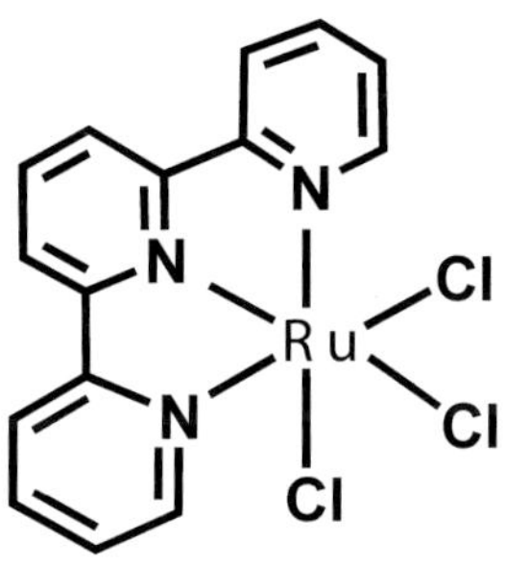

7

The cytotoxicity of three isomeric dichlororuthenium(II) complexes α-, β-, and γ-[Ru(II)(azpy)$_2$Cl$_2$] (azpy = 2-phenylazopyridine) has been investigated against a series of tumor cell lines (Hotze et al., 2005). The complex α-[Ru(II)(azpy)$_2$Cl$_2$] exhibits cytotoxicity higher than cisplatin. It is in contrast to much lower cytotoxicity of *trans*-dichloro complex γ-[Ru(II)(azpy)$_2$Cl$_2$] and the *cis*-dichloro isomer β-[Ru(II)(azpy)$_2$Cl$_2$]. The binding of the complex has been compared with previously observed results for bis(bipyridyl)ruthenium(II) complex (Hotze et al., 2005). The ligands 9-ethylguanine and guanosine form monofunctional adducts. The guanine derivatives in the azpy complexes can have more orientations than found for related *cis*-[Ru(II)(bpy)$_2$Cl$_2$] species. This versatility is considered important in the binding of α-[Ru(II)(azpy)$_2$Cl$_2$] complex to DNA. It is related to the cytotoxicity of this compound. Tris(ligand) complexes [RuL$_3$](PF$_6$)$_2$ (L = 2 phenylazopyridine or *o*-tolylazopyridine) and mixed ligand [RuL′2L″](PF$_6$)$_2$ (L′ and L″ are 2-phenylazopyridine or bpy) have been synthesized, structurally characterized and investigated for their cytotoxicity (Hotze et al., 2005). These complexes were designed to test the hypothesis that the compound α-[Ru(II)(azpy)$_2$Cl$_2$] exhibits a high cytotoxicity due to its two *cis* chloride ligands, which might be exchanged with biological targets such as DNA as in the case of cisplatin. The cytotoxicity of *mer*-[Ru(II)(azpy)$_3$](PF$_6$)$_2$ and *mer* [Ru(II)(tazpy)$_3$](PF$_6$)$_2$ (tazpy = *o*-tolylazopyridine) against human tumor cell lines (A498 (renal carcinoma cell lines), H226 (lung cancer cell line), IGROV (ovarian cancer cell lines), M19 (melanoma), EVSA-T, MCF-7 (human breast adenocarcinoma) and WiDR (colon adenocarcinoma cell line)) was found moderate. So, even though no chloride ligands are present in these tris(ligand) complexes, cytotoxic activity is observed. This implies that the 2-phenylazopyridine ruthenium (II) complexes act by a completely different mechanism than cisplatin. A series of monochloro-ruthenium complexes, [Ru(II)(terpy)(NN)Cl]$^+$ (NN, bidentate nitrogen ligand), containing different electron-donating groups were also prepared (Cheng et al., 2000). DNA binding and formation of Ru–DNA adducts were confirmed by gel mobility shift assay. The preferential DNA binding sites of [Ru(II)(terpy)(tmephen)Cl]$^+$ (tmephen = tetramethylphenanthroline) were purine residues. Surprisingly, [Ru(II)(terpy)(tmephen)Cl]$^+$ inhibited bacterial cell growth (*E. coli*) at lower concentrations than *cis*-[Ru(II)(bpy)$_2$Cl$_2$]. It was suggested that ruthenium complexes modified with electron-rich groups may represent a new class of anticancer ruthenium drugs. The interactions of a metal complex [Ru(II)(phen)$_2$ PMIP]$^{2+}$ {phen = 1,10-phenanthroline, PMIP = 2-(4-methylphenyl) imidazo[4,5-*f*]1,10-phenanthroline} (**8**) with yeast transfer RNA and calf thymus DNA have been investigated (Xu et al., 2005). Binding modes of these Ru(II) polypyridyl complexes to both nucleic acids involve intercalation. The results also suggest that interactions with these nucleic acids are enantioselective, with more binding of Δ-enantiomer than that of Δ-enantiomer.

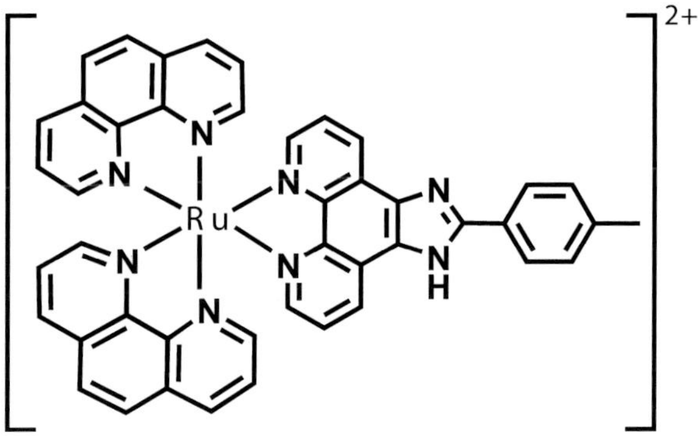

8

Another novel ruthenium(II)-complex containing 4-carboxy *N*-ethylbenzamide (CNEB) as (Ru(II)-CNEB) was found to interact with and inhibit M4-lactate dehydrogenase (M4-LDH), a tumor growth supportive enzyme, at the tissue level (Koiri et al., 2009). Modulation of M4-LDH by this compound in a T-cell lymphoma (Dalton's Lymphoma: DL) causes regression of the tumor in vivo. The compound showed a dose-dependent cytotoxicity to DL cells in vitro. It has also been observed that Ru(II)-CNEB could decline expression of the inducible form of 6-phosphofructo-2-kinase (iPFK2: PFKFB3), a regulator of glycolysis in DL cells. The complex also activated superoxide dismutase activity but declined the levels of catalase and glutathione peroxidase to impose oxidative stress in the DL cells. This was found consistent with enhanced p53 (a tumor supressor protein) level, decline in Bcl2/Bax (Bax (an apoptosis promoter) and Bcl-2 (an apoptosis inhibitor)) ratio and activation of caspase 9 in those DL cells. The findings suggest that Ru(II)-CNEB is able to activate oxidative stress-apoptosis pathway via p53-mediated repression of iPFK2, a key glycolytic regulator, in the DL cells in vivo. These findings provide a biochemical mechanism which can be utilized for defining pharmacological targets for the novel anticancer agents suitable for in vivo applications (Koiri et al., 2015).

8.3.4 Ruthenium(II) Arene Complexes

Organometallic ruthenium(II) complexes with arene ligands represent a relatively new group of ruthenium compounds with antitumor activity. The complexes of type $[(\eta^6\text{-arene})Ru(II)(en)Cl][PF_6]$ (en = ethylenediamine) constitute a group of anticancer compounds (Morris et al., 2001; Aird et al., 2002; Novakova et al., 2005) with a mechanism of action different from those of the ruthenium(III) complexes NAMI-A and $(HInd)[Ru(III)Cl_4(Ind)_2]$, which are currently on clinical trials (Alessio et al., 2004). The $(\eta^6\text{-arene})Ru(II)$ π-bonds in the monofunctional $[(\eta^6\text{-arene})Ru(II)(en)(Cl)]^+$ complexes are inert toward hydrolysis, but the chloride ligand is readily lost and the complex is transformed into correspondingly more reactive aquated species (Wang et al., 2003). It has also been shown that in cell-free media ethylenediamine Ru(II) arene compounds, in which arene = biphenyl (**9**), dihydroanthracene, tetrahydroanthracene **10**), *p*-cymene (**11**), or benzene, bind preferentially to guanine residues in natural double-helical DNA. DNA may be a favored reaction site for these ruthenium anticancer complexes since a recent study (Wang et al., 2005) demonstrated that the presence of cytochrome *c* or l-histidine had little effect on the course of the reaction with the short DNA fragment. In addition, DNA binding of the complexes containing biphenyl, dihydroanthracene, or tetrahydroanthracene ligands can involve combined coordination to guanine N7 and non-covalent, hydrophobic interactions between arene ligand and DNA, which may include arene intercalation and minor groove binding (Chen et al., 2002; Novakova et al., 2003). In contrast, the single hydrocarbon rings in the *p*-cymene and benzene complexes of ruthenium cannot interact with double-helical DNA by intercalation (Novakova et al., 2003). Interestingly, adducts of the complex containing *p*-cymene ligand, having methyl and isopropyl substituents, distort the conformation and thermally destabilize double-helical DNA distinctly more than the adducts of the tricyclic-ring Ru(II) arene compounds.

The activity of two Ru(II) arene complexes containing tetrahydroanthracene or *p*-cymene ligand has also been examined in two tumor cell lines A2780 (human ovarian cancer cell line) and HT29 (Human Colorectal Adenocarcinoma Cell Line).

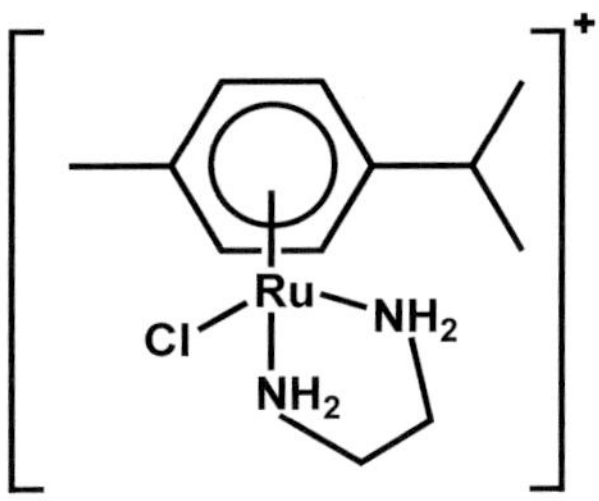

9

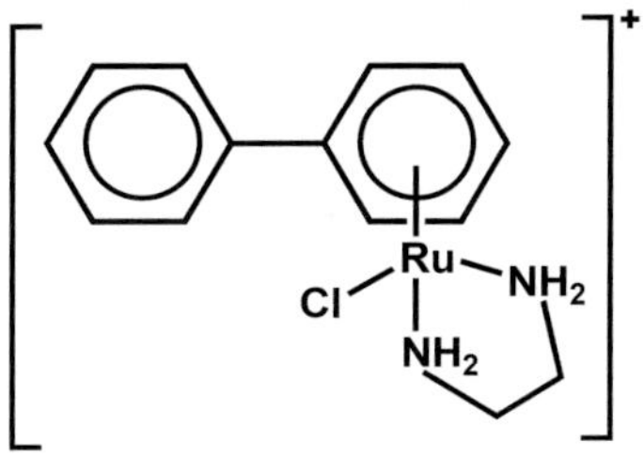

10

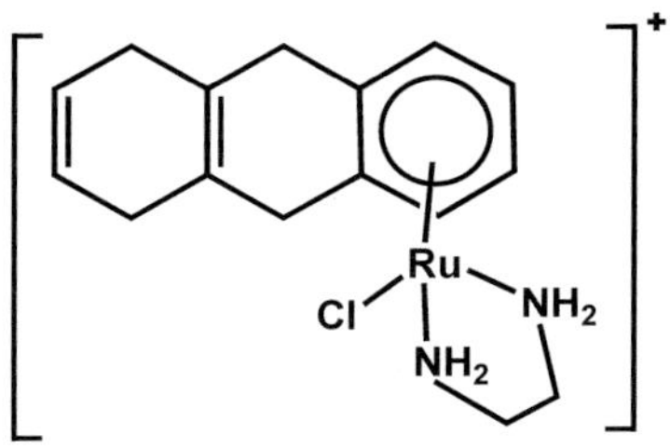

11

These complexes were chosen as representatives of two different classes of Ru(II) arene compounds for which initial studies of global modification of natural DNA was revealed (Novakova et al., 2003). They have different binding modes: one may involve DNA intercalation (tricyclic-ring Ru(II) complex containing tetra hydroanthracene ligand) and other (mono-ring Ru(II) complex containing the *p*-cymene ligand) cannot interact with double-helical DNA by intercalation. These ruthenium complexes are capable of non-covalent, hydrophobic interaction with DNA considerably with enhanced cytotoxicity against tumor cell lines owing to presence of arene ligand (Novakova et al., 2005). An analysis of DNA duplexes modified by mono-ring and tricyclic ring Ru(II) compounds revealed substantial differences in the impact of their monofunctional adducts on the conformation and thermodynamic stability of DNA and DNA polymerization *in vitro* (Novakova et al., 2005). The adducts of Ru(II) arene compounds are preferentially removed from DNA by mechanisms other than by nucleotide excision repair. This repair mechanism is believed to be the main process by which major cisplatin adducts are removed from DNA (Kartalou and Essigmann, 2001). The latter observation provides additional support for a mechanism underlying antitumor activity of Ru(II) arene compounds, different to that of cisplatin (Novakova et al., 2005). Thus, these results support the view that the different character of conformational alterations induced in DNA as a consequence of its global modification by Ru(II) arene compounds may damage DNA. Consequently, different biological effects may result in this new class of metal-based antitumor compounds (Novakova et al., 2005).

8.4 Conclusions

DNA is most important molecule in any cell and any structural change in it brought out by foreign bodies such as metallomolecules has a greater impact in the fields of molecular biology and medicine. Since the discovery of double-helical structure of B-DNA recognized in 1950s, many synthetic molecules have been shown to bind DNA in different binding modes and promoting a variety of structural changes in the DNA. Early metallo-drugs such as cisplatin, a leading chemotherapeutic agent, have been found to target DNA. The binding modes of metallomolecules can be divided into an external association with DNA, groove binding in either the major or the minor groove and intercalation between DNA base pairs. The precise binding mode and consequent DNA structural changes can be analyzed using a range of experimental techniques, such as UV visible, Emission and NMR spectra, finally by X-ray crystallography, gel electrophoresis, circular dichroism. Each of these techniques has different advantages and limitations too. Despite several restrictions, they have been proved reliable and accurate techniques when used appropriately for discerning the nature of DNA–metallomolecule interactions. Several ruthenium compounds are transported into cells relatively easy and bind to cellular DNA. A variety of cytotoxic ruthenium complexes induce different conformational distortions in DNA. In addition, several studies have demonstrated that ruthenium compounds discussed in this chapter are cytotoxic in tumor cell lines which exhibited inherent or acquired resistance to cisplatin. Thus, it suggested that DNA binding modes of ruthenium compounds could be correlated with their anticancer activity.

References

Aird R, Cummings J, Ritchie A, Muir M, Morris R, Chen H, Sadler P, Jodrell D (2002). In vitro and in vivo activity and cross resistance profiles of novel ruthenium (II) organometallic arene complexes in human ovarian cancer. *Br J Cancer*, **86**: 1652.

Alessio E, Mestroni G, Bergamo A, Sava G (2004). Ruthenium antimetastatic agents. *Curr Top Med Chem*, **4**: 1525.

Allardyce CS, Dyson PJ (2001). Ruthenium in medicine: Current clinical uses and future Prospects. *Platinum Met Rev*, **45**: 62.

Anagnostopoulou A, Moldrheim E, Katsaros N, Sletten E (1999). Interaction of *cis*- and *trans*-$RuCl_2(DMSO)_4$ with the nucleotides GpA, d(GpA), ApG, d(ApG) and d(CCTGGTCC): High-field NMR characterization of the reaction products. *J Biol Inorg Chem*, **4**: 199.

Ang WH, Dyson PJ (2006). Classical and non-classical ruthenium-based anticancer Drugs: Towards targeted chemotherapy. *Eur J Inorg Chem*, **2006**: 4003.

Arjmand F, Jamsheera A (2011). DNA binding studies of new valine derived chiral complexes of tin(IV) and zirconium(IV). *Spectrochim Acta A*, **78**: 45–51.

Barton JK (1986). Metals and DNA: molecular left-handed complements. *Science*, **233**: 727.

Barton JK, Lolis E (1985). Chiral discrimination in the covalent binding of is(phenanthroline)dichlororuthenium(II) to B-DNA. *J Am Chem Soc*, **107**: 708–709.

Berg JM, Tymoczko JL, Stryer L (2002). *Biochemistry*, Freeman, New York, 5th ed.

Bhadra K, Kumar GS (2011). Interaction of berberine, palmatine, coralyne, and sanguinarine to quadruplex DNA: A comparative spectroscopic and calorimetric study. *Biochim Biophys Acta*, **18**: 485–496.

Branden C, Tooze J (1999). *Introduction to Protein Structure*, Garland, New York, 2nd edn.

Berger MR, Garzon FT, Keppler BK, Schmähl D (1989). Efficacy of new ruthenium complexes against chemically induced autochthonous colorectal carcinoma in rats. *Anticancer Res*, **9**: 761.

Brindell M, Kulis E, Elmroth SKC, Urbanska K, Stochel G (2005). Light-induced anticancer activity of $[RuCl_2(DMSO)_4]$ complexes. *J Med Chem*, **48**: 7298–7304.

Brown JM (1999). The hypoxic cell: A target for selective cancer therapy--eighteenth Bruce F. Cain Memorial Award lecture. *Cancer Res*, **59**: 5863–5870.

Caspar R, Musatkina L, Tatosyan A, Amouri H, Gruselle M, Guyard-Duhayon C, Duval R, Chen HM, Parkinson, JA, Parsons S, Coxall RA, Gould RO, Sadler PJ (2002). Organometallic ruthenium(II) diamine anticancer complexes: Arene-nucleobase stacking and stereospecific hydrogenbonding in guanine adducts. *J Am Chem Soc*, **124**: 3064.

Cheng CC, Lee WL, Su JG, Liu CL (2000). Covalent interaction of Ru(terpy)(tmephen)Cl^+ with DNA: A potential ruthenium-based anticancer drug. *J Chin Chem Soc*, **47**: 213.

Clarke MJ (1980). Oncological implications of the chemistry of ruthenium, in *Metal Ions in Biological Systems*, ed., Sigel H, Marcel Dekker, Basel, vol. 11: pp. 231–283.

Clarke MJ (2003). Ruthenium metallopharmaceuticals. *Coord Chem Rev*, **236**: 209–233.

Clarke MJ, Zhu F, Frasca, DR (1999). Non-platinum chemotherapeutic metallopharmaceuticals. *Chem Rev*, **99**: 2511.

Coggan DZM, Baters PJ, Haworth IS, Rodger A, Robinson AJ (1999). DNA binding of ruthenium tris(1,10-phenanthroline): Evidence for the dependence of binding mode on metal complex concentration. *Inorg Chem*, **38**: 4486.

Cordier C (2004). Efficient DNA binding by optically pure ruthenium tris(bipyridyl) complexes incorporating carboxylic functionalities. Solution and structural analysis. *Inorg Chem*, **43**: 7987.

Da Ros T, Spalluto G, Prato M, Saison-Behmoaras T, Boutorine A, Cacciari B (2005). Oligonucleotides and oligonucleotide conjugates: A new approach for cancer treatment. *Curr Med Chem*, **12**: 71–88.

Depenbrock H, Schmelcher S, Peter R, Keppler BK, Weirich G, Block T, Rastetter J, Hanauske A-R (1997). Preclinical activity of trans indaz olium[tetrachlorobisindazoleruthenate(III)] (NSC 666158; IndCR; KP 1019) against tumour colony-forming units and haematopoietic progenitor cells. *Eur J Cancer*, **33**: 2404.

Dittrich C, Scheulen ME, Jaehde U, Kynast B, Gneist M, Richly H, Schaad S, Arion VB, Egger A, Arion VB, Reisner E, Cebrian-Losantos B, Shova S, Trettenhahn G, Keppler BK (2005). Reactions of potent antitumor complex trans-[Ru(III)Cl_4(indazole)$_2$]- with a DNA-relevant nucleobase and thioethers: Insight into biological action. *Inorg Chem*, **44**: 122.

Eichhorn GL, Shin YA (1968). Interaction of metal ions with polynucleotides and related compounds. XII. The relative effect of various metal ions on DNA helicity. *J Am Chem Soc*, **90**: 7323.

Frasca D, Ciampa J, Emerson J, Umans RS, Clarke M (1996). Effects of hypoxia and transferrin on toxicity and DNA binding of ruthenium antitumor agents in hela cells. *Met Based Drugs*, **3**: 197–209.

Frasca DR, Gehrig LE, Clarke MJ (2001). Cellular effects of transferring coordinated to $Cl(NH_3)_5RuCl_2$ and *cis*-$Cl_2(NH_3)_4RuCl$. *J Inorg Biochem*, **83**: 139.

Frausin F, Scarcia V, Cochietto M, Furlani A, Serli B, Alessio E, Sava G (2005). Antitumour metal compounds: More than theme and variations. *Pharmacol Exp Ther*, **313**: 227.

Galanski M, Arion VB, Jakupec MA, Keppler BK (2003). Recent developments in the field of tumor-inhibiting metal complexes. *Curr Pharm Des*, **9**: 2078.

Gale EF, Cundliffe E, Reynolds PE, Richmond MH, Waring MJ (1981). *The Molecular Basis of Antibiotic Action*, John Wiley & Sons Ltd., Chichester, 2nd edn.

Galeano A, Berger MR, Keppler BK (1992). Antitumor activity of some ruthenium derivatives in human colon cancer cell lines in vitro. *Arzneimittel Forschung/Drug Res*, **42**: 821.

Gelasco A, Lippard SJ (1998). NMR solution structure of a DNA dodecamer duplex containing a *cis*-diammineplatinum(II) d(GpG) intrastrand cross-link, the Major Adduct of the anticancer drug cisplatin. *Biochemistry*, **37**: 9230.

Goodsell D, Dicherson RE (1986). Isohelical analysis of DNA groove-binding drugs. *J Med Chem*, **29**: 727.

Gonzalez-Ruiz V, Olives AI, Martin MA, Ribelles P, Ramos MT, Menendez JC (2011). An overview of analytical techniques employed to evidence drug–DNA interactions. Applications to the design of genosensors, in: Komorowska MA, Olsztynska-Janus S, eds., *Biomedical Engineering, Trends, Research and Technologies*, pp. 65–90.

Gopal YNV, Jayaraju D, Kondapi AK (1999). Inhibition of topoisomerase II catalytic activity by two ruthenium compounds: A ligand dependent mode of action. *Biochemistry*, **38**: 4382.

Grover N, Welch TW, Fairley TA, Cory M, Thorp HH (1994). Covalent binding of aquaruthenium complexes to DNA. *Inorg Chem*, **33**: 3544.

Hartinger CG, Zorbas-Seifried S, Jakupec MA, Kynast B, Zorbas H, Keppler BK. (2006). From bench to bedside—preclinical and early clinical development of the anticancer agent indazolium *trans*-[tetrachlorobis

(1*H*-indazole) ruthenate(III)] (KP1019 or FFC14A). *J Inorg Biochem,* **100**: 891.

Haq I, Lincoln P, Suh D, Norden B, Chowdhry BZ, Chaires JB (1995). Interaction of DELTA- and LAMBDA-$[Ru(phen)_2DPPZ]^{2+}$ with DNA: A calorimetric and equilibrium binding study. *J Am Chem Soc,* **117**: 4788.

Hotze ACG, van der Geer EPL, Kooijman H, Spek AL, Haasnoot JG, Reedijk J (2005). Characterization by NMR spectroscopy, X-ray analysis and cytotoxic activity of the ruthenium(II) compounds $[RuL_3](PF_6)_2$ (L = 2-phenylazopyridine or *o*-tolylazopyridine) and $[RuL'_2L''](PF_6)_2$ (L′, L″ = 2-phenylazopyridine, 2,2′-bipyridine). *Eur J Inorg Chem,* 2648–2657.

Jaumot J, Gargallo R (2012). Experimental methods for studying the interactions between G-uadruplex structures and ligands. *Curr Pharmaceut Des,* **18**(14): 1900–1916.

Kamatchi TS, Chitrapriya N, Jamal Ahamed VS, Moon S-S, Fronczek FR, Natarajan K (2013). Ruthenium(II) complexes of 2,2′-bipyridine-5,5′-dicarboxylic acid: Synthesis, structure, DNA binding, cytotoxicity and antioxidant activity. *Inorg Chim Acta,* **404**: 58–67.

Kapitza S, Jakupec MA, Uhl M, Keppler BK, Marian B (2005a). The heterocyclic ruthenium (III) complex KP1019 (FFC14A) causes DNA damage and oxidative stress in colorectal tumor cells. *Cancer Lett,* **226**: 115.

Kapitza S, Pongratz M, Jakupec MA, Heffeter P, Berger W, Lackinger L, Keppler BK, Marian B (2005). Heterocyclic complexes of ruthenium(III) induce apoptosis in colorectal carcinoma cells. *J Cancer Res Clin Oncol,* **131**: 101.

Kartalou M, Essigmann JM (2001). Mechanisms of resistance to cisplatin. *Mutat Res,* **478**: 23.

Keppler BK (2005). *Proc Am Assoc Cancer Res,* **46**: P472.

Koiri RK, Trigun SK, Mishra L (2015). Activation of p53 mediated glycolytic inhibition-oxidative stressapoptosis pathway in Dalton's lymphoma by a ruthenium (II)-complex containing 4-carboxy N-ethylbenzamide. *Biochimie,* **110**: 52–61.

Koiri RK, Trigun SK, Mishra L, Pandey K, Dixit D, Dubey SK (2009). Regression of Dalton's lymphoma in vivo via declinein lactatedehydrogenase and induction of apoptosis by a ruthenium(II)-complex containing 4 carboxethylbenzamide as ligand. *Invest New Drug,* **27**: 503–516.

Kratz F, Beyer U (1998). Serum proteins as drug carriers of anticancer agents: A review. *Drug Deliv*, **5**: 281–299.

Kratz F, Hartmann M, Keppler B, Messori L (1994). The binding properties of two antitumor ruthenium(III) complexes to apotransferrin. *J Biol Chem,* **269**: 2581.

Kumar KA, Reddy KL, Vidhisha S, Satyanarayana S (2009). Synthesis, characterization and DNA binding and photocleavage studies of $[Ru(bpy)_2BDPPZ]^{2+}$, $[Ru(dmb)_2BDPPZ]^{2+}$ and $[Ru(phen)_2BDPPZ]^{2+}$ complexes and their antimicrobial activity. *Appl Organometal Chem,* **23**: 409–420.

Kung A, Pieper T, Keppler BK (2001). Investigations into the interaction between tumor-inhibiting ruthenium(III) complexes and nucleotides by capillary electrophoresis. *J Chromatogr B Biomed Appl,* **759**: 81.

Lakowicz JR (2006). *Principles of Fluorescence Spectroscopy,* 3rd ed., Springer.

Larsen TA, Goodsell DS, Cascio D, Grzeskowiak K, Dickerson RE (1989). The structure of DAPI bound to DNA. *J Biomol Struct Dyn,* **7**: 477–91.

Lerman L (1961). Structural considerations in the interaction of DNA and acridines. *J Mol Biol*, **3**: 18.

Lerman LS (1963). The structure of the dna-acridine complex. *Proc Natl Acad Sci U S A*, **49**: 94.

Liu J, Zhang T, Lu T, Qu L, Zhou H, Zhang Q, Ji L (2002). DNA-binding and cleavage studies of macrocyclic copper(II) complexes. *J Inorg Biochem,* **91**: 269–276.

Li N, Ma Y, Guo L, Yang X (2005). Interaction of anticancer drug mitoxantrone with DNA analyzed by electrochemical and spectroscopic methods. *Biophys Chem,* **116**: 199–205.

Levitt MH (2001). *Spin Dynamics: Basics of Nuclear Magnetic Resonance,* John Wiley & Sons Ltd, Chichester.

Loseto F, Alessio E, Mestroni G, Lacidogna G, Nassi A, Giordano D (1991). Interaction of $RuCl_2$ (dimethylsulphoxide)$_4$ isomers with DNA. *Anticancer Res,* **11**: 1549.

Mainwaring WIP, Parish JH, Pickering JD, Mann NH (1982). *Nucleic Acid Biochemistry and Molecular Biology,* Blackwell Scientific Publications, Oxford.

Malina J, Novakova O, Keppler BK, Alessio E, Brabec V (2001). Biophysical analysis of natural, double-helical DNA modified by anticancer heterocyclic complexes of ruthenium(III) in cell-free media. *J Biol Inorg Chem,* **6**: 435.

Mestroni G, Alessio E, Calligaris M, Attia WM, Quadrifoglio F, Cauci S, Sava G, Zorzet S, Morris RE, Aird RE, Murdoch PD, Chen HM, Cummings J, Hughes ND, Parsons S, Parkin A, Boyd G, Jodrell DI, Sadler PJ (2001). Inhibition of cancer cell growth by ruthenium(II) arene complexes. *J Med Chem,* **44**: 3616.

Neidle S (1994). *DNA Structure and Recognition*, OUP, Oxford.

Neyhart GA, Cheng CC, Thorp HH (1995). Kinetics and mechanism of the oxidation of sugars and nucleotides by oxoruthenium(IV): Model studies for predicting cleavage patterns in polymeric DNA and RNA. *J Am Chem Soc*, **117**: 1463.

Novakova O, Chen H, Vrana O, Rodger A, Sadler PJ, Brabec V (2003). DNA interactions of monofunctional organometallic ruthenium(II) antitumor complexes in cell-free media. *Biochemistry,* **42**: 11544.

Novakova O, Hofr C, Brabec V (2000). Modification of natural, doublehelical DNA by antitumor *cis*- and *trans*-$[Cl_2(Me_2SO)_4Ru]$ in cell-free media. *Biochem Pharmacol*, **60**: 1761.

Novakova O, Kasparkova J, Bursova V, Hofr C, Vojtiskova M, Chen H, Sadler PJ, Brabec V (2005). Conformation of DNA modified by monofunctional Ru(II) arene complexes: Recognition by DNA-binding proteins and repair. Relationship to cytotoxicity. *Chem Biol,* **12**: 121.

Novakova O, Kasparkova J, Vrana O, van Vliet PM, Reedijk J, Brabec V (1995). Correlation between cytotoxicity and DNA binding of polypyridyl ruthenium complexes. *Biochemistry*, **34**, 12369.

Pacor S, Monti-Bragadin C, Tamaro M, Dolzani L (1989). Chemical, biological and antitumor properties of ruthenium(II) complexes with dimethylsulfoxide. *Prog Clin Biochem Med*, **10**: 71.

Pacor S, Zorzet S, Cocchietto M, Bacac M, Vadori M, Turrin C, Gava B, Castellarin A, Sava G (2004). Intratumoral NAMI-A treatment triggers metastasis reduction, which correlates to CD44 regulation and tumor infiltrating lymphocyte recruitment. *Pharmacol Exp Ther,* **310**: 737.

Pluim D, van Waardenburg RCAM, Beijnen JH, Schellens JHM, (2004). Cytotoxicity of the organic ruthenium anticancer drug Nami-A is correlated with DNA binding in four different human tumor cell lines. *Cancer Chemother Pharmacol*, **54**: 71.

Pongratz M, Schluga P, Jakupec MA, Arion VB, Hartinger CG, Allmaier G, Keppler BK (2004). Transferrin binding and transferrin-mediated cellular uptake of the ruthenium coordination compound KP1019 studied by means of AAS, ESI-MS and CD spectroscopy. *J Anal Atomic Spectrom*, **19**: 46–51.

Pratviel G, Bernadou J, Meunier B (1998). DNA and RNA cleavage by metal complexes. *Adv Inorg Chem*, **45**: 251.

Pray L (2008). Discovery of DNA structure and function: Watson and Crick. *Nature Education*, **1**(1): 100.

Rademaker-Lakhai JM, van den Bongard D, Pluim D, Beijnen JH, Schellens JHM (2004). A Phase I and pharmacological study with imidazolium-trans-DMSO-imidazole-tetrachlororuthenate, a novel ruthenium anticancer agent. *Clin Cancer Res*, **10**: 3717.

Reisner E, Arion VB, Eichinger A, Kandler N, Giester G, Pombeiro AJL, Keppler BK (2005). Tuning of redox properties for the design of ruthenium anticancer drugs: Part 2. Syntheses, crystal structures and electrochemistry of potentially antitumor $[Ru^{III}/^{II}Cl_6\text{-}n(Azole)_n]_z$ (n = 3, 4, 6) Complexes. *Inorg Chem*, **44**: 6704–6716.

Reisner E, Arion VB, Guedes da Silva MFC, Lichtenecker R, Eichinger A, Keppler BK, Kukushkin VYu, Pombeiro AJL (2004). Tuning of redox potentials for the design of ruthenium anticancer drugs—An electrochemical study of [trans-$RuCl_4$L(DMSO)]– and [trans-$RuCl_4L_2]^-$ complexes, where L = imidazole, 1,2,4-triazole, indazole. *Inorg Chem*, **43**: 7083–7093.

Reisner E, Arion VB, Keppler BK, Pombeiro AJL (2008). Electron-transfer activated metal-based anticancer drugs. *Inorg Chim Acta*, **361**: 1569–1583.

Satyanarayana S, Dabrowiak JC, Chaires JB (1993). Tris(phenanthroline) ruthenium(II) enantiomer interactions with DNA: Mode and specificity of binding. *Biochemistry*, **32**: 2573.

Sava G, Alessio E, Bergano A, Mestroni G (1999). Sulfoxide ruthenium complexes: Non toxic tools for the selective treatment of solid tumour metastases, in: Clarke MJ, Sadler PJ, eds., *Topics in Biological Inorganic Chemistry. Metallopharmaceuticals*, Springer, Berlin, vol. **1**, pp. 143–169.

Sava G, Bergamo A (2000). Ruthenium-based compounds and tumour growth control. *Int J Oncol*, **17**: 353.

Sava G, Frausin F, Cocchietto M, Vita F, Podda E, Spessotto P, Furlani A, Scarcia V, Zabucchi G (2004). Actin-dependent tumour cell adhesion after short-term exposure to the antimetastasis ruthenium complex NAMI-A. *Eur J Cancer*, **40**: 1383.

Sava G, Zorzet S, Turrin C, Vita F, Soranzo MR, Zabucchi G, Cocchietto M, Bergamo A, DiGiovone S, Pezzoni G, Sartor L, Garbisa S (2003). Dual Action of NAMI-A in inhibition of solid tumor metastasis: Selective targeting of metastatic cells and binding to collagen. *Clin Cancer Res*, **9**: 1898.

Schleif R (1988). DNA binding by proteins. *Science*, **241**: 1182–1187.

Schluga P, Hartinger CG, Egger A, Reisner E, Galanski M, Jakupec MA, Keppler BK (2006). Redox behavior of tumor-inhibiting ruthenium(III) complexes and effects of physiological reductants on their binding to GMP. *Dalton Trans*, **14**: 1796–1802.

Seelig MH, Berger MR, Keppler BK (1992). Antineoplastic activity of three ruthenium derivatives against chemically induced colorectal carcinoma in rats. *J Cancer Res Clin Oncol*, **118**: 195–200.

Shahabadi N, Kashanian S, Khosravi M, Mahdavi M (2010). Multispectroscopic DNA interaction studies of a water-soluble nickel(II) complex containing different dinitrogen aromatic ligands. *Tran Meter Chem*, **35**: 699–705.

Sirajuddin M, Ali S, Haider A, Shah NA, Shah A, Khan MR (2012). Synthesis, characterization, biological screenings and interaction with calf thymus DNA as well as electrochemical studies of adducts formed by azomethine [2-((3,5-dimethylphenylimino)methyl)phenol] and organotin(IV) chlorides. *Polyhedron*, **40**(1): 19–31.

Sirajuddin M, Ali S, Shah NA, Khan MR, Tahir MN (2012). Synthesis, characterization, biological screenings and interaction with calf thymus DNA of a novel azomethine 3-((3,5-dimethylphenylimino) methyl)benzene-1,2-diol. *Spectrochim Acta A*, **94**: 134–142.

Smith CA, Sutherland-Smith AJ, Keppler BK, Kratz F, Baker EN (1996). Binding of ruthenium(III) anti-tumor drugs to human lactoferrin probed by high resolution X-ray crystallographic structure analyses. *J Biol Inorg Chem*, **1**: 424.

Suh D, Chaires JB (1995). Criteria for the mode of binding of DNA binding agents. *Biorg Med Chem*, **3**(6): 723–728.

Sun H, Li H, Sadler PJ (1999). Transferrin as a metal ion mediator. *Chem Rev*, **99**: 2817–2842.

Sun H, Xiang J, Liu Y, Li L, Li Q, Xu G, Tang Y (2011). A stabilizing and denaturing dual-effect for natural polyamines interacting with G-quadruplexes depending on concentration. *Biochimie*, **93**: 1351–1356.

Tacchini L, Bianchi L, Bernelli-Zazzera A, Cairo G (1999). Transferrin receptor induction by hypoxia. *J Biol Chem*, **274**: 24142.

Tan C, Liu J, Chen L, Shi S, Ji L (2008). Synthesis, structural characteristics, DNA binding properties and cytotoxicity studies of a series of Ru(III) complexes. *J Inorg Biochem* **102**: 1644–1653.

Timerbaev AR, Hartinger CG, Aleksenko SS, Keppler BK (2006). Interactions of antitumor metallodrugs with serum proteins:

Advances in characterization using modern analytical methodology. *Chem Rev,* **106**: 2224.

Trynda-Lemiesz L (2004). interaction of an anticancer ruthenium complex HInd[$RuInd_2Cl_4$] with cytochrome c. *Acta Biochim Pol,* **51**: 199.

Vacca A, Bruno M, Boccarelli A, Coluccia M, Ribatti D, Bergamo A, Garbisa S, Sartor L, Sava G (2002). Inhibition of endothelial cell functions and of angiogenesis by the metastasis inhibitor NAMI-A. *Br J Cancer,* **86**: 993.

Van Vliet PM, Toekimin SMS, Haasnoot JG, Reedijk J, Novakova O, Vrana O, Brabec V (1995). *mer*-[$Ru(terpy)Cl_3$] (terpy = 2,2′:6′,2″-terpyridine) shows biological activity, forms interstrand cross-links in DNA and binds two guanine derivatives in a *trans* configuration. *Inorg Chim Acta,* **231**: 57.

Wang F, Bella J, Parkinson JA, Sadler PJ (2005). Competitive reactions of a ruthenium arene anticancer complex with histidine, cytochrome c and an oligonucleotide. *J Biol Inorg Chem,* **10**: 147.

Wang F, Chen H, Parsons S, Oswald IDH, Davidson JE, Sadler PJ (2003). Kinetics of aquation and anation of Ruthenium(II) arene anticancer complexes, acidity and X-ray structures of aqua adducts. *Chem Eur J,* **9**: 5810.

Wardman P (2001). Electron transfer and oxidative stress as key factors in the design of drugs selectively active in hypoxia. *Curr Med Chem,* **8**(7): 739–761.

Watson JD, Crick FHC (1953). A structure for deoxyribose nucleic acid. *Nature,* **171**: 737.

Wei C, Wang J, Zhang M (2010). Spectroscopic study on the binding of porphyrins to (G4T4G4)4 parallel G-quadruplex. *Biophys Chem,* **148**: 51–55.

Xu H, Liang Y, Zhang P, Du F, Zhou B R, Wu J, Liu J-H, Liu Z-G, Ji L-N (2005). Biophysical studies of a ruthenium(II) polypyridyl complex binding to DNA and RNA prove that nucleic acid structure has significant effects on binding behaviors. *J Biol Inorg Chem,* **10**: 529.

Yan YK, Melchart M, Habtemariam A, Sadler PJ (2005). Organometallic chemistry, biology and medicine: Ruthenium arene anticancer complexes. *Chem Commun,* **38**: 4764.

Chapter 9

Ruthenium Aryl Sulfides Complexes

Minu Gupta Bhowon and Sabina Jhaumeer Laulloo

Department of Chemistry, Faculty of Science, University of Mauritius, Reduit, Mauritius

mbhowon@uom.ac.mu

Ruthenium metal complexes involving organic sulfur donor ligands are of considerable interest in coordination chemistry and can serve as potential models to the biological systems. They are known to exhibit a wide range of reactivity and chelating ability at the sulfur atoms. In this chapter, ruthenium complexes derived from aryl thiols, sulfides, and disulfides will be explored. Ruthenium with aryl thiols leads to the formation of binuclear and trinuclear complexes via-sulfur bridging. Diarylsulfides are prone to rearrangement and sulfur redistribution in the presence of ruthenium due to lability of S-S bond. In such cases, in situ sulfur-sulfur cleavage followed by reductive oxidation to sulfoxide or sulfinate anion prior to coordination occurs. In some cases, the sulfur atom is bonded to ruthenium without any S-S cleavage.

Ruthenium Chemistry
Edited by Ajay Kumar Mishra and Lallan Mishra

ISBN 978-981-4774-39-0 (Hardcover), 978-1-315-11058-5 (eBook)
www.panstanford.com

9.1 Introduction

Aryl sulfides are fundamental building blocks, which have the capacity to coordinate with metal atoms to form different structural frameworks, exhibiting interesting functional properties, including biological and catalytic activities. Their synthetic procedures for complexation show great advantages in obtaining coordination architectures that are unattainable by direct methods. The importance of sulfur as a single atom between adjacent transition metal sites allows the delocalization of electron-density toward the bridging atoms giving rise to attractive supramolecular architectures. Among the d-block transition metals, ruthenium, which occupies the central position, is known for its rich coordination with the sulfur donor ligands. These ruthenium sulfur compounds are useful in many important fields and can serve as potential models to the biological systems. Subsequent studies of ruthenium complexes include a wide range of reactivities and chelating abilities at the sulfur atoms. In this chapter, ruthenium complexes derived from aryl thiols, sulfides and disulfides will be explored to develop a thorough knowledge and understanding of the coordinating properties of ruthenium toward sulfur.

9.2 Aryl Thiols

The aryl thiolate anion $C_6H_5S^-$ is a fundamental ligand and can be classified as soft Lewis base and coordinate strongly with metals that behave as Lewis acids. The sulfide can be considered as a handle while the aryl group can be manipulated and adjusted to effect steric and electronic control of ligation ability. There have been various reports on the structural determination of ruthenium aryl thiolate metal complexes with multi-dimensional properties.

Ruthenium complexes with neutral aryl thiols are relatively uncommon due to the high acidity of SH. In 1994, Tocher synthesized mononuclear thiol complex $[Ru(\eta^3{:}\eta^3\text{-}C_{10}H_{16})Cl_2(HSC_6H_5)]$ where presence of neutral SH was confirmed by the ν_{SH} at 2460 cm^{-1} in the IR spectrum. In ^1H NMR spectrum, the broadness of the aromatic signals suggested the dynamic equilibrium in the complex which was due to the weak binding of the aryl thiols to the metal center leading to partial disproportionation of the complex into free thiol and starting ruthenium(IV) chloro bridge dimer $[\{Ru(\eta^3{:}\eta^3\text{-}C_{10}H_{16})Cl(\mu\text{-}Cl)\}_2]$ (Belchem, 1994).

Bis(2,6-dimethylbenzenethiolato) ruthenium complexes containing mono, bi and tri-phosphine ligands are known. $Ru(\eta^4$-1,5-COD)(η^6-1,3,5-COT) (COD = cyclooctadiene, COT = cyclooctatetraene) reacted with 2,6-dimethylbenzenethiol in the presence of PMe_3 to yield the mononuclear ruthenium complex $[Ru(\eta^5\text{-}C_8H_{11})(SC_6H_3Me_2\text{-}2,6)(PMe_3)_2]$, liberating 1,5-COD (Fig. 9.1). On heating the mononuclear complex with PMe_3, the thiaruthenacycle *cis*-$Ru[SC_6H_3(2\text{-}CH_2)(6\text{-}Me)\text{-}{}^2_{\kappa}S,C](PMe_3)_4$ complex is isolated with concomitant formation of 1,3-COD (Hirano et al., 2005).

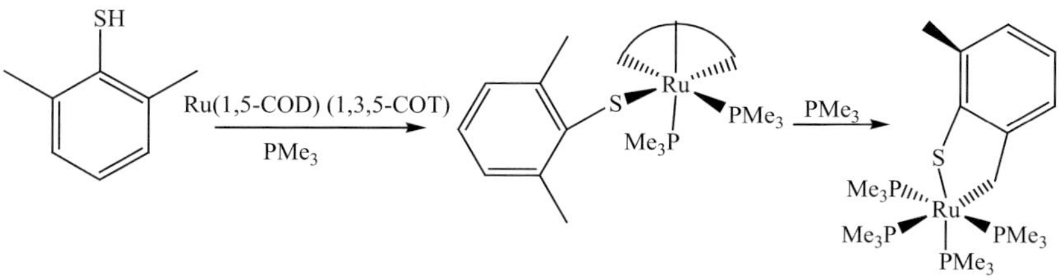

Figure 9.1 Formation of thiaruthenacycle complex.

In these reactions, sp^3-C-H bond activation of one of the *ortho*-methyl group occurs. The sp^3-C-H bond cleavage proceeds by the concerted mechanism and the coordinative unsaturation is an intrinsic factor for C-H bond cleavage reaction (Fig. 9.2). It is proposed that the methyl C-H activation occurs via the interaction of the sulfur atom of the arene thiolate group, which also acts as a good acceptor of the cleaved proton (Hirano et al., 2010).

The aryl thiolate anion can also act as a bridging ligand between ruthenium centers, leading to the formation of mono, bis, and tris-thiolate bridge complexes. The reaction of aryl thiols with dimeric arene ruthenium dichloride complexes has been extensively studied. The first examples are the cationic tristhiolato complexes of the type $[(\eta^6\text{-}C_6Me_6)Ru_2(\mu\text{-}SC_6H_4R)_3]^+$ (Schacht et al., 1992) and the *p*-cymene derivative $[(p\text{-}MeC_6H_4{}^iPr)_2Ru(SC_6H_5)_3]^+$ (Mashima, 1992) where both complexes contain three thiophenolato bridges between two ruthenium centers. Since then, a number of dithiolates $[(\eta^6\text{-}L_2)Ru_2(\mu\text{-}SC_6H_4R)_2]^+$ (Ibao et al., 2012) and trithiolates $[(\eta^6\text{-}L)_2Ru_2(\mu\text{-}SC_6H_4R)_3]^+$ (Cherioux et al., 2002, 2003a,b, 2004; Tschan et al., 2004; Fowe et al., 2008; Giannini et al., 2012, 2013; Gras et al., 2010) (L = C_6H_6, C_6Me_6, $Me_4C_6H_2$, $MeC_6H_4{}^iPr$; R = H, Br, CH_3, OH, *t*-Bu, ^{i}Pr, F, *m*-$C_6H_4C_6H_5$) ruthenium complexes have been reported.

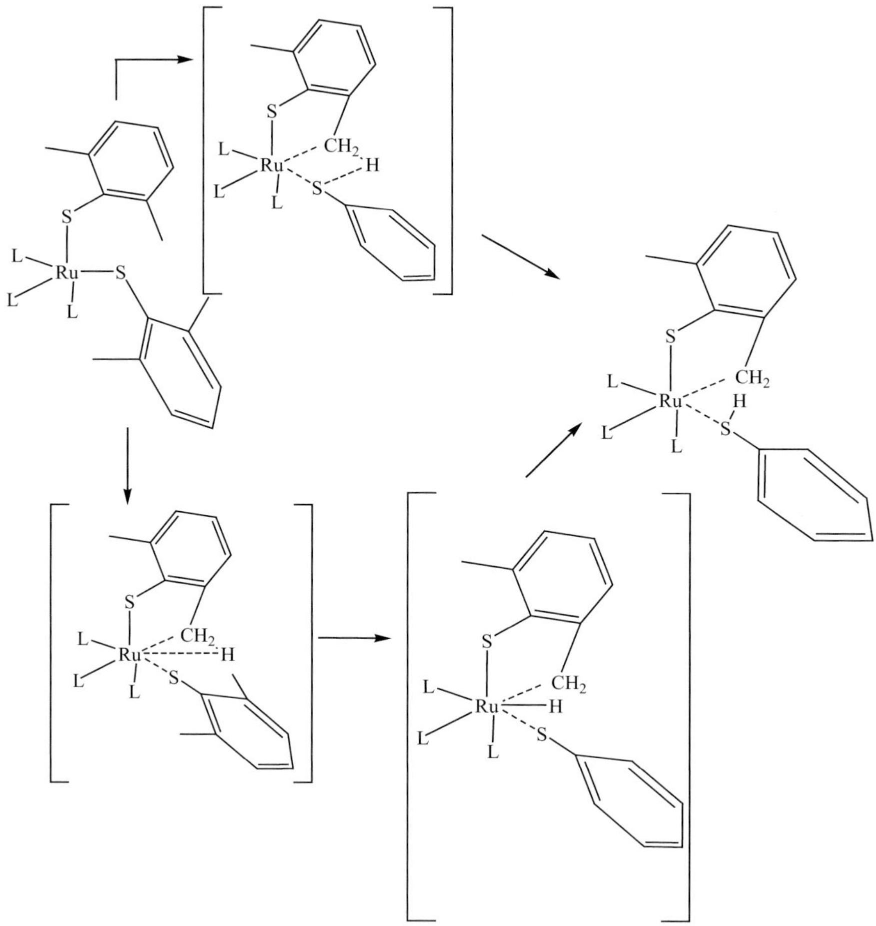

Figure 9.2 Proposed mechanistic details of C-H bond cleavage of *ortho*-methyl group.

These reactions proceed through the partial chloro bridge cleavage and coordination of the monodentate thiolate anion as a two electron donor bridging ligand giving an intermediate species which further bridge another thiolate anion leading to the formation of bis and tris bridge thiolate binuclear ruthenium complexes (Fig. 9.3). In these complexes, no metal-metal bond is present and binuclear backbone is stabilized by the bridging thiolate ligand. This has also been confirmed using DFT calculations for the complexes $[(\eta^6\text{-L})_2Ru_2(\mu_2\text{-}SC_6H_4R)_n(\mu_2\text{-}H)_x]$ (n = 1, 2, 3; x = 0, 1, 2) (R = Br) (Fowe et al., 2008).

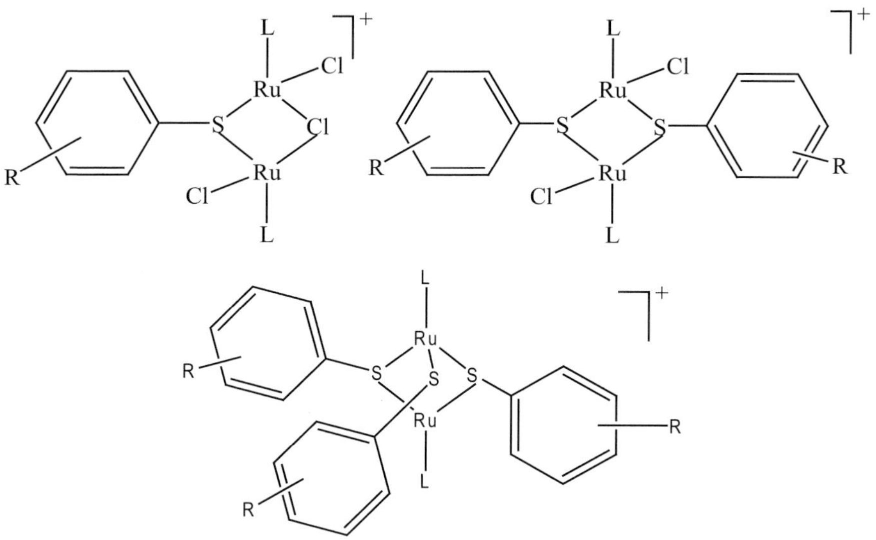

Figure 9.3 General structures of mono, bis, and tris thiolato ruthenium complexes.

The dithiolates compounds are often contaminated with either the monothiolates or trithiolates derivatives. Reaction of $[p\text{-}MeC_6H_4{}^i Pr]_2Ru_2Cl_4]$ with $p\text{-}{}^t BuC_6H_4CH_2SH$ gives a mixture of trithiolate, $[p\text{-}MeC_6H_4{}^i Pr]_2Ru_2(p\text{-}{}^t BuC_6H_4CH_2S)_3]^+$ and dithiolate, $[p\text{-}MeC_6H_4{}^i Pr]_2Ru_2(p\text{-}{}^t BuC_6H_4CH_2S)_2Cl_2]$ complexes which are difficult to separate. When the reaction is carried out under milder conditions (room temperature, 0°C), the neutral dithiolate complexes $[(p\text{-}MeC_6H_4{}^i Pr)_2Ru_2(SR)_2Cl_2]$ (R = $CH_2C_6H_5$, $CH_2CH_2C_6H_5$, $CH_2C_6H_4p\text{-}{}^t Bu$, C_6H_{11}, $CH_2C_6H_4\text{-}OCH_3$) are isolated (Ibao et al., 2012; Stilbal et al., 2015).

Electron-withdrawing groups on the aryl ring favor the formation of the monothiolato complex. The monothiolato complex

$[(\eta^6\text{-p-MeC}_6H_4i\text{Pr})_2Ru_2Cl_2(\mu\text{-Cl})(\mu\text{-SC}_6H_4R)]]$ (L = MeC_6H_4iPr; R = NO_2) is formed as the major product and the dithiolato complex is formed in trace amount (Stílbal et al., 2014). When the sterically demanding substituent such as admantyl thiol is used, the pure monothiolato complex is isolated.

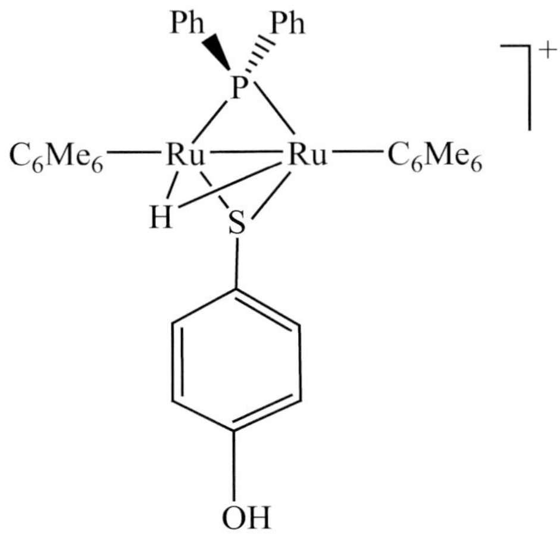

Figure 9.4 Structure of $[(\eta^6\text{-C}_6Me_6)_2Ru_2(\mu_2\text{-SC}_6H_4OH)(\mu_2\text{-PPh}_2)(\mu_2\text{-H})]^+$.

These types of arene ruthenium thiolato complexes have been reported to exhibit anticancer properties. Tristhiolato complexes show better activities compared to the dithiolato and monothiolato complexes. The tristhiolato complexes exhibited an IC_{50} value in the range of nanomolar for A2780 human ovarian cancer cells and the *cis*-platin-resistant mutant A2780cisR (Stilbal et al., 2014; Giannini et al., 2013, 2015).

Another triple bridge dinuclear cation $[(\eta^6\text{-C}_6Me_6)_2\ Ru_2(\mu_2\text{-SC}_6H_4OH)(\mu_2\text{-PPh}_2)(\mu_2\text{-H})]^+$ (Fig. 9.4) is formed by the reaction of $[(\eta^6\text{-C}_6Me_6)_2Ru_2(\mu_2\text{-PPh}_2)(\mu_2\text{-H})_2]^+$ and *p*-hydroxy thiophenol, (Tschan et al., 2006). The presence of two phenyl groups on the phosphorus bridging ligand forces the Ru-Ru system to adopt a distorted geometry, with no chiral centers and Ru-Ru bond length of 2.852 Å.

The thiolato bridge trinuclear ruthenium cluster $[HRu_3(C_6H_6)(C_6Me_6)_2(\mu_3\text{-SC}_6H_4CH_3)(\mu_2\text{-X})]^{2+}$ (Fig. 9.5) are obtained when dinuclear thiolato complex $[H_2Ru_2(C_6Me_6)_2(\mu_2\text{-SC}_6H_4CH_3)]^+$ reacts with $[Ru(C_6H_6)(H_2O)_3]^{2+}$. The two complexes adopt a *nido* Ru_3 framework. The two Ru-Ru single bonds are in the range of

2.926–2.947 Å. The non-bonding Ru-Ru distance (3.599 Å) of the μ_2-S is slightly longer than the μ_2-O (3.437 Å) (Tschan et al., 2005).

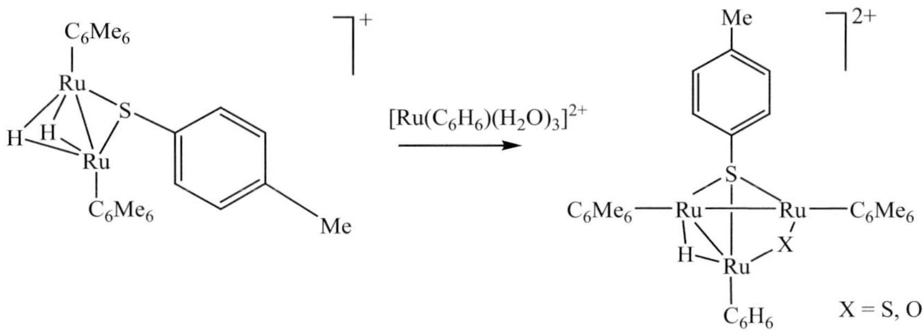

Figure 9.5 Structure of trinuclear cluster $[HRu_3(C_6H_6)(C_6Me_6)_2(\mu_3\text{-}SC_6H_4CH_3)(\mu_2\text{-}X)]^{2+}$.

Another μ_2-S trinuclear cluster $[H_2Ru_3(1,2,4,5\text{-}Me_4C_6H_2)_3)_2(\mu_2\text{-}S)(Cl)]^+$ (Fig. 9.6) was isolated when *p*-bromo thiophenol was reacted with tetra-nuclear hydrido complex $[H_4Ru_4(1,2,4,5\text{-}(Me)_4C_6H_2)_4]^{2+}$. In this reaction, C-S bond cleavage occurred (Tschan et al., 2005) together with the desulfurization of the bromo thiophenol into bromobenzene were observed. The Ru-Ru distance (2.784 Å) of the μ_2-S bridge is of the range of metal-metal single bond and the complex adopts a *closo* framework.

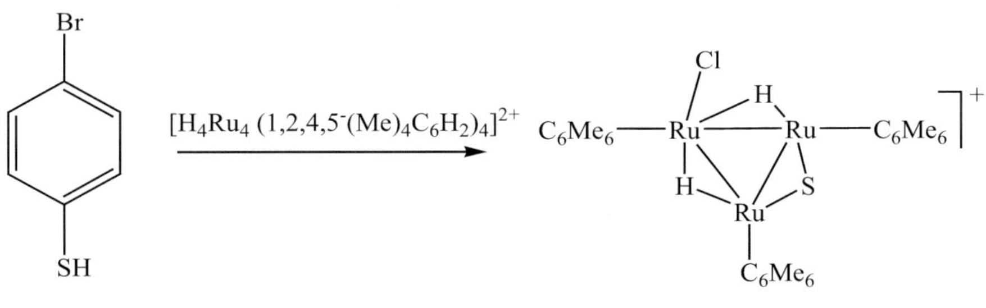

Figure 9.6 Structure of trinuclear cluster $[H_2Ru_3(1,2,4,5\text{-}Me_4C_6H_2)_3)_2(\mu_2\text{-}S)(Cl)]^+$.

The tetra-nuclear $[(\mu\text{-}SC_6H_3(Me)_2)_2\{Ru_2(CO)_4(\mu\text{-}Cl)_2(\mu\text{-}SC_6H_3(Me)_2)\}_2$, or hexa-nuclear $[(\mu\text{-}SC_6H_2(Me)_3)_3\{Ru_2(CO)_4(\mu\text{-}Cl)_2(\mu\text{-}SC_6H_2(Me)_3)\}_3$ clusters are obtained when $[Ru(CO)_2(\mu\text{-}Cl)_2]_n$ is reacted with 2,6-dimethylthiophenol or 2,4,6-trimethylthiophenol, respectively, displaying unique 8 and 12 metallocycle rings with

Ru_4S_4 and Ru_6S_6 cores (Fig. 9.7) (Ye et al., 2013). In the tetranuclear complex, the two dinuclear ruthenium species $\{[[Ru(CO)_2(\mu\text{-}Cl)]_2(\mu\text{-}SC_6H_3(Me)_2]\}^+$ linked through two thiolate anions doubly bridging the two ruthenium(II) atoms, forming an eight atom planar Ru_4S_4 planar ring. The structure of hexanuclear complex contains three dinuclear ruthenium $\{[Ru(CO)_2(\mu\text{-}Cl)]_2(\mu\text{-}SC_6H_2(Me)_3]\}^+$ and three double bridging thiolate anions giving Ru_6S_6 with alternating chloride and thiolate bridge between ruthenium(II) atoms. The average distances from the ruthenium to the bridging chloro atoms are non-equivalent (2.082 and 2.289 Å for tetra; 2.438 Å and 2.478 Å for hexa) and average Ru-S bond lengths are 2.415 Å (tetra) and 2.414 Å (hexa).

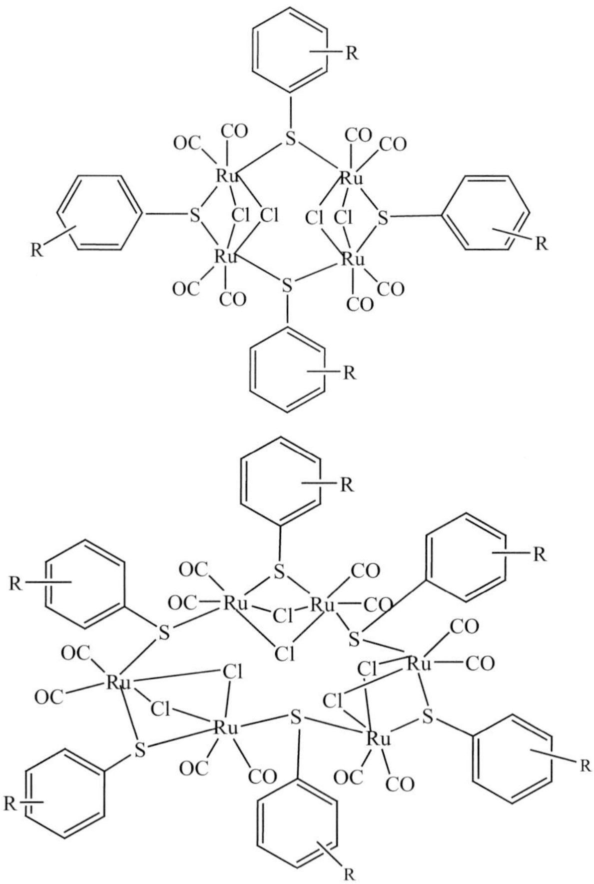

Figure 9.7 Structures of tetra and hexanuclear ruthenium complexes.

Treatment of $Ru_3(CO)_{12}$ with C_6H_5SH/RC_6H_4SH (R = Me, ^{i}Pr) resulted in the formation of the hexamer $[Ru(SC_6H_5R)_2(CO)_2]_6$ and octamer $[Ru(SC_6H_5)_2(CO)_2]_8$ (Fig. 9.8) complexes, respectively. These complexes adopt a M_6/M_8 wheel where two thiolate ligands are bridged between two metal centers. These molecular wheels have planar M_6/M_8 arrangement constituting of regular hexagons/octagons (Chan et al., 2012).

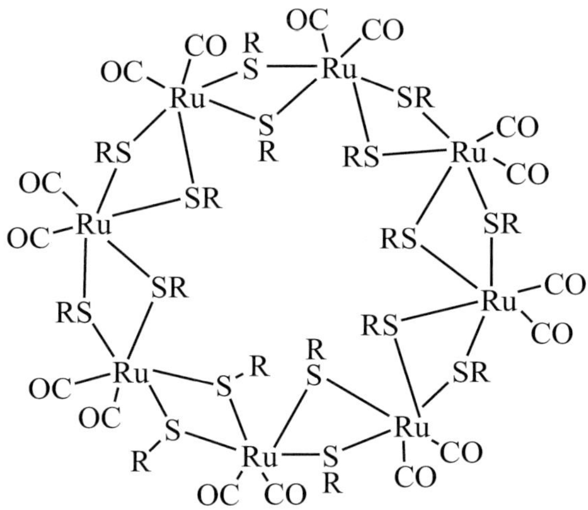

Figure 9.8 Structure of M_8 molecular wheel.

The hexamer $[Ru(SC_6H_5R)_2(CO)_2]_6$ (R = ^{i}Pr) exhibited in vitro cytotoxic effect against various number of human cells with $IC_{50} >$ 100 M using the Mosmann-based cytotoxic MTT assay (Chan et al., 2014).

Reaction of the substituted thiol, (N_3S_3 = trianion of 1,4,7-tris(4-tert butyl-2-mercaptobenzyl)-1,4,7-triazacyclononane) with $RuCl_2(DMSO)_4$ yields an air sensitive mononuclear $[Ru^{III}N_3S_3]$ complex. Further chemical or electrochemical oxidation of mononuclear complex gives dinuclear $[Ru_2(N_3S_3)_2](PF_6)_4$ complex containing two reduced Ru(II) ions and a neutral *tris*(disulfide). The dinuclear complex having a C_2 symmetry and the two ruthenium centers are connected by three coordinated disulfide group with the average S-S bond at 2.209 Å. Further reaction of mononuclear with $RuCl_2(DMSO)_4$ gave trinuclear complex $[Ru_3(N_3S_3)_2](PF_6)_2$ where two terminal Ru ions have C_3 symmetry, (*cis*N_3S_3) donor set and the central Ru ion is in a distorted octahedral environment comprising of 6 bridging sulfur atoms (Fig. 9.9). The Ru-Ru distance (2.775 Å) is indicative of a metal-metal bond with a formal bond order of 0.5. The Ru-S distances of terminal ruthenium ion are slightly shorter (2.321 Å) than those of central ion (2.390 Å) (Albela et al., 1999).

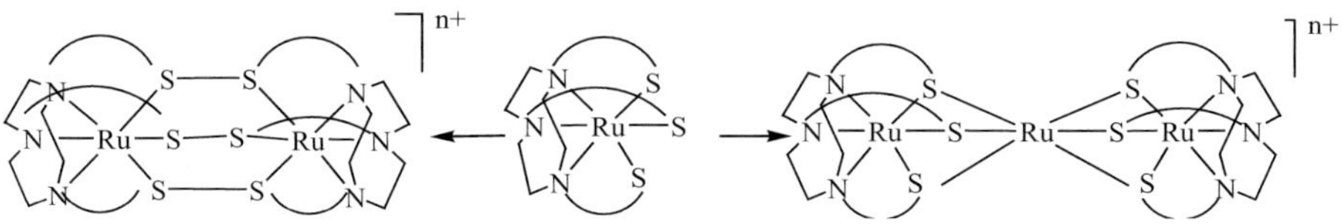

Figure 9.9 Structures of mono-, di-, and trinuclear ruthenium complexes.

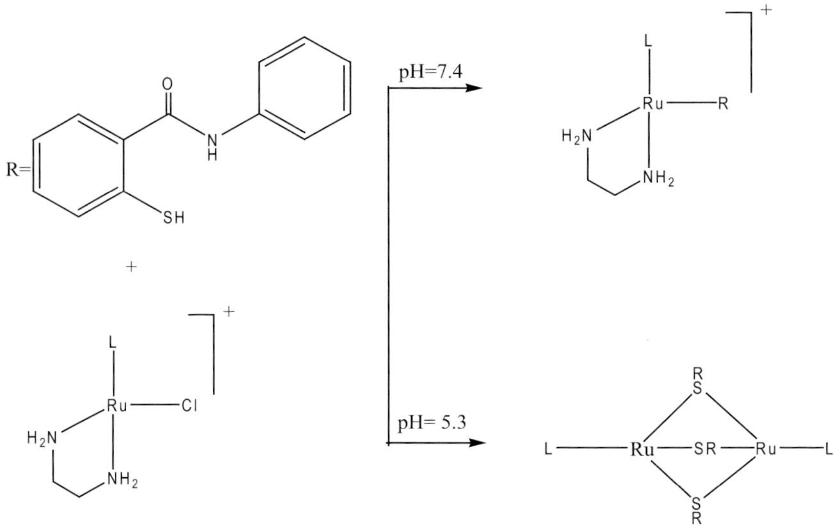

Figure 9.10 Structure of mononuclear and triply bridged ruthenium complexes.

Reaction of $[(\eta^6\text{-L})Ru(en)Cl]PF_6$ with 2-mercaptobenzanilide at pH 7.4 or 5.3 yielded the mononuclear$[\eta^6\text{-L})Ru_2(en)(SR)$ (L = *p*-cymene; R = $(C_6H_4)CONH(C_6H_5)$, en = ethylene diamine) or dinuclear triply bridged $[\eta^6\text{-L})Ru_2(en)(\mu\text{-SR})_3]$ ruthenium complexes (Fig. 9.10). The (NH-OC) bond length of 2.02 Å indicates the presence of strong H-bond between the NH of the en ligand and the carbonyl oxygen in the mononuclear complex. The triply bridged complex adopts a sandwich configuration with an intermolecular H-bond between the amide NH and the adjacent sulfur atom in the thiolate ligand. The bond lengths of the three thiolato ligands and the six Ru-S bonds differ slightly from each other, while the angles of the three thiolato differ markedly from each other (Han et al., 2011).

9.3 Ortho-Substituted Aryl Thiols

O-Substituted aryl thiols can be described to have a rigid structure, with a constant S and other donor atoms and can bring accommodating changes by variations in the angles at the sulfur and other donor chelating atoms. The presence of hard donor together with soft thiolate donors makes them versatile potential ligands to form ruthenium complexes of different nuclearities.

9.3.1 Thiosalicylic Acid

Thiosalicylic acid (TSA), often known as 2-mercaptobenzoic acid or 2-sulfanylbenzoic acid, is commercially available as an odorless and off-white solid, which is readily soluble in lower alcohols and water under alkaline conditions. In acidic condition, TSA remains unchanged, while at pH 6, 50% of the TSA is oxidized to its disulfide derivative ($HOOCC_6H_4SSC_6H_4COOH$) while-at pH 7, complete oxidation of TSA is observed (Rowland, 2011).

TSA is known to show a variety of coordination modes toward metals due to the presence of both hard (O) and soft (S) donor atoms, which have the ability to bridge between the ruthenium centers, giving rise to binuclear and trinuclear complexes.

The reaction of TSA with $[(\sigma^6\text{-L})RuCl(\mu\text{-Cl})]_2$ (L = MeC_6H_4iPr) in the presence of Et_3N (triethylamine) gives the binuclear complex $[(\sigma^6\text{-L})Ru(TSA)]_2$ (Fig. 9.11). The metal is found to adopt a distorted octahedral geometry with ruthenium lying almost directly under the center of the cymene ring. The thiosalicylate forms a six-membered ring bonded to ruthenium via sulfur and one of the oxygen atoms of the carboxyl group, with sulfur bridging to another Ru center forming a dimer having a Ru-S-Ru-S four membered ring core. X-ray diffraction studies show that ruthenium does not bind equidistantly to the benzene ring of the cymene ligand (Henderson et al., 2001). This complex proved to be quite stable as no cleavage of the thiolate bridge was observed when dimer was made to react with CO, PPh_3 or *cis*-[$PtCl_2(PPh_3)_2$]. Addition of TSA to the dimer gives rise to the tris-(thiosalicylate) species (Fig. 9.11), having m/z value of 931. In this complex, the carboxylic acid did not coordinate and remain protonated.

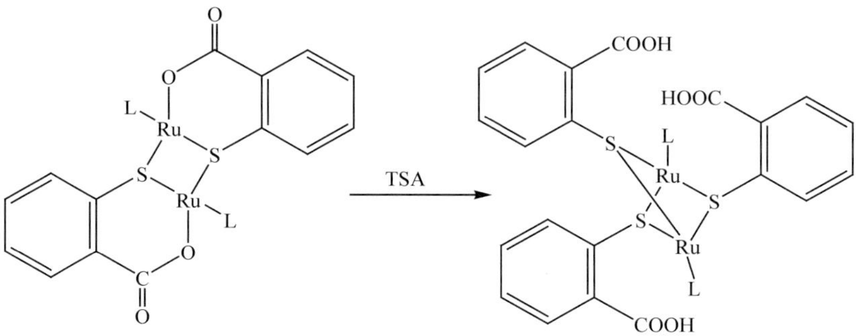

Figure 9.11 Structures of bis and tris-bridge thiolate complexes.

A trinuclear complex [NEt_4] [Ru_3^{III}(TSA-H)$_2$(TSA)$_4$] is formed when the TSA reacts with $Ru(acac)_3$ (acac=acetyl acetonate) in the presence of NEt_4(OAc). $6H_2O$ in one pot reaction. In the complex, each TSA is bonded to two Ru atoms, one via the μ-S bridging and one via the S,O-chelating mode (Chan et al., 2011, 2012). The structure is a linear S-bridged face sharing tri-octahedral [Ru_3] unit with Ru-Ru distance of 2.736 Å, which is shorter than the sum of van der Waals radii (2.8 Å) but is comparable to the reported Ru-Ru distances of 2.805 Å. The complex has a $Ru^{III}(\mu\text{-S})_3$ $Ru^{III}(\mu\text{-S})_3Ru^{III}$ core with Ru-Ru bonding interaction and six carboxylate groups, each coordinated to Ru via one oxygen atom. The Ru-S distances of the complex varied with two Ru-S being 2.271–2.280 Å and one Ru-S being 2.393–2.395 Å.

The trinuclear ruthenium complex [Ru_3^{III}(TSA-H)$_2$(TSA)$_4$][NEt_4] (Fig. 9.12) is stable under physiological conditions, forming adducts with biological reductant glutathione. The complex shows cytotoxic activities toward a panel of human cancer cell lines including cervical epithelioid carcinoma (HeLa), breast cancer (MDA-MB-231), nasopharyngeal carcinoma (SUNE1) and hepatocellular carcinoma (HepG2) using MTT assay which are comparable to the clinically used *cis*-platin. This is the first ruthenium complex reported to significantly attenuate the Wnt-β-catenin signaling in both transcriptomic and proteomic levels.

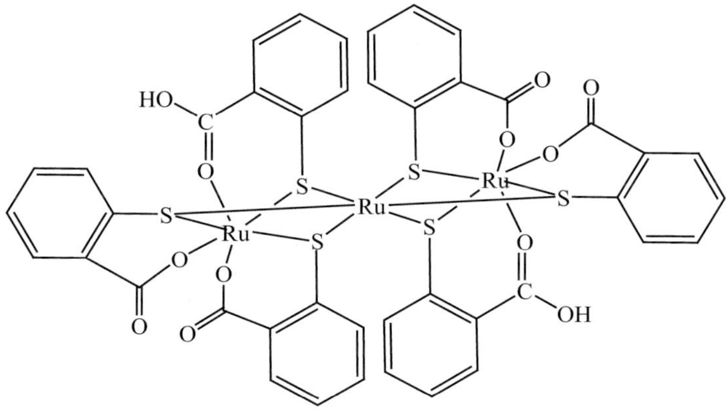

Figure 9.12 Structure of $[Ru_3^{III}(TSA\text{-}H)_2(TSA)_4]$.

2-(Pyridine-2-methylsulfanyl)-benzoic acid, formed by the reaction of TSA with 2-chloromethylpyridine, reacts with [CpRu $(\eta^6\text{-}C_6H_6)Cl(\mu\text{-}Cl]_2$ to give the water soluble pseudo octahedral half sandwich piano stool ruthenium(II) complex $[Ru(\eta^6\text{-}C_6H_6)$ (2-(Pyridine-2-methylsulfanyl)-benzoic acid)Cl]PF_6 (Fig. 9.13) with Ru-S bond length of 2.407 Å (Prakash et al., 2014). This complex has been reported to catalyze a series of aromatic aldehydes into alcohols in water using glycerol as hydrogen donor.

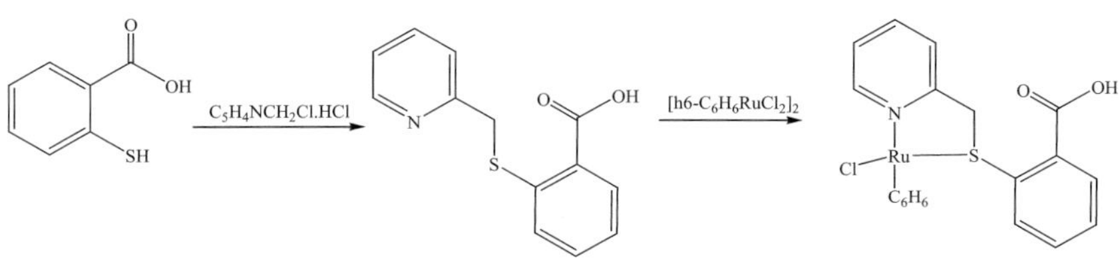

Figure 9.13 Synthesis of $[Ru(\eta^6\text{-}C_6H_6)$ (2-(Pyridine-2-methylsulfanyl)-benzoic acid)Cl].

9.3.2 Hydroxythiophenol

Ortho-hydroxythiophenol (Htp) has two coordination sites, the O and S donor atoms behaving as a bidentate ligand with ruthenium metal ion.

The reaction of $Ru(CO)(Cl_2)(trpy)$ (trpy = terpyridine) with Htp gave [Ru(CO)(Htp)(trpy)]. The IR spectrum displays two strong bands at 1925 and 1911 cm^{-1} with peak intensity of 1:1 indicating the formation of two isomers (Fig. 9.14) due to the difference in the orientation of the chelate ring of the Htp (Sugimoto and Tanaka, 2001).

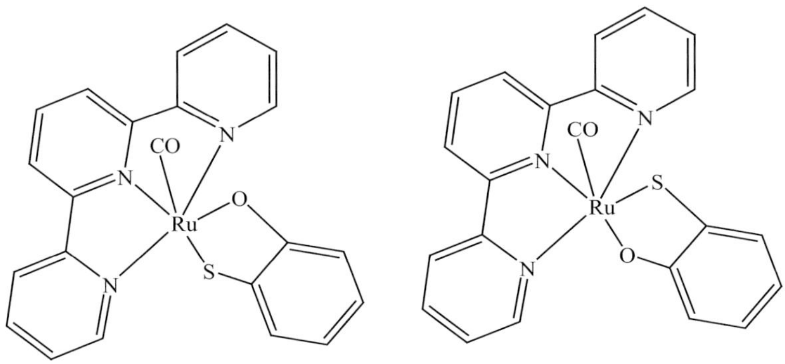

Figure 9.14 Structure of isomers of [Ru(CO)(Htp)(trpy)].

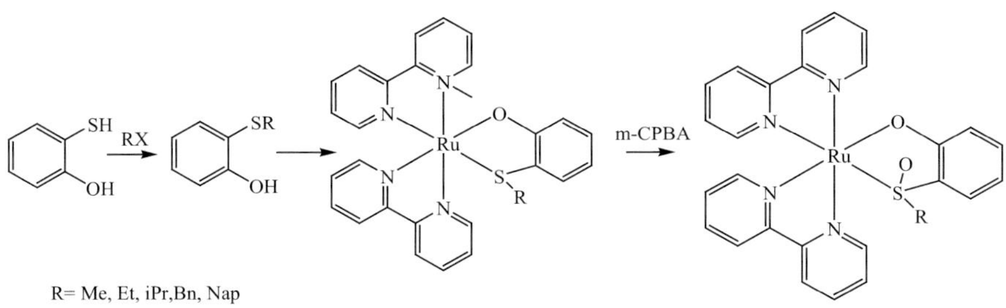

Figure 9.15 Synthesis of $[Ru(bipy)_2(C_6H_4(SR)O)]$ and $[Ru(bipy)_2(C_6H_4(S(O)R)O)]$.

Reaction of Htp with alkylbromide in the presence of $KHCO_3$ as base yielded 2-alkylthiophenol $(C_6H_4(SR)O)$ which reacts with $Ru(bipy)_2Cl_2/[Ru(bipy)_2py_2]^{2+}$ to form the complex, $[Ru(bipy)_2(C_6H_4(SR)O)]$, where oxygen and sulfur of thiophenol are bonded to ruthenium (Fig. 9.15). In the presence of m-CPBA (CPBA = chloro perbenzoic acid) as oxidant, the obtained complex is oxidized to sulfoxides derivative $[Ru(bipy)_2(C_6H_4(S(O)R)O)]$ (Fig. 9.15) where the sulfoxide ligand is generated in situ and forms hydrogen bonding with the bipy (bipy = 2,2′-bipyridine). The Ru-S bond length is 2.250 Å and Ru-N (2.105 Å) is slightly longer than those of other Ru-N (2.058–2.070 Å), indicating the stronger *trans* effect of coordinated sulfur atom. The configurations of the complexes are stable during the coordination and oxidation reactions (Li et al., 2014).

9.3.3 Aminothiophenol

Ortho-aminothiophenol (Atp) is known to act as either uninegative bidentate ligand or can behave as a redox active chelating ligand existing in three oxidation states (i) fully reduced catecholate dianion (tQ^{2-}), fully oxidized neutral quinone (tQ^0) and partially oxidized semi-quionone monoradical ($tSQ^{\bullet -}$) (Fig. 9.16).

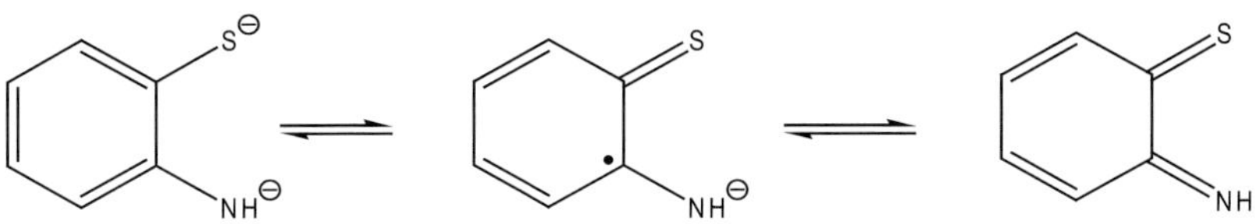

Figure 9.16 Resonance structures of *o*-aminothiophenol.

Kingston reported the first octahedral ruthenium complex $[Ru\{C_6H_4(S)NH_2\}_2(CO)_2]$ of Atp (Kingston et al., 1967). Later, Velasquez synthesized $[Ru(CO)_2(SC_6H_4NH_2)_2]$ complex by the reaction of $RuCl_3$ with *o*-aminothiophenol in ethanol in the presence of CO (Velasquez et al., 2012). Similar type of ruthenium complexes $RuCO(L)_2(S\text{-}C_6H_4NH_2)Cl]$, $[Ru(L)_2(S\text{-}C_6H_4NH_2)_2]$ and $[RuX(L)(S\text{-}C_6H_4NH_2)_2]$ are reported when Atp reacts with $RuCl_2(L_4)$, $RuH(CO)L_3$ and RuX_3L_3 ($L = PPh_3$, $AsPh_3$; $X = Cl$, Br), respectively (Fig. 9.17). In these complexes Atp binds through S and N in a bidentate manner (Pandey and Sharma, 2013).

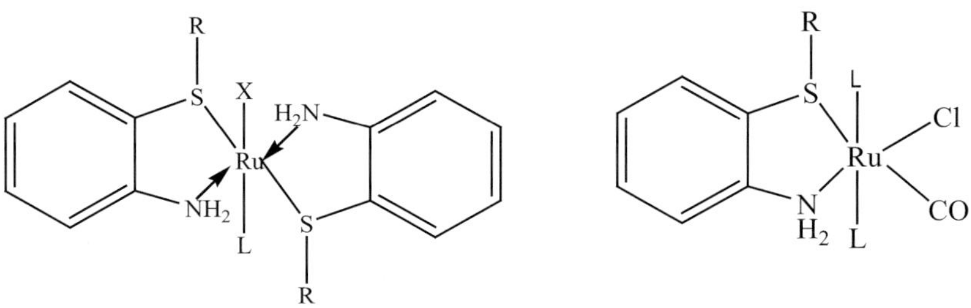

Figure 9.17 Structures of ruthenium aminothiophenol complexes.

When $CpRuCl(PPh_3)_2$ reacts with Atp, the mononuclear ruthenium complex $[CpRu(PPh_3)(NHC_6H_4S)]Cl$ is formed where Atp coordinates via fully oxidized neutral quinone in a bidentate manner (Fig. 9.18).

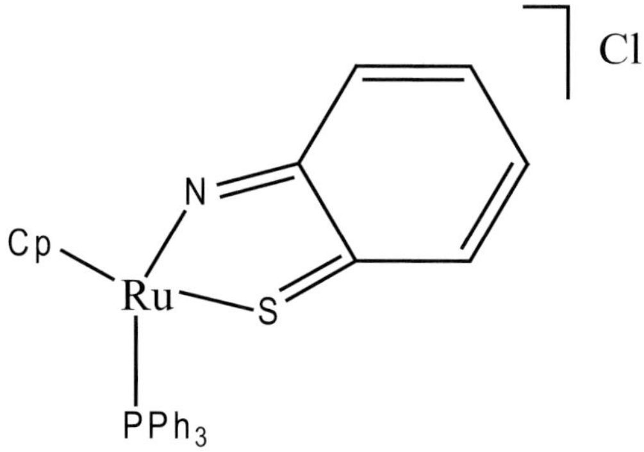

Figure 9.18 Structure of $[CpRu(PPh_3)(NHC_6H_4S)]Cl$.

When the thiol proton is replaced by either methyl or phenyl group, then a dicationic dinuclear complex $[Cp_2Ru_2(PPh_3)_2(NHC_6H_4S)_2]Cl_2$ is isolated where the metal coordinates via the semi-quionone monoradical form of Atp (Fig. 9.19). In this reaction, a bischelated NH-$(CH_3S)C_6H_3$-$C_6H_3(SCH_3)$-NH bridging ligand is generated in situ via an unusual C-C bond formation which occurs between two Atp that act as bridge between two $[CpRu(PPh_3)]$ moieties (Ghosh et al., 2015).

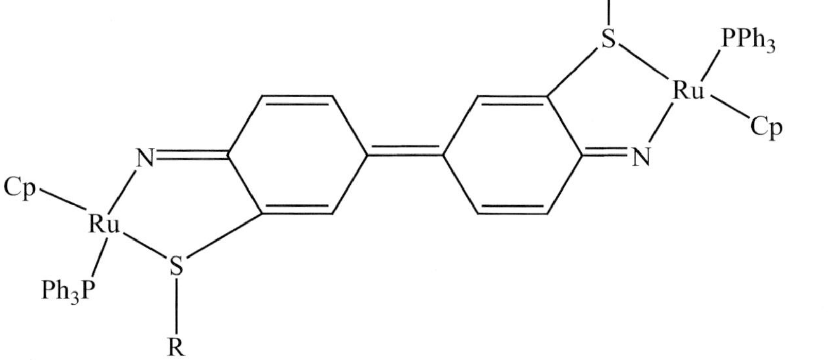

Figure 9.19 Structure of $[Cp_2Ru_2(PPh_3)_2(NHC_6H_4S)_2]^{2+}$.

In the reaction of $RuCl_2(PPh_3)_3$ or $Ru(acac)_2(CH_3CN)_2$ with Atp, the ruthenium(III) iminosemiquinone complexes are isolated (Bhattacharya et al., 2002; Patra et al., 2003).

There are many reports of ruthenium Schiff base complexes derived from Atp which gives mononuclear or binuclear complexes. In these complexes, the ruthenium binds via S and N donor atoms of the Schiff base together with other ligands (Fig. 9.20) (Arunachalam et al., 2009, 2011; Bhattacharya et al., 1999; Prabhakaran et al., 2004; Thangadurai and Natarajan, 2001; Roy et al., 2009). These ruthenium Schiff base complexes are known to act as potential catalyst for the hydrogenation reactions of methoxybenzene and benzaldehyde, oxidation of alcohols and C-C coupling reactions and the products are obtained in good yield. These complexes also exhibits good antibacterial activity against a panel of bacterial strains such as *E. feacalis, A. hydrophilla, E. Coli, S aureus, S. Typhi, V. cholera*, and *P. aeruginosa* (Thangadurai and Natarajan, 2001; Prabhakaran et al., 2004).

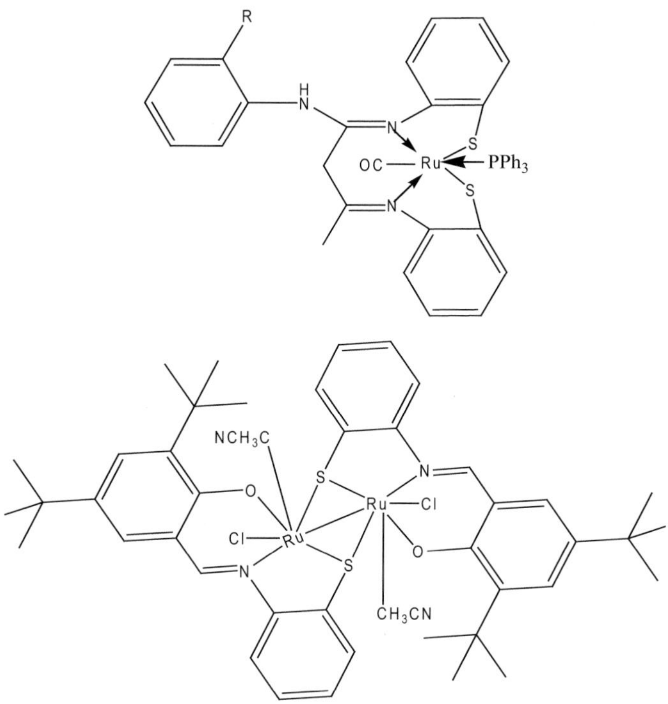

Figure 9.20 Examples of ruthenium Schiff base complexes.

9.4 Diaryl Sulfides

Diaryl sulfide ligands possess flexible S_n spacer groups between the two aryl rings and can adopt different conformation when coordinated to ruthenium metal giving rise to mononuclear and binuclear complexes with sulfur or no sulfur co-ordination. In the presence of other donor atoms, the sulfur atom can remain uncoordinated or take part in weak coordination or undergo reductive cleavage.

9.4.1 Diaryl Monosulfide Complexes

Reaction of $(C_6H_5)_2S$ with Ru(η^5-C_5H_5)(dppe)Cl (dppe = 1,2-bis (diphenylphosphino)ethane) in the presence of $AgBF_4$ led to the formation of cationic mononuclear complex [Ru(η^5-C_5H_5) ($Ar_2PCH_2CH_2PAr_2$)($(C_6H_5)_2S$)]BF_4 (Ohkita et al., 1994) (Fig. 9.21).

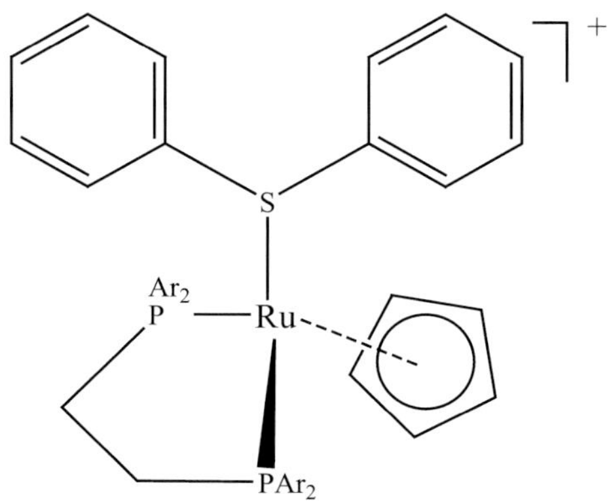

Figure 9.21 Structure of [Ru(η^5-C_5H_5)($Ar_2PCH_2CH_2PAr_2$)($(C_6H_5)_2S$)]$^+$.

Another mononuclear complex [$RuCl_2(CO)_2(S(C_6H_5)_2)_2$] is formed by the reaction of [$RuCl_2(CO)_3$]$_2$ with $(C_6H_5)_2S$ in THF (Fig. 9.22). In this complex, the ruthenium is in an octahedral geometry with CO and Cl atoms in *cis* position while the two diarylsulfides are *trans* to each other. The two Ru-S bond lengths are 2.387 Å and 2.406 Å and the complex is joined into stacks by a weak three dimensional H-Cl hydrogen bonding network. When CH_2Cl_2 is used as solvent, a mixture of two products [$RuCl_2(CO)_3H_2O$]

and $[RuCl_3(CO)_3]_2 \cdot 2H_2O$ are isolated where no coordination with diaryl sulfide occurred and on prolonged refluxing only trace amount of the $[RuCl_2(CO)_2(S(C_6H_5)_2)_2]$ complex is obtained (Taimisto et al., 2003).

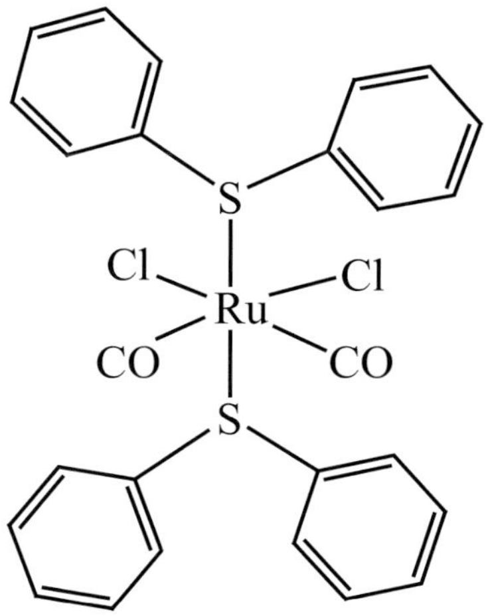

Figure 9.22 Structure of $[RuCl_2(CO)_2(S(C_6H_5)_2)_2]$.

Diaryl monosulfides containing additional donor groups are known to coordinate with ruthenium via tridentate, bidentate or monodentate fashion.

$[Ru(\eta^6\text{-}L)Cl_2]_2$ (L = $MeC_6H_4{}^iPr$) reacted with $[(p\text{-}Ph_2POC_6H_4)_2S]$ in a non-coordinating solvent such as toluene to afford the binuclear complex $[Ru(\eta^6\text{-}L)Cl_2]_2[\mu\text{-}(p\text{-}Ph_2POC_6H_4)_2S]$ with no sulfur coordination and with cymene group in a *trans* position with respect to Ru-Ru axis (Fig. 9.23) (Arena et al., 1999). The ^{31}P NMR shows a single resonance peak at 115.4 ppm indicating chemical equivalence of the phosphorus atoms. In the presence of the coordinating solvent, CH_3CN, substitution of one terminal Cl with CH_3CN is observed leading to the formation of cationic complex $[Ru(\eta^6\text{-}L)Cl\ (CH_3CN)]_2\ [\mu\text{-}(p\text{-}Ph_2POC_6H_4)_2S][PF_6]_2$. In these complexes, the two stereogenic ruthenium atoms are held together by the bridging phosphinito ligands. Bubbling of CO in the cationic complex gives $[Ru(\eta^6\text{-}p\text{-cymene})Cl(CO)]_2\ [\mu\text{-}(p\text{-}Ph_2POC_6H_4)_2S][PF_6]_2$ where CH_3CN is replaced by CO and the complex is only detected in the CO atmosphere and have limited stability.

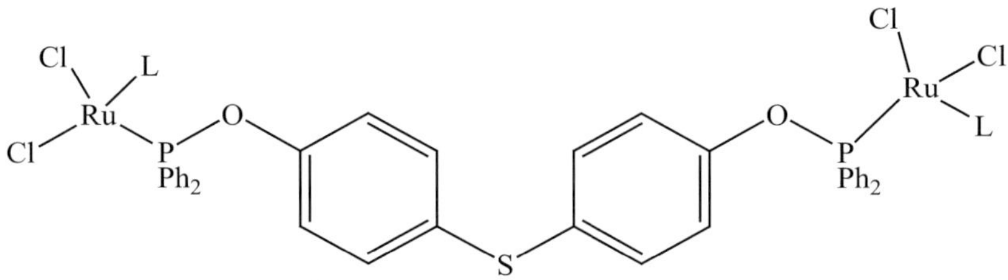

Figure 9.23 Structure of $[Ru(\eta^6\text{-}L)Cl_2]_2[\mu\text{-}(p\text{-}Ph_2POC_6H_4)_2S]$.

The binol-based diphosphate ligand (PSP) is not capable of chelating a transition metal since the *p*-thiophenol spacer is too large. However, it readily bridges two ruthenium centers to give the dinuclear complex $[Ru_2Cl_4(L)_2(PSP)]$ (L = C_6H_6, $C_6H_4(CH_3)(^i Pr)$, (Fig. 9.24) upon the reaction of the dimeric complex $[(\eta^6\text{-}C_6H_6)RuCl_2]_2$ with PSP in CH_2Cl_2/benzene. In this complex PSP act as a bidentate bridging ligand bonded to two ruthenium atoms via phosphorous atoms (Chen et al., 2002).

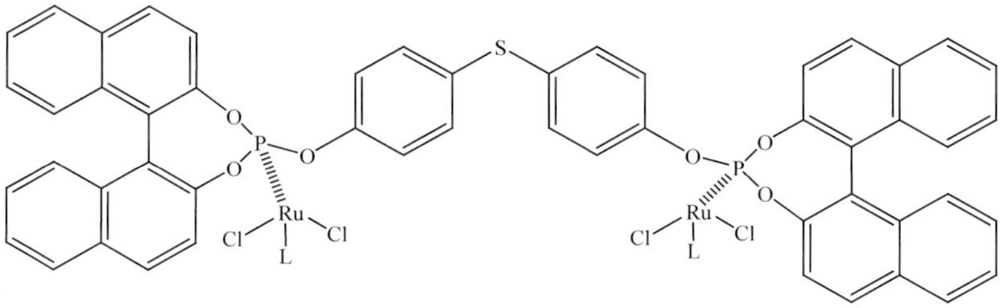

Figure 9.24 Structure of $[Ru_2Cl_4(L)_2(PSP)]$.

Similar type of complexes with no sulfur coordination to ruthenium is also reported (Punji and Balakrishna, 2007). The diphosphate ligand is prepared by the reaction of chlorophosphate with 2,6-pyridinedimethanol in the presence of NEt_3 which on reaction with $[Ru(\eta^6\text{-}L)(\mu\text{-}Cl)Cl_2]$ gives $[RuCl((\eta^6\text{-}L)(\eta^2\text{-}2,6\text{-}C_6H_3N\{CH_2OP(\mu\text{-}OC_{10}H_6)(\mu\text{-}S)(C_{10}H_6O\text{-})\}_2\text{-}\kappa P,\kappa P]Cl$ (L = *p*-cymene) (Fig. 9.25).

The reaction of the dimeric complex $[(\eta^6\text{-}C_6H_6)RuCl_2]_2$ with bis(diarylamido)/thioether, led to the formation of the mononuclear complex $[(\eta^6\text{-}C_6H_6)Ru(X_2NSN)]$(X = xylene, 3,5-$(CF_3)_2C_6H_3$), where the ligand co-ordinates in a tridentate manner (NSN) with the ruthenium center. The arene group can be replaced by PMe_3, PhCN, MeCN, N_2H_4 to yield $[(PMe_3)_3Ru(X_2NSN)]$ and $[(PMe_3)_2LRu(X_2NSN)]$ (L = PhCN, MeCN, N_2H_4), respectively (Fig. 9.26) (Takemoto et al., 2002).

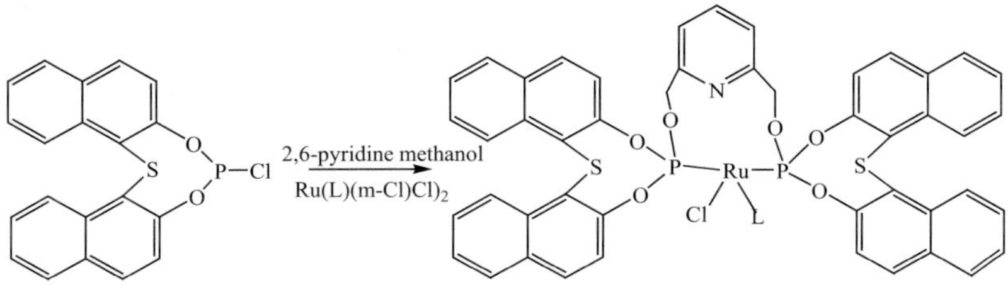

Figure 9.25 Synthesis of $RuCl((\eta^6\text{-}L)(\eta^2\text{-}2,6\text{-}C_6H_3N\{CH_2OP(\text{-}OC_{10}H_6)(\mu\text{-}S)(C_{10}H_6O\text{-})\}_2\text{-}kP,kP]Cl$.

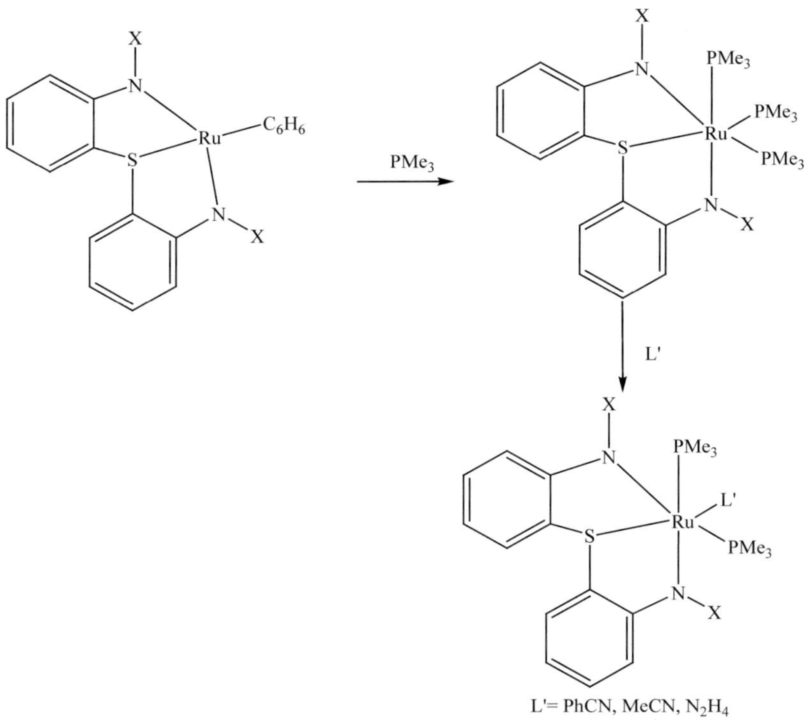

Figure 9.26 Structures of bis(diarylamido)/thioether ruthenium complexes.

When $[(PMe_3)_3Ru(X_2NSN)]$ reacts with benzoylhydrazine, the complex $[(PMe_3)_3Ru(k^2\text{-}(O\text{-}N)PhC(O) = NNH)\ H(X_2NSN)]$ (Fig. 9.27) is isolated, where the ruthenium is bonded to bidentate amido/thioether (NS) with a dangling $XNHC_6H_4$. The η^2 benzoylhydrazido ligand is bonded to ruthenium via oxygen and terminal N atom forming a five-membered chelate ring.

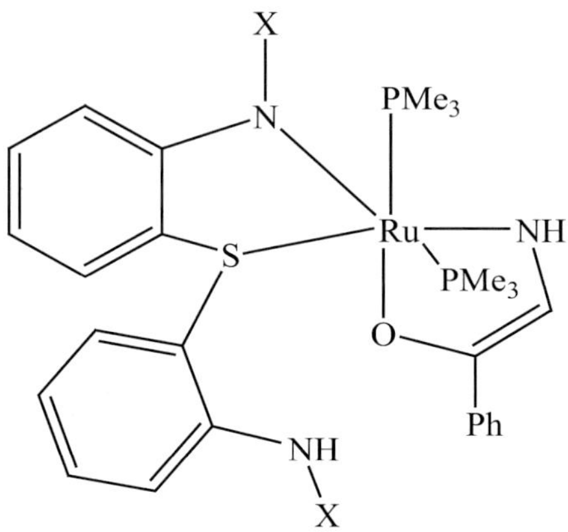

Figure 9.27 Structure of $[(PMe_3)_3Ru(k^2\text{-}(O\text{-}N)PhC(O){=}NNH)\,H(X_2NSN)]$.

9.4.2 Diaryl Disulfide Complexes

Diaz et al. reported the ruthenium diaryl disulfide complex $[CpRu(PPh_3)_2(C_6H_5SSC_6H_5)]PF_6$ by the reaction of $CpRu(PPh_3)_2Cl$ with $(C_6H_5S)_2$ in methanol in the presence of PF_6 where only one S atom is coordinated to the metal (Diaz and Gómez, 2001).

The reaction of $(C_6H_5S)_2$ with the ruthenium azobenzene complex $[Ru_2\{\eta^6{:}\eta1{:}\mu\text{-}C_6H_5(CH_2)_3O\}_2(Y)](BF_4)_2$ (Y = (PhN=NPh, $PhN{=}NC_6H_4\text{-}4\text{-}OCH_3$) in CH_2Cl_2 under irradiation (>510 nm) gives the complex $[Ru_2\{\eta^6{:}\eta1{:}\ \mu\text{-}C_6H_5(CH_2)_3O\}_2(\mu\text{-}C_6H_5SSC_6H_5)]^{2+}$ where both sulfur atoms are coordinated to two ruthenium metals (Fig. 9.28) (Kitaura et al., 2002). The S-S bond length (2.110 Å) is slightly longer compared to the free disulfide ligand (1.900–2.000 Å). The S atoms adopt a chiral non-planar configuration in order to avoid repulsions from the azobenzene fragment.

A striking feature of compounds containing S-S linkers is that these ligands can undergo diversified changes such as oxidative or reductive cleavage or extrusion of one sulfur atom. The cleavage of the S-S bond occurs via thermal, chemical, or photochemical conditions, in the presence of metal ions. The mechanism involves (i) in situ S-S function reactions (ii) in situ cleavage and

(iii) in situ thiol atom reactions. The activation of the S-S bond probably involves coordination of $C_6H_5SSC_6H_5$ to RuL_n leading to an intermediate producing a back-donation (electron donation) from metal to sulfur atom $(\pi) \rightarrow S(\pi)$. This leads to a destabilization of the S-S bond, resulting into its cleavage to give rise to either mononuclear or dinuclear complexes with bridging thiolate ligands.

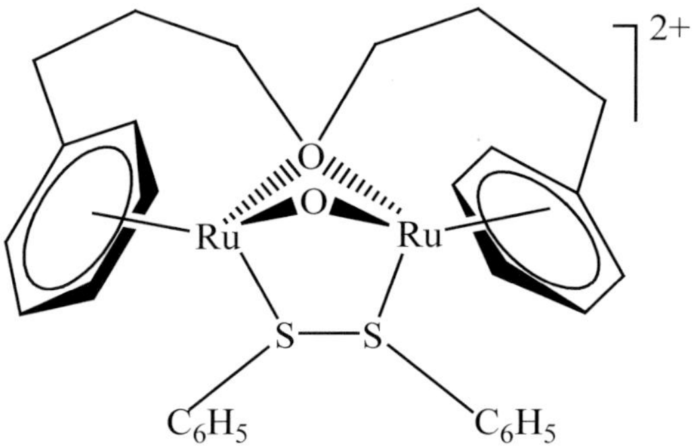

Figure 9.28 Structure of $[Ru_2\{\eta^6:\eta 1:\mu\text{-}C_6H_5(CH_2)_3O\}_2(C_6H_5SSC_6H_5)]^{2+}$.

The reaction of the tetranuclear complex $[CpRu(\mu\text{-Cl})]_4$ with $(C_6H_5S)_2$ yields an *anti*-isomer phenyl thiolate bridged diruthenium complex $[CpRuCl(\mu_2\text{-}SC_6H_5)]_2$, in which the S-S bond is cleaved (Fig. 9.29). The reaction of the neutral phenyl thiolate bridged ruthenium complex with AgOTf in THF yields the cationic phenyl thiolate bridged ruthenium complex $[CpRu(\mu_2\text{-}Cl)(\mu_2\text{-}SC_6H_5)_2\ RuCp]^+$ (Fig. 9.29) (Nishibayashi et al., 2004). The X-ray structure suggests that the ruthenium atoms are bridged by one chloride and two benzene thiolate groups in a *cis*-configuration.

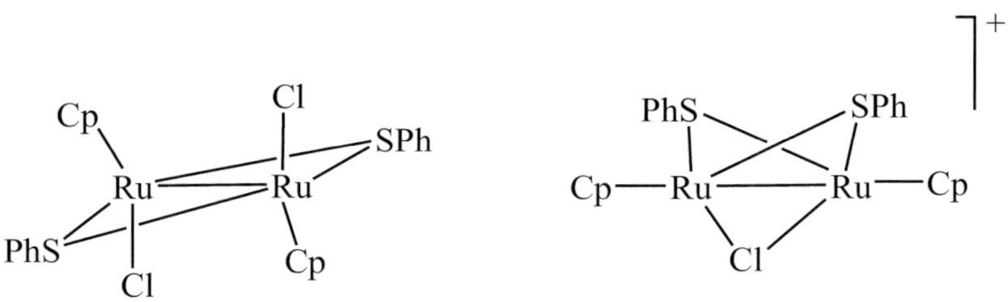

Figure 9.29 Structures of $[CpRuCl(\mu_2\text{-}SC_6H_5)]_2$ and $[CpRu(\mu_2\text{-}Cl)(\mu_2\text{-}SC_6H_5)_2$ $RuCp]^+$.

Treatment of RuTp(COD)Cl (Tp = hydridotris(pyrazalyl) borate anion) with $RC_6H_4SSC_6H_4R$ (R = H, Me) in DMF gives the complexes of $RuTp_K{}^2(S,S)\text{-}SC_6H_3R\text{-}S\text{-}C_6H_4R)(CO)$ (Fig. 9.30). In these reactions the conversion of $(C_6H_5S)_2$ into 2-(arylmercapto)aryl mercaptan is observed in the coordination sphere of the ruthenium (IV) center. This may be due the presence of the steric shielding of Tp ligand which inhibits the formation of binuclear bridging complex. The CO ligand in the product stems from the decomposition of the solvent, DMF. The coordination geometry around the ruthenium atom is slightly distorted octahedral (Standfest-Hauser et al., 2004).

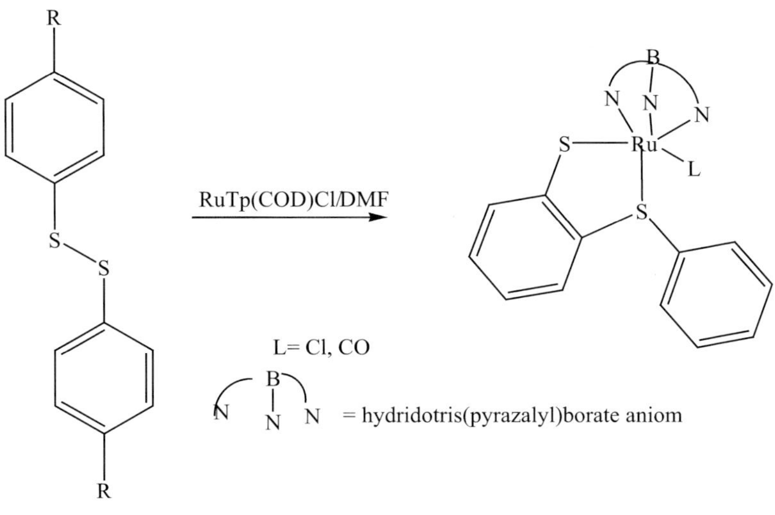

Figure 9.30 Synthesis of $RuTp_K{}^2(S,S)\text{-}SC_6H_3R\text{-}S\text{-}C_6H_4R)(CO)$.

The reaction of 2-aminophenyl disulfide with $RuCl_3$, gives the mononuclear $[Ru(C_6H_5SSC_6H_5)Cl_3(H_2O)]$ (Bhowon et al., 2007) where metal is bonded to NH_2 group and no S coordination is observed.

2,2′-Dithiobenzoic acid, a multifunctional ligand containing both thio and carboxylic groups, adopts unique structural topologies and exhibits interesting properties upon coordination with metals. The dithiodibenzoic acid undergoes S-S mediated bond reaction followed by S-oxidation leading to the formation of sulfinato-benzoate ligand $(C_6H_4(SO_2)COO^-)$ in the presence of $RuCl_3$ and phen/bipy under slow diffusion conditions. The sulfinato-benzoate ligand coordinates with ruthenium where the metal ion coordinates with oxygen, the sulfur atom of the ligand and 4 nitrogen atoms of phen/bipy (Fig. 9.31). This is a rare example where the ruthenium is coordinated to the sulfur atom rather than the oxygen of sulfinate (Moosun et al., 2015).

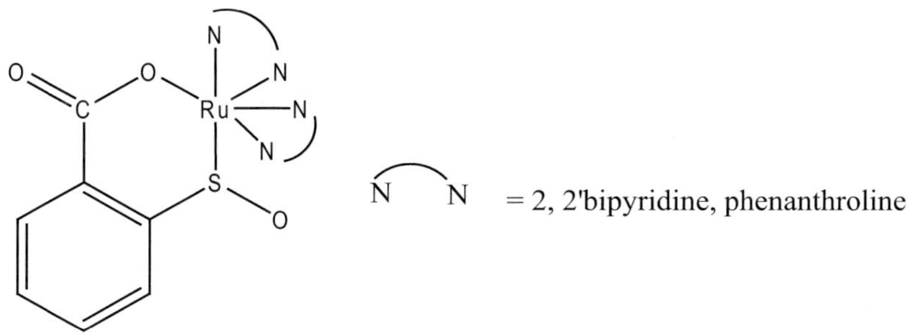

Figure 9.31 Structures of ruthenium sulfinato complex.

When Schiff bases of 2-aminophenyl disulfide are reacted with $RuCl_3$ in the presence of bipy, phen and PPh_3, binuclear complexes are isolated where one ruthenium is coordinated to an imine nitrogen, one terminal chlorine, one bipyridine and two bridging chlorine atoms while the other ruthenium is coordinated to the other imine nitrogen, one terminal Cl, two bridging Cl and two water molecules with (Fig. 9.32) (Bhowon et al., 2005) or without sulfur coordination (Bhowon et al., 2007).

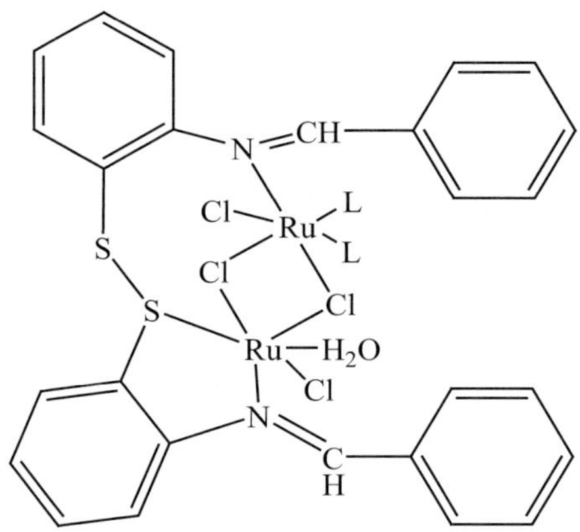

Figure 9.32 Structure of ruthenium Schiff base complex of 2-aminophenyl disulfide.

When the imine group is replaced by amide group, the resulting 2,2′-dithiobisbenzoylbenzamide ligand reacts with $RuCl_3$ in the presence of PPh_3 and N-donor ligands such as phen, bipy, pyridine and diaminotoluene to yield the binuclear complexes with no sulfur coordination (Bhowon et al., 2001; Jhaumeer Laulloo et al., 2003).

These ruthenium disulfide complexes possess interesting catalytic and biological properties. Ruthenium disulfide diamine complexes (Bhowon et al., 2007) are found to catalyze the oxidation of primary alcohols into corresponding aldehyde and cyclohexene into cyclohexene-oxide, cyclohexenone and cyclohexenol derivatives.

Ruthenium diaryl disulfides containing diimines or diamides moieties show antibacterial activities against a panel of bacteria such as *E. Coli, S aureus, S. Typhi, P. vulgaris, B. subtilus,* and *P. aeroginosa* (Bhowon et al., 2007, 2001; Jhaumeer-Laulloo et al., 2002).

9.5 Conclusion

This chapter presents an overview of the coordination chemistry of diaryl sulfide ligands with ruthenium, where the useful data has been selected for the readers who are not familiar with this topic and to detail the most promising systems for future development. In all well-characterized cases, ruthenium with aryl thiols leads to the formation of mononuclear and polynuclear complexes via μ-sulfur bridging. Thiosalicylic acid gives rise to the formation of binuclear and trinuclear complexes with μ-sulfur bridging while the derivatives such as *o*-hydroxythiophenol and *o*-aminothiophenol gives rise to only mononuclear complexes. Diarylsulfides are prone to rearrangement and sulfur redistribution in the presence of ruthenium due to lability of S-S bond. In addition, after careful review of these ruthenium compounds, it becomes apparent that, given the molecular design and the ease with which the structure of these complexes can be modified, they offer a promising approach in the development of key compounds in both medical and catalytic fields.

References

Albela B, Bothe E, Brosch O, Mochizuki K, Weyhermüller T, Wieghardt K (1999). Redox chemistry of (1,4,7-tris(4-tert butyl 2-mercaptobenzyl) 1,4,7-triazacyclononane) ruthenium(III), $[Ru^{III}L]$: Synthesis and characterization of $[Ru^{II}_2(L\text{-}L)](BPh_4)_4 \cdot 10CH_3CN$ and $[LRuRuRuL](PF_6)_2 \cdot H_2O$. *Inorg Chem*, **38**: 5131–5138.

Arena CG, Drago D, Panzalorto M, Bruno G, Faraone F (1999). Diphosphinito-bridged ruthenium(II) and rhodium(III) complexes with stereogenic metal centers. *Inorg Chim Acta*, **292**: 84–95.

Arunachalam S, PadmaPriya N, Jayabalakrishnan, Chinnusamy V (2011). Synthesis and characterization of organosoluble ruthenium complexes bearing Schiff base ligands: Efficient reusable catalyst for the hydrogenation reactions. *Int J Appl Biol Pharm Technol*, **2**: 352–365.

Belchem G, Steed JW, Tocher DA (1994). Organometallic thiol and thiolato complexes of ruthenium(IV). *J Chem Soc Dalton Trans*, **13**: 1949–1962.

Bhattacharyya D, Chakraborty S, Munshi P, Lahiri GK (1999). Ruthenium(II/III) bipyridine complexes incorporating thiol based imine functions synthesis, spectroscopic and redox properties. *Polyhedron*, **18**(23): 2951–2959.

Bhattacharya S, Gupta P, Basuli F (2002). Structural systematics for *o*-C_6H_4XY ligands with X,Y = O, NH and S donor atoms. *O*-Iminoquinone and *o*-iminothioquinone complexes of ruthenium and osmium. *Inorg Chem*, **41**: 5810–5816.

Bhowon MG, Jhaumeer-Laulloo S, Dowlut M, Curpen S, Jumnoodoo V (2005). Synthesis and characterization of bis(2-aminophenyl)disulfide diamine derivatives and their ruthenium complexes. *Trans Met Chem*, **30**: 35–39.

Bhowon MG, Jhaumeer-Laulloo S, Ramnial T (2001). Antibacterial properties of ruthenium(II) benzamide complexes. *Trans Met Chem*, **26**: 329–332.

Bhowon MG, Jhaumeer-Laulloo S, Soukhee N, Allibacus A, Shiboo V (2007). Synthesis, catalytic and antibacterial activity of 2-amino-phenyldisulfide. *J Coord Chem*, **60**(12): 1335–1343.

Chan SL-F, Gao S, Chui SS-Y, Shek L, Huang J-S, Che C-M (2012). Supramolecular self-assembly of 1D and 3D heterometallic coordination polymers with triruthenium building blocks. *Chem Eur J*, **18**: 11228–11237.

Chan SL-F, Shek L, Huang J-S, Chui SS-Y, Sun RW-Y, Che C-M (2012). Molecular wheels of ruthenium and osmium with bridging chalcogenolate ligands: Edge-shared-octahedron structures and metal-ion binding. *Angew Chem Int Ed*, **51**: 2614–2617.

Chan SL-F, Sun RW-Y, Choi M-Y, Zeng Y, Shek L, Chui SS-Y, Che C-M (2011). An anti-cancer trinuclear ruthenium(III) complex with 2-thiosalicylate ligands attenuates Wnt-β-catenin signaling. *Chem Sci*, **2**: 1788–1792.

Chen Y, Valentini M, Pregosin PS, Albinati A (2002). Synthesis of Ru-arene complexes of Me-Duphos. X-ray, PGSE NMR diffusion and catalytic studies. *Inorg Chim Acta*, **327**: 4–14.

Chérioux F, Therrien B, Süss-Fink G (2003). First star-like oligophenylene molecules containing a dinuclear organometallic core. *Eur J Inorg Chem*, **6**: 1043–1047.

Chérioux F, Therrien B, Süss-Fink G (2004). Synthesis and structural characterisation of new cationic dinuclear ruthenium(II) thiolato complexes of the type $[Ru_2(\eta^6\text{-arene})_2(\mu\text{-}p\text{-S–}C_6H_4\text{–Br})_3]^+$. *Inorg Chim Acta*, **357**: 834–838.

Chérioux F, Thomas CM, Monnier T, Süss-Fink G (2003). Specific reactivity of SH versus OH functions towards dinuclear arene ruthenium units: Synthesis of cationic complexes of the type $[(arene)_2Ru_2(SR)_3]^+$. *Polyhedron*, **22**(4): 543–548.

Chérioux F, Thomas CM, Therrien B, Süss-Fink G (2002). Dendritic systems based on dinuclear ruthenium or rhodium units generating peripheral catalytic sites. *Chem Eur J*, **8**(19): 4377–4382.

Diaz C, Gomez A (2001). New ruthenium(II) complexes containing organodisulfide ligands. *J Coord Chem*, **54**(3–4): 261–266.

Fowe EP, Therrien B, Süss-Fink G, Daul C (2008). Electron-structure calculations and bond order analysis using density functional theory of cationic dinuclear arene ruthenium complexes. *Inorg Chem*, **47**(1): 42–48.

Ghosh P, Mandal S, Chattarjee I, Mondal TK, Goswami S (2015). Comparison of redox activity between 2-aminothioether and 2-aminothiophenol: Redox-induced dimerization of 2-aminothioether via C-C coupling. *Inorg Chem*, **54**(13): 6235–6244.

Giannini F, Bartoloni M, Paul LEH, Süss-Fink G, Reymond JL, Furrer J (2015). Cytotoxic peptide conjugates of dinuclear arene ruthenium trithiolato complexes. *MedChemComm*, **6**: 347–350.

Giannini F, Furrer J, Ibao A-F, Süss-Fink G, Therrien B, Zava O, Baquie M, Dyson PJ, Stěpnička P (2013). Highly cytotoxic trithiophenolatodiruthenium complexes of the type $[(\eta^6\text{-}p\text{-MeC}_6H_4Pr^i)_2Ru_2(SC_6H_4\text{-}p\text{-X})_3]^+$: Synthesis, molecular structure, electrochemistry, cytotoxicity, and glutathione oxidation potential. *J Biol Inorg Chem*, **17**: 951–960.

Giannini F, Furrer J, Süss-Fink G, Clavel CM, Dyson PJ (2013). Synthesis, characterization and in vitro anticancer activity of highly cytotoxic trithiolato diruthenium complexes of the type $[(\eta^6\text{-}p\text{-MeC}_6H_4{}^iPr)_2$ $Ru_2(\eta_2\text{-SR}^1)2(\eta_2\text{-SR}^2)$ containing different thiolato bridges. *J Organomet Chem*, **744**: 41–48.

Giannini F, Paul LEH, Furrer J, Therrien B, Süss-Fink G (2013). Highly cytotoxic diruthenium trithiolato complexes of the type $[(\eta^6\text{-}p\text{-MeC}_6H_4Pr^i)_2Ru_2(\mu_2\text{-SR})_3]^+$: Synthesis, characterization, molecular structure and in vitro anticancer activity. *New J Chem*, **37**: 3503–3511.

Gras M, Therrien B, Süss-Fink G, Zava O, Dyson PJ (2010). Thiophenolato-bridged dinuclear arene ruthenium complexes: A new family of highly cytotoxic anticancer agents. *Dalton Trans*, **39**: 10305–10313.

Han Y, Luo Q, Hao X, Li X, Wang F, Hu W, Wu K, Lü S, Sadler PJ (2011). Reactions of an organoruthenium anticancer complex with 2-mercaptobenzanilide-a model for the active site cysteine of protein tyrosine phosphatase 1B. *Dalton Trans*, **40**: 11519–11529.

Henderson W, Nicholson BK, Oliver AG, Rickard CEF (2001). Dimeric rhodium(III), iridium(III) and ruthenium(II) thiosalicylate complexes. *J Organomet Chem*, **625**: 40–46.

Hirano M, Sakaguchi Y, Yajima T, Kurata N, Komine N, Komiya S (2005). Stiochiometric and catalytic sp^3 C-H/D_2 exchange reactions of ortho-substituted benzenethiol and phenols by a ruthenium(II) complex. Effect of a chalcogen anchor on the bond cleavage reaction. *Organometallics*, **24**: 4799–4809.

Hirano M, Togashi S, Ito M, Sakaguchi Y, Komine N, Komiya S (2010). Carbon-hydrogen bond cleavage reaction in 5-coordinate bis(2,6-dim ethylbenzenethiolato)ruthenium(II) complexes. *Organometallics*, **29**: 3146–3159.

Ibao A-F, Gras M, Therrien B, Süss-Fink G, Zava O, Dyson PJ (2012). Thiolato-bridged arene–ruthenium complexes: Synthesis, molecular structure, reactivity, and anticancer activity of the dinuclear complexes $[(\text{arene})_2Ru_2(SR)_2Cl_2]^+$. *Eur J Inorg Chem*, **2012**(9): 1531–1535.

Jhaumeer-Laulloo S, Bhowon MG (2003). Synthesis, biological and catalytic properties of Ru(II) benzamides Schiff base complexes. *Ind J Chem*, **42A**: 2536–2540.

Jhaumeer-Laulloo S, Bhowon MG, Hosany A (2002). Antibacterial & catalytic activities of 2,2′-dithio(2-hydroxyphenyl)benzamide (DNBH) and its ruthenium metal complexes. *Afr J Sci Technol*, **3**(2): 52–55.

Kingston JV, Jamieson JWS, Wilkinson G (1967). Octahedral ruthenium(II) carbonyl complexes with nitrogen, Sulphur and tin donor ligands. *J Inorg Nucl Chem*, **29**: 133–138.

Kitaura R, Miyaki Y, Onishi T, Kurosawa H (2002). Bridge coordination of bidentate ligands to a dinuclear η^6-arene ruthenium(II) unit constructed by a chelating and bridging alkoxo ligand. *Inorg Chim Acta*, **334**: 142–148.

Li Z-Z, Wen A-H, Yao S-Y, Ye B-H (2015). Enantioselective synthesis of sulfoxides in octahedral ruthenium complexes via a chiral-at-metal strategy. *Inorg Chem*, **54**(6): 2726–2733.

Mashima K, Mikami A, Nakamura A (1992). Preparation and structural characterization of cationic dinuclear ruthenium (II). Thiolate complexes, (Ru_2 $(SPh)_3$ (η^6-*p*-cymene)$_2$) Y (Y = Cl and PF_6). *Chem Lett*, **21**: 1795–1798.

Moosun SB, Jhaumeer-Laulloo S, Coles SJ, Blair LH, Hosten EC, Bhowon MG (2015). Slow diffusion in situ ruthenium/ligand reaction: Crystal structures, fluorescence and biological properties. *Inorg Chem Commun*, **62**: 71–76.

Nishibayashi Y, Imajima H, Onodera G, Inada Y, Hidai M, Uemura S (2004). Preparation of alkanechalcogenolate and benzenechalcogenolate-bridged diruthenium complexes and their catalytic activity toward propargylation of acetone with propargylic alcohol. *Organometallics*, **23**: 5100–5103.

Ohkita K, Kurosawa H, Hirao T, Ikeda I (1994). Relative stability of η^5-cyclopentadienyl(olefin)ruthenium cations: The first estimation of the nature of the Ru(II) olefin bond. *J Organomet Chem*, **470**: 189–190.

Pandey RN, Sharma P (2013). Synthesis and spectral characterization of organometallic complexes of ruthenium(II and III) with mixed ligands. *Int J Chem Environ Pharm Res*, **4**(1): 8–12.

Patra S, Sarkar B, Mobin SM, Kaim W, Lahiri GK (2003). Separating innocence and non-innocence of ligands and metals in complexes $[(L)Ru(acac)_2]^n$ (n = −1, 0, +1; L = *o*-iminoquinone or *o*-iminothioquinone. *Inorg Chem*, **42**: 6469–6473.

Prabhakaran R, Geetha A, Thilagavathi M, Karvembu R, Krishnan V, Bertagnolli H, Natarajan K (2004). Synthesis, characterization, EXAFS investigation and antibacterial activities of new ruthenium(III) complexes containing tetradentate Schiff base. *J Inorg Biochem*, **98**: 2131–2140.

Prakash O, Joshi H, Sharma KN, Gupta PL, Singh AK (2014). Transfer hydrogenation (pH independent) of ketones and aldehydes in water with glycerol: Ru, Rh and Ir catalysts with a COOH group near the metal on a (phenylthio)methyl-2-pyridine scaffold. *Organometallics*, **33**: 3804–3812.

Punji B, Balakrishna MS (2007). Large bite bisphosphate, 2,6-$C_5H_3N\{CH_2OP(-OC_{10}H_6)(\mu\text{-}S)(C_{10}H_6O\text{-})\}_2$: Synthesis, derivatization, transition metal chemistry and application towards hydrogenation of olefins. *J Organomet Chem*, **692**: 1683–1689.

Rowland CE, Cantos PM, Toby BH, Frisch M, Deschamps JR, Cahill CL (2011). Controlling disulfide bond formation and crystal growth from 2-mercaptobenzoic acid. *Cryst Growth Des*, **11**: 1370–1374.

Roy N, Sproules S, Weyhermüller T, Wieghardt K (2009). Trivalent iron and ruthenium complex with a redox noninnocent(2-mercaptophenylimino)-methyl-4,6-di-tert-butylphenolate(2-)ligand. *Inorg Chem*, **48**: 3783–3791.

Schacht HT, Haltiwanger RC, Rakovoski M (1992). Structural characterization of 1:2 tris(μ-benzenethiolato)bis[(hexamethylbenzeneruthenium] chloride-chloroform. *Inorg Chem*, **31**: 1728–1730.

Standfest-Hauser CM, Mereiter K, Schmid R, Kirchner K (2004). Reaction of diaryl disulfides with RuTp(COD)Cl in N,N-dimethylformamide: Formation of 2-(arylmercapto)aryl mercaptan complexes. *Organometallics*, **23**: 2194–2196.

Stibal D, Süss-Fink G, Bruno T (2015). Crystal structure of bis[μ-(4-methoxyphenyl)methanethiolato-κ^2 S:S]bis[chloride(μ_6-1-isopropyl-4-methylbenzene)ruthenium(II)] chloroform disolvate. *Acta Cryst*, **71**: 1216–1218.

Stibal D, Therrien B, Giannini F, Paul LEH, Furrer J, Süss-Fink G (2014). Monothiolato-bridged dinuclear arene ruthenium complexes: The missing link in the reaction of arene ruthenium dichloride dimers with thiols. *Eur J Inorg Chem*, **2014**: 5925–5931.

Sugimoto H, Tanaka K (2001). Syntheses of new ruthenium carbonyl terpyridine *o*-phenylene complexes: Strong interaction between carbonyl and *o*-phenylene ligands. *J Organomet Chem*, **622**: 280–285.

Taimisto M, Oilunkaniemi R, Laitinen RS, Ahlgrén M (2003). Formation and structural characterization of $[RuCl_2(CO)_2(SPh_2)_2]$, $[RuCl_2(CO)_3(OH_2)]$, and $[Ru(OH_2)_6][RuCl_3(CO)_3]_2 \cdot 2H_2O$. *Z Naturforsch*, **58b**: 959–964.

Takemoto S, Kawamura H, Yamada Y, Okada T, Ono A, Yoshikawa E, Mizobe Y, Hidai M (2002). Ruthenium complexes containing bis(diarylamido)/thioether ligands: Synthesis and their catalysis for the hydrogenation of benzonitrile. *Organometallics*, **21**: 3897–3904.

Thangadurai TD, Natarajan K (2001). Synthesis and characterization of new ruthenium(III) complexes containing tetradentate Schiff bases. *Synth React Inorg Met Org Chem*, **31**(4): 549–567.

Tschan MJ-L, Chérioux F, Therrien B, Brelot LK, Süss-Fink G (2006). (μ-Diphenylphoaphido-$_\kappa$P:P)-μ-hydido-(μ-4-hydroxybenzene-thiolato-$_\kappa^2$S:S)bis[η^6-hexamethylbenzene)ruthenium(II)]tetrafluorobo rate. *Acta Cryst*, **E62**: 2916–2918.

Tschan MJ-L, Chérioux F, Therrien B, Süss-Fink G (2004). Subsequent hydride substitution in (arene)trihydridodiruthenium complexes: Synthesis and structure of thiolato-bridged diruthenium cations. *Eur J Inorg Chem*, 2405–2411.

Tschan M, Therrien B, Chérioux F, Süss-Fink G (2005). Sulfur containing trinuclear arene ruthenium clusters. *J Mol Struct*, **743**: 177–181.

Velásquez VPL, Jazzazi TMA, Malassa A, Görls H, Gessner G, Heinemann SH, Westerhausen M (2012). Derivatives of photosensitive CORM-S1-CO complexes of iron and ruthenium with the $(OC)_2M(S\text{-}C\text{-}C\text{-}NH_2)_2$ fragment. *Eur J Inorg Chem*, DOI: 10.1002/ejic.201101230, 1072–1078.

Ye R, Qin Yi, Jia A-Q, Chen Q, Zhang Q-F (2013). Thiolato-bridged tetra- and hexa-nuclear ruthenium(II) carbonyl chloro complexes. *Inorg Chim Acta*, **405**: 218–221.

Chapter 10

Tuning of Ruthenium–DMSO Complexes for Search of New Anticancer Agents

Ruchi Gaur[a] and Lallan Mishra[b]

[a]*Department of Chemistry, Indian Institute of Technology, Kanpur 208016, Uttar Pradesh, India*
[b]*Department of Chemistry, Institute of Science, Banaras Hindu University, Varanasi 221005, Uttar Pradesh, India*

ruchiuno@gmail.com

In the search of tumor-inhibiting coordination compounds with improved clinical effectiveness, reduced toxicity and broader spectrum of activity, ruthenium–DMSO complexes are found of growing significance. Recently, the focus has been centered on coordination complexes and mechanisms as "activation by reduction," "ligand exchange kinetics," and "transferrin-targeted delivery," which account for the excellent cytotoxicity of these complexes. Despite differences between coordination compounds and organometallic compounds, similar modes of action are suggested. Among ruthenium–DMSO complexes investigated so far, ruthenium-halide-DMSO-based antitumor drugs like *cis*-$Ru(DMSO)_4Cl_2$, *trans*-$Ru(DMSO)_4Cl_2$, NAMI $Na[RuImCl_4(DMSO)]$, NAMI-A and several arene derivatives have enthused the chemical

Ruthenium Chemistry
Edited by Ajay Kumar Mishra and Lallan Mishra

ISBN 978-981-4774-39-0 (Hardcover), 978-1-315-11058-5 (eBook)
www.panstanford.com

community to get involved in the design and development of new and novel ruthenium-based drugs. One of the classic targets for the cancer chemotherapy is DNA. Its binding with metal-based drugs is believed to be the dominant mechanism for their cytotoxicity. However, non-classical targets such as topoisomerases are thought to be more important. Present chapter deals with the development of novel ruthenium–DMSO-based drugs. Due to larger list of metallo-drugs, this is limited only to ruthenium–DMSO complexes with greater emphasis on their DNA binding, nuclease activity, bioactivities and their mode of actions.

10.1 Introduction

There is an enormous rise in the number of cancer patients, which constantly demands more efficient and potent anticancer agent for clinical purpose. To meet the demand, a variety of ruthenium complexes have been developed in the laboratory of chemists and they have of course shown activity against wider range of cancers, but most of them caused serious side effects. Against this backdrop, selective targeting of a chemotherapeutic agent specifically at the tumor site has been the most challenging task in the area of cancer therapy. The deeper understanding of cancer biology triggered the introduction of targeted chemotherapies, using metal-based agents/drugs, able to address specific cancer physiology or diseased states. Metal-based agents are used in about 50% of all tumor therapies and display tremendous therapeutic activity in a series of cancerous cells (Jakupec, 2008). Metal-based chemotherapeutic agents offer remarkable versatility owing to a variety of building blocks, different interactions (H-bond, π-π-stacking, coordinative bond, spatial recognition), redox properties, flexibility to coordinate with ligands and different reaction kinetics when interacted with biomolecules.

The interest in metal-based chemotherapeutic agents originated in the early 1970s with the serendipitous discovery of cisplatin by Rosenberg (Rosenberg, 1969), and it became world's first selling anticancer drug. It showed the inhibition of the division of bacterial cells and mainly used in the treatment of metastatic testicular, ovarian, and transitional bladder cancer

(Kelland, 2007). Several promising metal-based chemotherapeutic agents have been extensively studied after the discovery of cisplatin (Allardyce, 2016). After cisplatin, nearly 40 platinum-based agents have been designed and investigated clinically as anticancer agents. These include carboplatin, oxaliplatin, and satraplatin. Generally, platinum-based complexes (cisplatin and carboplatin) inhibit DNA synthesis through covalent binding of DNA molecules to form intrastrand and interstrand DNA cross-links. However, oxaliplatin showed a different mechanism of action as its bulky diaminocyclohexane carrier ligand activates oxaliplatin and form platinum-DNA adduct, which blocks the DNA replication and turns out to be more cytotoxic (Alcindor, 2011). At present, octahedral platinum(IV) complex satraplatin is the most promising as applied orally due to its kinetic inertness in advanced clinical stage (Bhargava, 2009). Despite several efforts of current platinum-based drugs, they were found effective only to a limited range of cancerous cells. Some cancerous cells attain intrinsic resistance and these drugs have severe side effects such as gastrointestinal symptoms (nausea, vomiting, diarrhea, and abdominal pain), renal tubular injury, neuromuscular complications, and ototoxicity. Therefore, there is still a need to design new approaches to outwit these drawbacks (Jakupec, 2003). Among transition metal complexes, octahedral ruthenium complexes are found appealing candidates in the search for new diagnostic and therapeutic agents (Erkkila, 1999).

A large number of Ru(II) and Ru(III) compounds were studied for their DNA binding, cleavage, and cytotoxic properties and have emerged as most promising antitumor and antimetastatic agents after platinum complexes (Antonarakis, 2010) and are being extensively studied in clinical trials to fight metastases and colon cancer (Ang, 2006). Theoretically, ruthenium complexes possess unique biochemical features allowing them to be exploited as chemotherapeutic agent. These features include (1) long span of variable accessible oxidation states to tune ability of electron transfer rates and redox potentials under biochemical processes, (2) slow rate of ligand exchange, (3) well-developed coordination chemistry with reliable approaches, and (4) less toxic effects with easy absorption and excretion by body (Jakupec, 2008). It is believed that active ruthenium complexes can mimic iron in binding to several proteins (e.g., transferrin, albumins) in order to exert low toxicity and transport iron in plasma. Owing to high stability of the ruthenium(II) complexes also proved to be good models for the study of in vitro interactions of octahedral complexes with DNA and related molecules (Alessio, 1989). The first systematic investigation of ruthenium complexes and their antitumor properties against the platinum resistant murine P388 leukemia (P388/DDP) was studied on *fac*-$[RuCl_3(NH_3)_3]$ and *cis*-$RuCl_2(NH_3)_4]Cl$ in the early 1980s by Clarke et al. (Clarke, 1980). With this, am(m)ine complexes of Ru(II) and Ru(III) are approaching as novel innovative new metallopharmaceuticals. The am(m)ine complexes of Ru(II) and Ru(III) tend to bind selectively histidyl imidazole nitrogens on proteins and N7 site on the imidazole ring of purine nucleotides (Clarke et al., 1999; Messori et al., 2000; Bailey et al., 1997) and considered to be specific target for specific tissues (Gray et al., 1996; Messori et al., 2000; Messori et al., 2000; Bailey et al., 1997). Wang et al. also reported ethylenediamine Ru(II) arene compounds, in which arene = biphenyl, dihydroanthracene, tetrahydroanthracene, *p*-cymene, or benzene, bind preferentially to guanine residues in natural double-helical DNA in cell-free media (Wang, 2005). However, the real breakthrough in clinic trial of ruthenium complexes was demonstrated by Keppler et al. in bis-imidazoletetrachlororuthenate $[RuCl_4(Im)_2]$, Im = imidazole for autochthonous colorectal cancers (Keppler, 1987) and in the early 1990s, Sava et al. Na[*trans*-$RuCl_4$(Im)(DMSO-*S*)] (NAMI) found it active specifically against solid metastasizing tumors in mice (Sava et al., 1994–1996). In the search of more active anti-metastatic drug NAMI-A (ImH) [*trans*-$RuCl_4$(Im)(DMSO-*S*)] was developed by the modification in NAMI and has completed phase 1 clinical trials (Bratsos et al., 2007; Bergamo et al., 2007). It is observed that ruthenium complexes, characterized by the presence of at least one sulfoxide ligand, display the capacity to interfere with the process of lung metastasis formation of solid metastasizing tumors of the mouse.

For the past five decades, a steady growth of ruthenium complexes as chemotherapeutic agent is reflected by the accelerating growth of publications in this area. Therefore, it was found interesting to have an overlook of ruthenium–DMSO complexes embodied in the form of a chapter. This chapter describes the historical prospects, mechanism of action, targets, and present scenario of ruthenium–DMSO complexes.

10.2 Historical Background

During the past five decades, ruthenium–DMSO complexes have been interesting starting materials for the synthesis of new coordination compounds (Evans, 1973) and exhibit antitumor and antimetastatic activities (Allessio et al., 2004; Hesek et al., 1999). They also unveiled their applications in medicinal chemistry particularly radiosensitizers (Mandal et al., 2004; Chan et al., 1989) as well as catalytic property for the organic transformations. Dimethyl sulfoxide displayed ambidentate nature, coordinating as *S*-DMSO and *O*-DMSO depending on the nature and characteristics of transition metal ions. As sulfynil group, it provides a good acceptor site for π-electron donor species (soft metal atoms) such as low spin iron (II) and ruthenium (II) ions depending on the electronic and steric factors (Calligaris, 1996). In most of ruthenium(II) complexes, ruthenium center favored coordination through sulfur (*S*-DMSO) over coordination through oxygen (*O*-DMSO) to get stable species owing to steric hindrance, unless ligand overcrowding occurs (Alessio, 1988) or DMSO is *trans* to strong π-acceptors like CO (Alessio, 1995) and NO (Serli, 2002). So, Ru(II) center usually prefers S-bonding site while Ru(III) favor inverted O bonding donor site of DMSO. This ambidentate nature of DMSO is responsible for linkage isomerism that often accompanies changes in oxidation states of ruthenium metal. Tomita et al. described a classic example of linkage isomerism in $[Ru(NH_3)_5(DMSO)]^{2+}$ as shown in Scheme 10.1 (Tomita, 1994). The *S*/*O*-DMSO linkage isomerization is mostly the oxidation state dependent on the metal center and has a significant effect on the redox nature, kinetics, and thermodynamics of ligand-exchange reactions with biological molecules (Heffeter, 2005). This potential importance of linkage isomerization of metal complexes and showed diverse biological environments which enabled them as prodrugs.

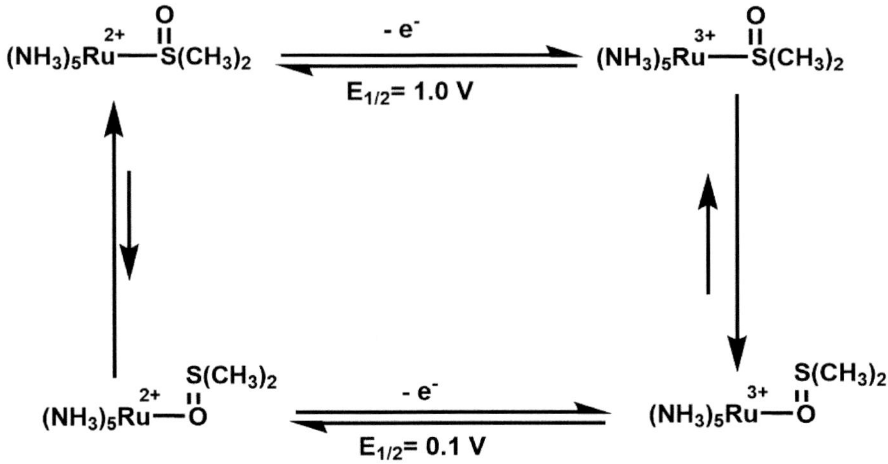

Scheme 10.1 Linkage isomerization of ruthenium DMSO complex (Tomita, 1994).

The pioneering work began in 1988, when Alessio et al. screened *cis*- and *trans*-[Ru(DMSO)$_4$X$_2$] (where X = Br, Cl) isomers for high activity in tumor systems, including the metastasizing Lewis lung tumors (Alessio et al., 1988; Sava et al., 1989). Examination of their effects on the primary tumor and on metastasis revealed that *trans*-Ru(DMSO)$_4$Cl$_2$ is 20-fold more active against metastasis than its *cis* analogue. Both *cis*- and *trans*-Ru(DMSO)$_4$Cl$_2$ isomers have been shown to react with nucleoside and nucleotide components in water (Davey, 1998). In water, *cis* isomer immediately form *trans,cis,cis*-[RuCl$_2$(DMSO)$_2$(H$_2$O)$_2$] by loss of the *O*-bonded DMSO whereas the *trans* compound rapidly loses two *S*-bonded DMSO ligands yielding *cis*-diaqua complex, i.e., *cis,fac*-[RuCl(DMSO)$_3$(H$_2$O)$_2$], respectively, as shown in Scheme 10.2. These hydrolyzed isomers then form cationic compounds undergoing slow reversible chloride dissociation. So, the *trans* compound contains three reactive groups, while the *cis* isomer contains only two, which makes the *cis* aqua species inert relative to the *trans* isomer (Mestroni, 1989). This difference correlates with a higher potency of the *trans* isomer as an antitumor agent (Alessio, 1991).

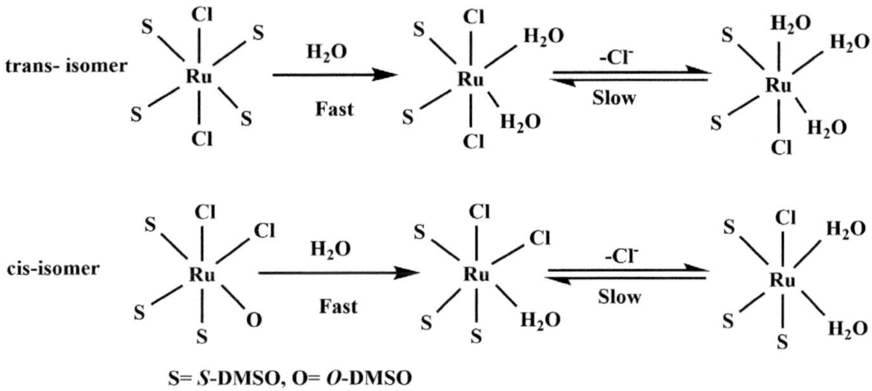

Scheme 10.2 Hydrolysis of *cis* and *trans* isomer of $Ru(DMSO)_4Cl_2$ (Mestroni, 1989).

The *cis*- and *trans*-$RuCl_2(DMSO)_4$ interacted with DNA coordinating via intrastrand cross-linking between two adjacent purines to the Ru center and displayed remarkable anticancer activity (Coluccia, 1993). Earlier studies based on circular dichroism (CD) suggested that the coordination of the *cis* isomer to DNA does not significantly alter the conformation of B-DNA, whereas *trans* isomer binds to DNA more rapidly with some changes in the CD spectra indicating conformational alterations in DNA of non-denaturational character (Mestroni, 1989). The *trans* isomer binds with DNA more effectively with bifunctional lesion to neighboring guanine residues at their N7 atoms whereas *cis* isomer forms mainly monofunctional lesions only on natural DNA (Novakova, 2000). These findings demonstrated considerably slower binding of the *cis* isomer to the dinucleotide d(GpG) compared with *trans*-[Ru(II)$(DMSO)_4Cl_2$] (Anagnostopoulou, 1999). The *cis*-$Ru(DMSO)_4Cl_2$ has also been used to prepare heterodinuclear Pt–Ru compound (Fig. 10.1) cross-linked by 1,4-diaminobutane, i.e., [{*cis,fac*-$RuCl_2$-(DMSO)$_3$}($NH_2(CH_2)_4NH_2$){*cis*-Pt(NH_3)Cl_2}] with a view to developing a compound that could cross-link DNA–DNA or DNA–protein structures. On the basis of DNA-binding studies, it was demonstrated that the heterodinuclear compound could form specific DNA lesions that could cross-link proteins to DNA, but its sensitivity to light and rapid hydrolysis prevented further evaluation (Van Houten, 1993).

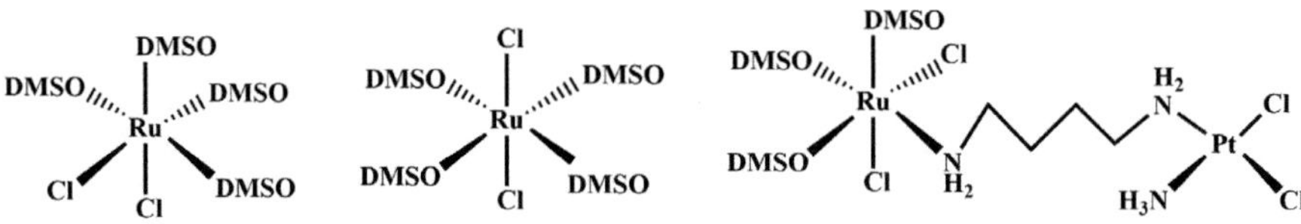

cis-$Ru^{II}(DMSO)_4Cl_2$ *trans*-$Ru^{II}(DMSO)_4Cl_2$

1 2 3

Figure 10.1 Examples for ruthenium–DMSO complexes evaluated for antitumor activity (Van Houten, 1993).

Yapp and coworkers reported in vitro studies of analogues of *cis*- and *trans* Ru-$(DMSO)_4Cl_2$ containing chelating DMSO ligands such as bis(methylsulfinyl)ethane, bis(ethylsulfinyl)ethane (BESE), bis(propylsulfinyl)ethane and bis(methylsulfinyl)propane (Yapp, 1997). The in vitro experiments described that *trans*-complexes accumulate in the cells and bind to DNA to a greater degree than the *cis*-complexes without any cytotoxicity under aerobic and hypoxic conditions. Arene–ruthenium complexes containing BESE, which is essentially a bidentate DMSO ligand, have been tested in vitro (human breast cancer cells: MDA-MB-435s), but their cytotoxicities were more than fivefold higher than that of cisplatin, and no further investigation was reported (Huxham, 2003).

10.3 Primary Targets for the Development of Metal-Based Therapeutic Agents

In the search of new and novel metal complexes for their anticancer activity, they are targeted initially at DNA, proteins (BSA and HSA), enzymes, i.e., topoisomerase, kinase, metalloproteinases (MMPs), telomerase etc. Here, we focus mainly on DNA and topoisomerase enzyme. DNA is an attractive target for the development of therapeutic agents owing to the availability of several binding sites on its skeleton. DNA stores genetic information and distribute it into the cell. Metal complexes can bind and interact with specific sites of DNA which can enable an access to information at cellular level. The interaction of transition metal complexes with DNA is found interesting with their applications in molecular biology, cancer therapy (Lippert, 2000) and photodynamic therapy (PDT) (Levina, 2009). In the development of anticancer activity, primary factors affecting the interaction of DNA with the metal complexes are binding behavior, binding mode (covalent, noncovalent binding: hydrogen bonding, groove binding and intercalation), reaction kinetics and cleavage (Metcalfe, 2003). Several metal complexes were bound with DNA and delivered into the cells. The binding behavior of ruthenium complexes with DNA reinforced by hydrophobic, base π–π stacking interactions, large number of aromatic rings, large surface area, H-bonding interactions, etc. The first technique for monitoring the direct DNA metal interaction was explored by Barton and Lolis using spectrophotometric covalent binding assay (Barton, 1985). Metal complexes bind with nucleobase or phosphate through *intra-* or *inter*-strand cross-link covalent bond and form metal DNA adduct. As a consequence, there are different biochemical and analytic techniques which have been gaining an insight into the mechanisms by which these molecules interact with DNA and assist to explore new therapeutic and diagnostic agents. These techniques include absorption spectrophotometry (Liu, 1998), nuclear magnetic resonance (NMR) spectroscopy (Haq, 1995), X-ray crystallography, gel electrophoresis, molecular modeling, fluorescence spectroscopy (Mahadevan, 1997), isothermal titration calorimetry, viscosity measurements (Mitsopoulou, 2008), electrochemiluminescence (Hu, 2009), inductively coupled plasma mass spectrometry (ICP-MS) (Egger, 2009), atomic absorption spectroscopy (AAS) (Kirin, 2008), electrochemical methods (Johnston, 1995), electrochemiluminescence (Cao, 2006), and circular dichromism spectroscopy (Carcell, 1995). Metal complexes displaying luminescence allow ready characterization of their uptake characteristics. They can be examined by fluorometry, confocal microscopy (Montgomery, 2009), and flow cytometry (Puckett, 2007).

The complex NAMI, Na[*trans*-$RuCl_4$(Im)(DMSO-*S*)], interacts with DNA irreversibly and construct double-helical bifunctional intrastrand DNA adducts efficient for terminating RNA synthesis in vitro (Malina, 2001). Whereas, NAMI-A was less cytotoxic than cisplatin as correlated to DNA binding and intracellular metal accumulation. NAMI-A, (ImH)[*trans*-$RuCl_4$(Im)(DMSO-*S*)] binds only with DNA without any intracellular ruthenium accumulation and attributed lower cytotoxicity than the cisplatin (Pluim, 2004) (Fig. 10.2). Mura et al. reported two less cytotoxic ruthenium complexes, thiazolium (bisthiazole)tetrachlororuthenate and thiazolium (thiazole, DMSO) tetrachlororuthenate, closely related to imidazole containing NAMI (Mura, 2004). Replacement of imidazole ligand by less basic thiazole exhibited slow rate of ligand exchange but displayed high reactivity of serum albumin compared to calf thymus DNA. Thus, low cytotoxicity of these complexes was associated with slow rate of ligand exchange and low reactivity with DNA. Aquation reaction followed by the binding of two adjacent guanine nucleobases with metal complexes can induce deformation in DNA structure and stimulate the cell death via apoptosis (Alderden et al., 2006; Reedijk et al., 2009).

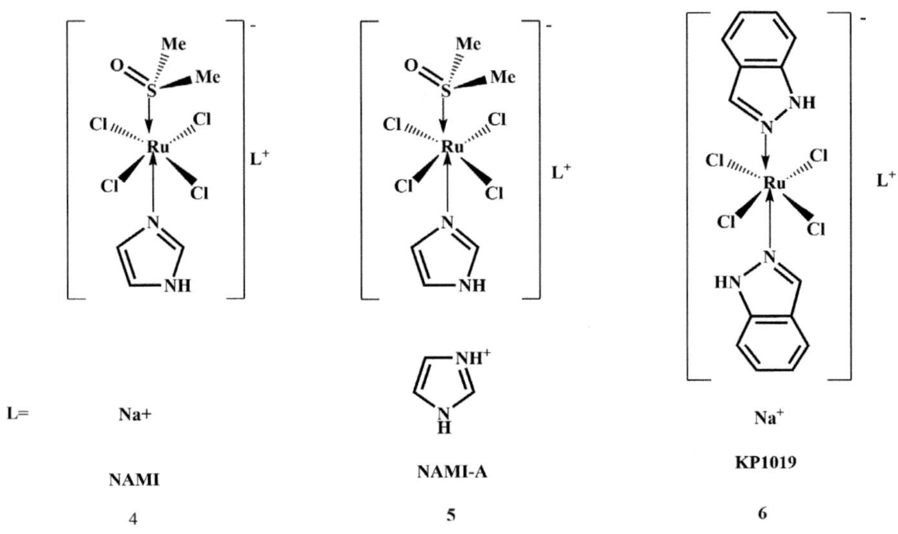

Figure 10.2 Structure of some ruthenium Ru (III) complexes (Kratz et al., 1994; Sava et al., 1999; Bergamo et al., 2000).

The study of DNA interaction with ruthenium–DMSO complexes and ruthenium-mediated DNA damage are important area of chemical research because of their therapeutic and anticancer research interest. DNA or RNA cleavage is elementary reaction in gene therapy as well as gene regulations and leads to apoptosis

(Hernandez-Gil, 2011). DNA cleavage generally occurs because of the break of basic units of DNA-heterocyclic base, phosphate, and sugar. Many complexes showed potential DNA-cleavage activity via hydrolytic or oxidative pathway. The hydrolytic cleavage mainly proceeds via targeting phosphodiester linkage of DNA. However, oxidative cleavage involves the oxidation of nucleobase and nucleosidic bond breakage. Guanine base is the most susceptible site for oxidation in all nucleobases.

Cell nucleus is composed by highly compact state of DNA strands and wrap-around scaffolding histone proteins to construct the coiled condensed form chromatin. Various proteins/enzyme regulate the mechanism of DNA accessibility and chromatin formation. DNA topoisomerases are a prominent class of enzymes which manipulate DNA topology as well as nuclear processes together with the unwinding and rewinding of the DNA helix during replication, recombination, repair, transcription and chromatin transformation (Chen, 2013). DNA topoisomerases are categorized in monomeric type I and multimeric type II, which are further divided according to their structures and mechanisms of action into IA/B and IIA/B (Chikamori et al., 2010; Wang et al., 2009). Most of the therapeutic drugs target mainly topoisomerase I and topoisomerase II for chemotherapy as they block the ligation phase of cell cycle and destruct the integrity of the genome. Topoisomerase I (topo I) changes the degree of supercoiling by unwinding duplex DNA and break phosphodiester bond of one strand. When one strand of supercoiled DNA is cleaved by topoisomerase I, the subsequent unwound DNA is more relaxed and easily detectable by gel electrophoresis technique (Gaur, 2013). The supercoiled double-stranded DNA reacts with topoisomerase I and forms a covalent 3′-phosphorotyrosyl adduct denoted as cleavable complex. Several drugs such as irinotecan, topotecan, and camptothecin (CPT) stabilize the cleavable complex by blocking the re-ligation step and enhance the formation of lethal DNA strand breaks. Topoisomerase I causes single strand cleavage and is nontoxic for cells because of efficient and rapid repair mechanism (Tsao et al., 1993; Wu et al., 1997).

On the other hand, topoisomerase II enzyme is crucial for the cellular division of rapidly growing proliferating tumor cells. Therefore, the inhibition of topo II has also been primary targeted for numerous antitumor agents (Larsen et al., 2003; Nitiss et al., 2009). The nuclear enzyme topoisomerase II is referred to as a "molecular engineer," which is essential for DNA replication, repair, transcription, topological changes, and chromosomal segregation at mitosis under physiological condition (Spence, 2005). The topoisomerase catalyzes the transient double strand of DNA, transport it into an intact fragment of DNA and relegate cleaved strands. The enzyme is a combination of three domains: (1) N-terminal ATP-binding domain: various catalytic inhibitors reduce ATPase activity by blocking ATP from its binding site, (2) DNA-binding/cleaving domain: catalytic active site necessary for construction of covalent complex, and (3) C-terminal tail (Bailly, 2012). To understand the mechanism of action, two categories of Topo II inhibitors were studied in detail: (1) those that form DNA–Topo-II cleavable complex by binding with the topo II and stimulate the cleavage of double standard DNA (etoposide) (Zeglis, 2011), (2) other class, which includes catalytic inhibitors which antagonize the activity of enzyme to implement catalysis (merbarone) Larsen (2003). The wide range of topoisomerase inhibitors, including etoposide, mitoxantrone, amsacrine, idarubicin, and doxorubicin mainly destroy all cells in DNA replication and sensing of DNA in protein production or DNA-damage repair (Li, 2001). The topoisomerase inhibition is fundamentally influenced by the nature of complexes, ligands, and available uncoordinated sites in the skeleton of coordinated ligands. In this connection, Jayaraju et al. reported topo II inhibitor salicylaldoxime cobalt complex (CoSAL), which results in cleavable complex formation by interacting oxime moiety of the salicylaldoxime ligand with the topo II (Jayaraju, 1999). Furthermore, (η^6-benzene)Ru(DMSO)Cl_2 (Fig. 10.3) displayed strong DNA-binding affinity together with cross-linking with topoisomerase II and inhibited the activity of topoisomerase II by cleavage complex formation (Gopal, 2002). Gopal et al. suggested that ruthenium complex interacts with DNA and forms cross-links with topoisomerase II. The complex exhibited antiproliferative activity in two human cancer cell lines Colo-205 (colon adenocarcinoma) and ZR-75-1 (breast carcinoma) in vitro, but it is inconclusive if there is a direct link to its ability to inhibit topoisomerase II activity (Gopal, 2002).

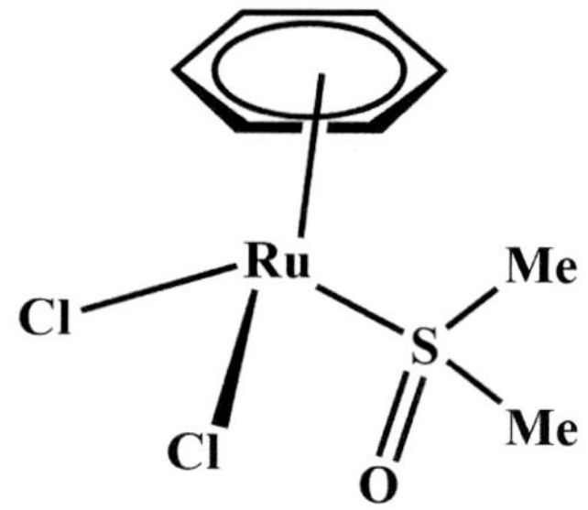

(η⁶-benzene)Ru(DMSO)Cl₂

7

Figure 10.3 Structure of (η^6-benzene) Ru (DMSO)Cl_2 complex (Gopal, 2002).

10.4 Types of Ruthenium–DMSO Complexes

The activity of metal complexes in therapeutic action usually depends upon the nature of ligand and metal center. Therefore, suitable combination of biologically active ligand with ruthenium metal complexes provides important prerequisite for the construction of a highly efficient drug. Ruthenium–DMSO complexes have been explored in the construction of other metal-based active drugs owing to the presence of labile DMSO molecule which provides its facile substitution with biomolecules. The activity of ligand is facilitated by the potency of overall metal complex and serves to modulate this activity in the development of anticancer agent. To follow this approach, we classified Ru–DMSO complexes according to coordinating ligands.

10.4.1 Ruthenium–DMSO Complexes Containing N-Coordinated Ligands

After the successful clinical trial of NAMI-A, its analogues have been studied for potential antimetastatic activity as a means to ascertain the structure–activity relationships in this class of compounds. Similar to NAMI-A, Hdmtp[*trans*-$RuCl_4$(DMSO)dmtp] (dmtp = 5,7-dimethyl[1,2,4]triazolo[1,5-*a*]pyridmidine) exhibited significant cytotoxicity (Velders, 2004) whereas [*mer*-$RuCl_3$(DMSO)(acv)

$(H_2O)]$ and [*mer*-$RuCl_3$(DMSO)(acv)-(MeOH)] and [*trans*-$RuCl_4$(DMSO) guaH] (where acv = acyclovir and gua = guanine in Fig. 10.4) displayed very low cytotoxicity in in vitro studies (Turel, 2004).

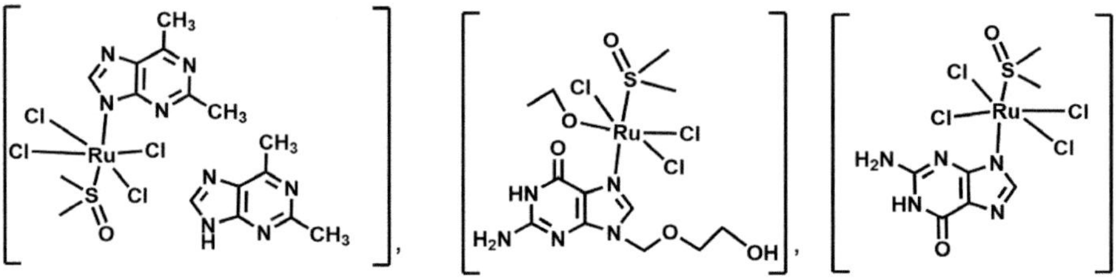

8 9 10

Figure 10.4 Structure of ruthenium dmtp, acyclovir, and guanine DMSO complexes (Velders, 2004; Turel, 2004).

Some ruthenium nitrosyl DMSO complexes were explored as controlled NO-releasing agents for biomedical applications, particularly anticancer agents owing to release of cytotoxic NO within tumor cells which leads to cell death (Lang et al., 2000; Slocik et al., 2001). It is well established that diamagnetic Ru(II) center binds to NO^+ and features the linear $\{Ru(NO)\}_6$ moiety (Wescott, 1999). Generally, modulation of NO release is induced by one electron reduction at NO^+ or by photolysis to yield coordinated $NO^•$. Based on these facts, Serli et al. reported the nitrosyl-nitro derivative *trans,cis,cis*-$RuCl_2(DMSO)_2(NO)(NO_2)$ from the precursors *trans*-$RuCl_2$(DMSO-S)$_4$ or *cis*-$RuCl_2(DMSO)_4$, which is redox active (Scheme 10.3) (Serli, 2003). The site of reduction in *trans,cis,cis*-$RuCl_2(DMSO)_2(NO)(NO_2)$ is the NO^+ moiety. The reduced complex readily liberates a NO_2^- followed by $NO^•$ free radical. These observations reveal that the complex might be implicated in some pharmacological applications as potential NO-releasing agents by in vivo reduction.

Griffith et al. reported three new NAMI-A type analogues of pyridinehydroxamic acid (pyha) complexes, $[NBu_4]$[*trans*-$RuCl_4$(DMSO-*S*)(4-pyha)]. CH_3COCH_3 **12**, [4-pyhaH][*trans*-$RuCl_4$(dmso-*S*)(4-pyha)] **13**, and [3-pyhaH][*trans*-$RuCl_4$(dmso-*S*)(3-pyha)] **14**, for pharmacological evaluation of their in vitro cytotoxicity, inhibition of cell invasion, and their gelatinase activity as depicted in Fig. 10.5. Complex **14** showed higher extent of inhibition of the highly invasive human breast adenocarcinoma cells (MDA-MB-231) than complex **13**. However, both complexes did not display any remarkable inhibitory effect on matrix metalloproteinase (MMP) enzyme production and activity (Griffith, 2008).

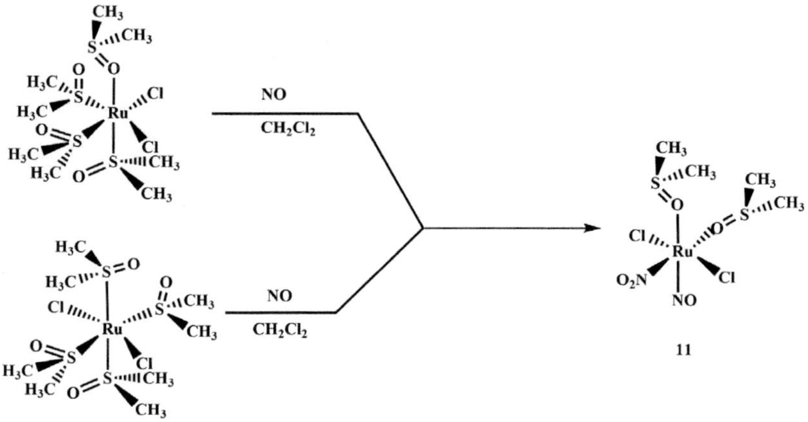

Scheme 10.3 The ruthenium nitrosyl-nitro derivative *trans,cis,cis*-$RuCl_2(DMSO\text{-}O)_2(NO)(NO_2)$ (Serli, 2003).

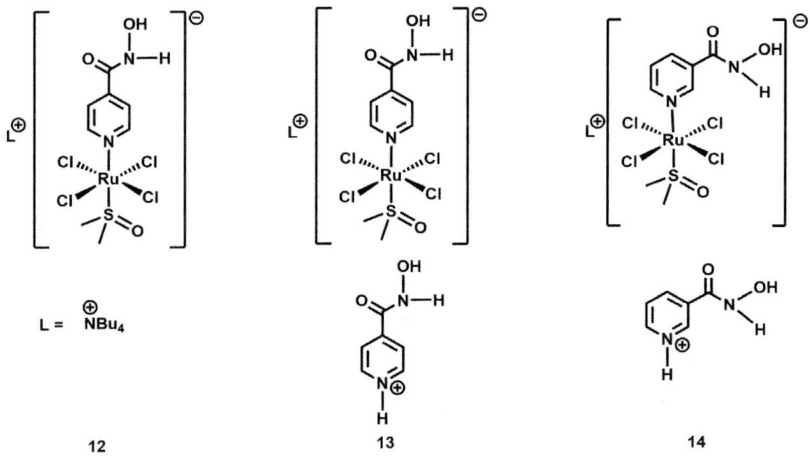

Figure 10.5 NAMI-A type analogues of pyridinehydroxamic acid complexes (Griffith, 2008).

The compound Hdmtp[*trans*-$RuCl_4$(DMSO)dmtp] strongly inhibits lung metastases formation without affecting primary tumors (Ang, 2006). The same traits were also observed in dinuclear analogues of NAMI-A with pyrazine, pyrazole, and bidentate *N*-heterocyclic ligands, e.g., 4,4′-bipyridine, 1,2-bis (4,4′-pyridyl)ethane, which displayed a mild cytotoxic activity against human and murine tumor cell lines and modify the cell cycle distribution of murine adenocarcinoma cells (Alessio et al., 2000; Bergamo et al., 2003; Bergamo et al., 2003; Bergamo et al., 2004) as shown in Fig. 10.6. These dinuclear ruthenium complexes form faster interstrand cross-links with DNA compared with NAMI-A.

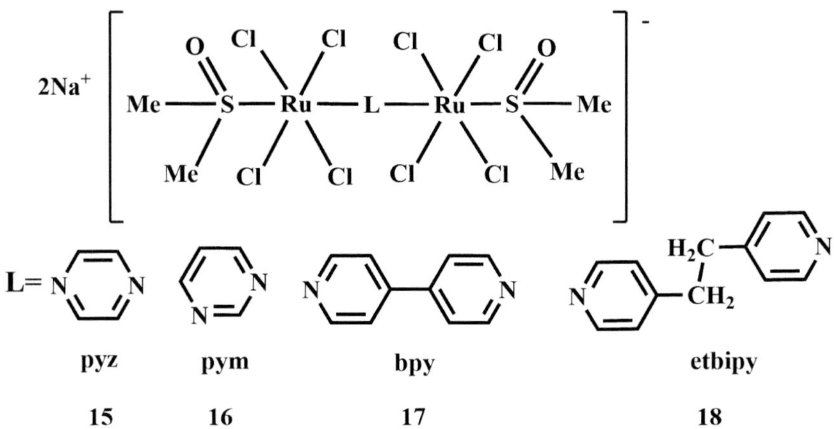

Figure 10.6 Ruthenium dimers with bridged ditopic non-chelating aromatic N-ligands (Alessio et al., 2000; Bergamo et al., 2003; Bergamo et al., 2003; Bergamo et al., 2004).

On the other hand, ruthenium–DMSO–thiolactam-modified kenpaullone derivatives **19, 20** displayed remarkable cytotoxicity against three different cell lines A549 (human alveolar epithelial cell line), CH1 (Lymphoma cancer cell line) and SW480 (colon cancer cell line), as shown in Fig. 10.7 (Schmid, 2007). Complex **19** easily solubilizes in aqueous media and exhibits hydrolysis compared to complex **20**. The binding of GMP, along with the inhibition of ethidium bromide, provides the evidence that the cytotoxicity of complexes owes to intact mono cation intercalation into DNA.

For further development of ruthenium–DMSO complexes, Caiping et al. established a relationship between the chemical properties and the biological activity of four related ruthenium(III) complexes, with the formulas mer-[$RuCl_3$(DMSO)(N-N)] (N-N = 2,2′-bipyridine **21**, 1,10-phenanthroline **22**, dipyrido[3,2-f:2′,3′-h] quinoxaline **23**, and dipyrido[3,2-a:2′,3′-c]phenazine **24**) as shown in Fig. 10.8 (Tan, 2008). They established the hydrolysis process and DNA binding affinity of complexes which showed that the extension of the N-N ligands can increase the stability, DNA binding affinity as well as cytotoxic studies in the order **21** < **22** < **23** < **24**. The lack of correlation between cell growth inhibition, DNA binding affinity, and hydrolysis stability suggests that differences in cellular uptake, subcellular distribution, and additional biological targets may play even more important roles in their antitumor activity.

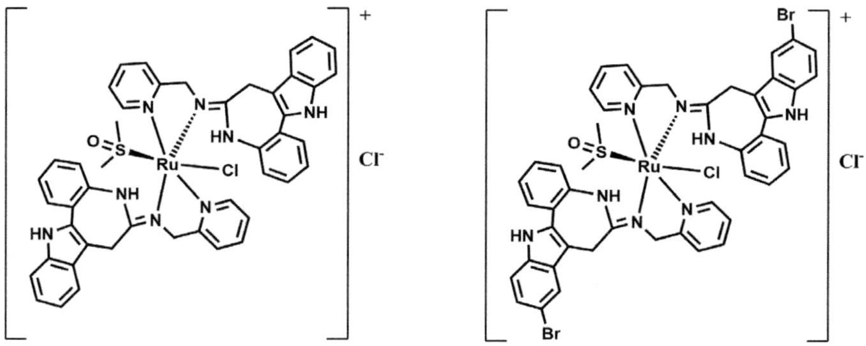

19 20

Figure 10.7 Structure of some ruthenium DMSO kenpaullone complexes (Schmid, 2007).

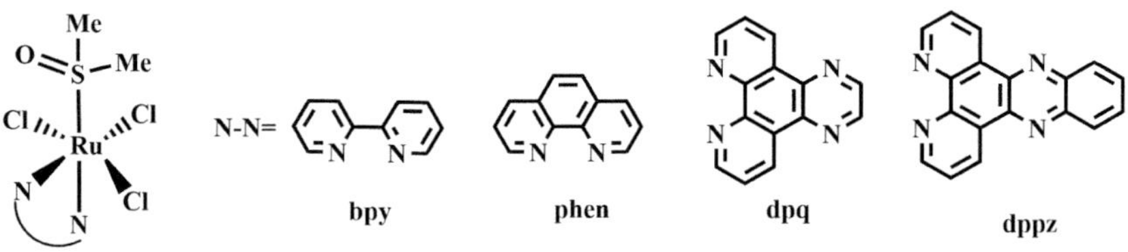

21 22 23 24

Figure 10.8 Structures of some ruthenium–DMSO complexes containing polypyridyl ligands (Tan, 2008).

Furthermore, a well-established specific inhibitor of epidermal growth factor receptor (EGFR), i.e., ethylenediamine or imidazole group modified 4-anilinoquinazoline ligand, was combined with Ru(II)/Ru(III)–DMSO complexes (**25–28** in Fig. 10.9) and explored in the treatment of metastatic non-small cell lung cancer (NSCLC) (Ji, 2014). The complex **26** with 4-(3′-chloro-4′-fluoroanilino)-6-(2-(2-aminoethyl)-aminoethoxy)-7-methoxyquinazoline ligand exhibited more potency specifically on the EGF-induced proliferation of breast cancer cell line MCF-7 than its ligand. The complex also actively influence the early-stage apoptosis and inhibit proliferation of MCF-7 via blocking the EGFR signaling, DNA synthesis and replication.

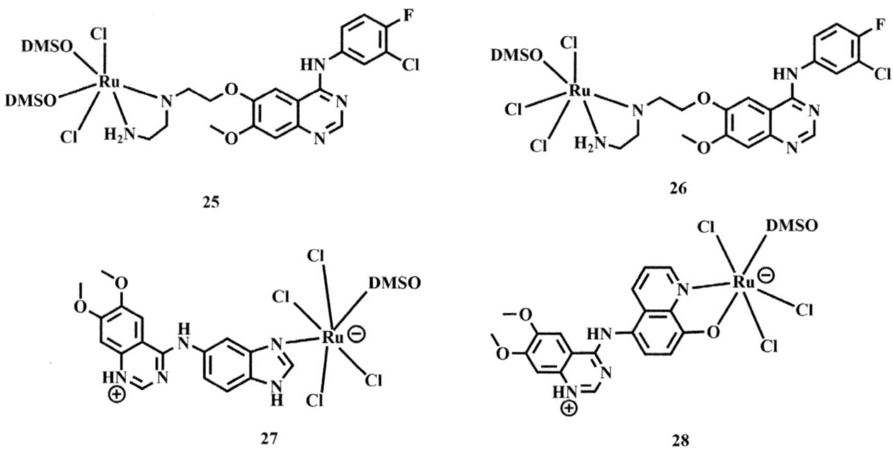

Figure 10.9 The structures of Ru(II) (**25**) and Ru(III) (**26–28**) complexes containing 4-anilinoquinazoline derivatives (Ji, 2014).

10.4.2 Ruthenium–DMSO Complexes Containing O-Coordinated Ligands

The introduction of chelating ligands generally increases the stability of Ru(II)–DMSO complexes against aquation and hydrolysis in aqueous solutions (Bratsos, 2008). Based on these facts, Bratsos et al. have reported a series of new Ru(II)-DMSO complexes containing dicarboxylate ligands (dicarb), namely, malonate (mal), methylmalonate (mmal), dimethyl-malonate (dmmal), succinate (suc), and oxalate (ox), 1,1-cyclobutane dicarboxylate (cbdc) and structurally characterized for their biological evaluations (Bratsos, 2008). These complexes were kinetically stable in an aqueous solution. Among all dicarboxylates, only dimer [*fac*-Ru(DMSO-S)$_3$ (H_2O)(μ-cbdc)]$_2$ (complex **35** in Fig. 10.10), exhibited moderate activity against both the human oral carcinoma (KB) and the murine B16-F10 melanoma cell lines. Unfortunately, studies in vivo displayed no activity against both the primary tumor growth and the formation of spontaneous metastases on the MCa mammary carcinoma model.

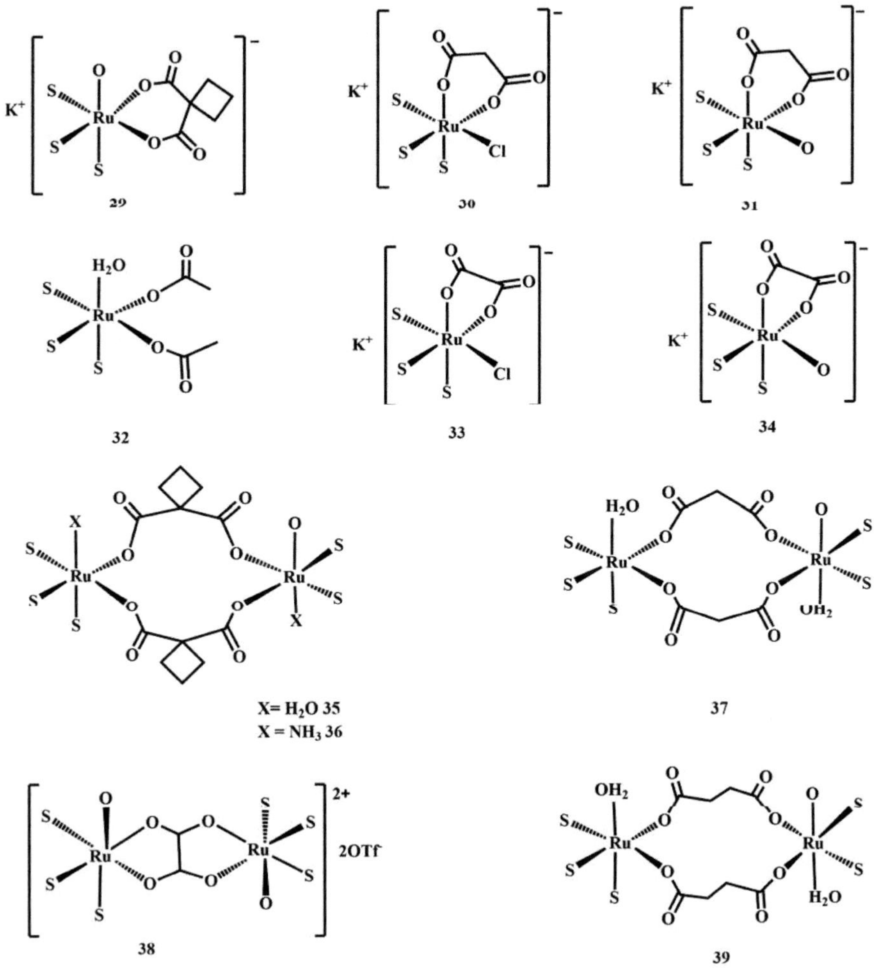

S= *S*-DMSO, O= *O*-DMSO

Figure 10.10 Structures of some Ru(II)–DMSO carboxylate complexes (Bratsos, 2008).

For the past several years, we have been working on ruthenium–DMSO complexes and their DNA binding, cleavage and anticancer activities. In 2010, we reported two ruthenium–DMSO complexes **40, 41** (Fig. 10.11) containing biologically active 3-(4-benzyloxyphenyl)-1-(2-hydroxylphenyl)-prop-2-en-1-one and 2-(4-benzyloxyphenyl)-3-hydroxy-chromen-4-one ligand and their anticancer activity on Dalton Lymphoma (DL) cells. Both complexes enhance their anticancer activity in vitro compared to cisplatin and they owe it to biologically active ligands (Prajapati, 2010).

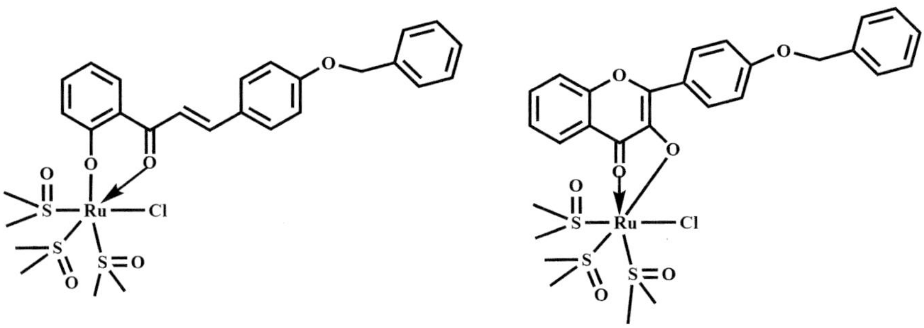

40 **41**

Figure 10.11 Structure of ruthenium–DMSO complexes with chalcone and flavanol ligands (Prajapati, 2010).

In our continuation to investigate the induced effect of biologically active chalcone ligand, a new ruthenium complex containing electron donating chloro group attached to the skeleton of a chalcone was structurally characterized (complex **42** in Scheme 10.4). Its binding with DNA sequence $d(ACCGACGTCGGT)_2$ and ruthenium guanine adduct reveals that the complex is stabilized by additional electrostatic and hydrogen bonding interaction with DNA besides the probable displacement of a labile DMSO by the N7 of guanine. The complex cleaved DNA via its major groove and oxidative pathway (Gaur, 2011). Further, five new ruthenium complexes (**43–47**) bearing different substituted chalcones were synthesized and structurally characterized. Their binding affinity to CT-DNA increased with increase in the conjugation of aromatic ring owing to larger surface area and facilitate its partial intercalation with base pair through DNA groove. Complex **46** and **47** bearing thiophene derivatives in their structural frameworks not only bind to DNA strongly but also show self-cleavage of DNA via oxidative pathway. The strong binding of complex **47** owes to electron releasing methyl group attached to a thiophene ring in its ligand. It reinforces electron density on the surface of chalcone, which strongly overlaps between symmetrical orbital of DNA bases. Further, both complexes **46** and **47** also inhibited topoisomerase II enzyme activity (Gaur, 2012). Both complexes **46** and **47** exerted their cytotoxic effect through G2 phase cell cycle arrest. The complex **47** inhibited MMP-2 activity of HeLa cells and exhibited greater cytotoxicity and apoptosis induction (Jovanovic, 2016).

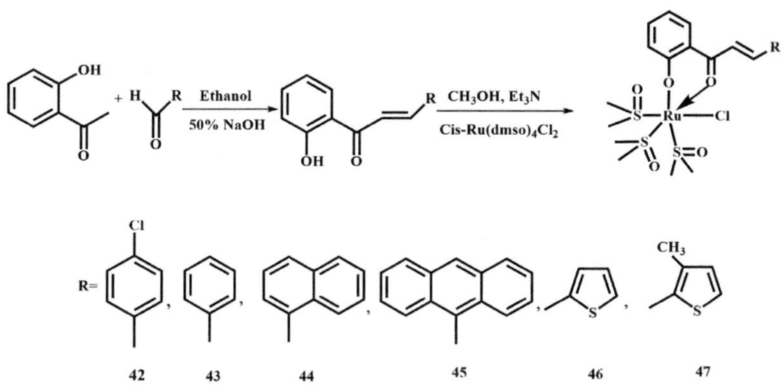

Scheme 10.4 Synthesis of ruthenium–DMSO chalcone complexes (Gaur, 2011; Gaur, 2012; Jovanovic, 2016).

As an extension, two dinuclear ruthenium complexes [Ru(II,II)] were synthesized from bis-chalcone and its flavonol (Fig. 10.12). Complexes transformed supercoiled DNA to nicked circular form under photo-induced conditions in visible region at λ_{max} 560 nm. Highly conjugated structural framework of complex **48** showed its strong binding with DNA through intercalation and minor groove compared to complex **49**. Results supported that both complexes follow type I pathway by the formation of $^{\bullet}OH$, which involves light-induced excitation of the photosensitizer promoting an electron to a higher energy state. In addition, both complexes inhibited topoisomerase I enzyme activity (Gaur, 2013).

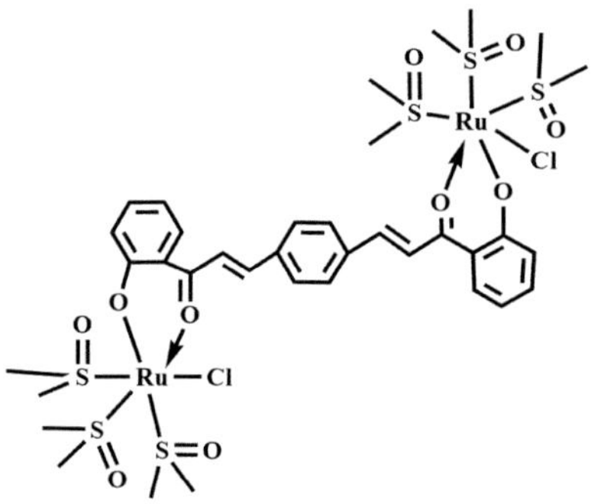

48

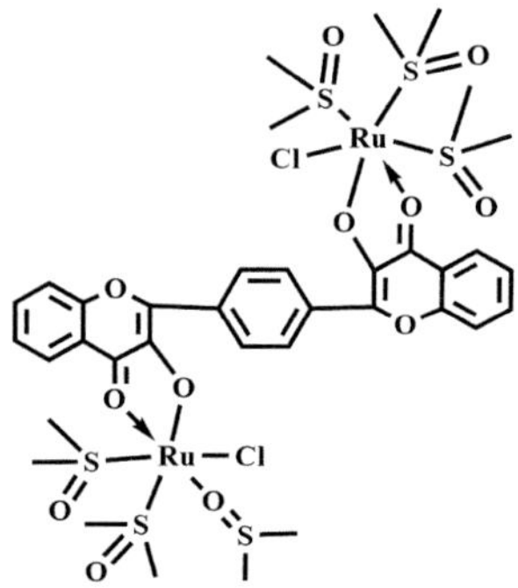

49

Figure 10.12 The structure of ruthenium–DMSO bis-chalcone and bis-flavanol complexes (Gaur, 2013).

Recently, Singh et al. also synthesized four Ru(II)-DMSO complexes **50–53** with substituted chalcone ligands viz. (E)-1-(2-hydroxyphenyl)-3-(4-methoxyphenyl)prop-2-en-1-one, (E)-1-(2-hydroxyphenyl)-3-(4-nitrophenyl)prop-2-en-1-one, (E)-3-(4-(dimethylamino)phenyl)-1-(2-hydroxyphenyl)prop-2-en-1-one, and (E)-1-(2-hydroxyphenyl)-3-(4-chlorophenyl)prop-2-en-1-one (Fig. 10.13) and screened for anti-cancer activity against breast cancer cell lines (MCF-7 and MDA MB-231). The complexes **50–53** displayed significant enhanced anticancer activity against MCF-7 topoisomerase I inhibition (Singh, 2016).

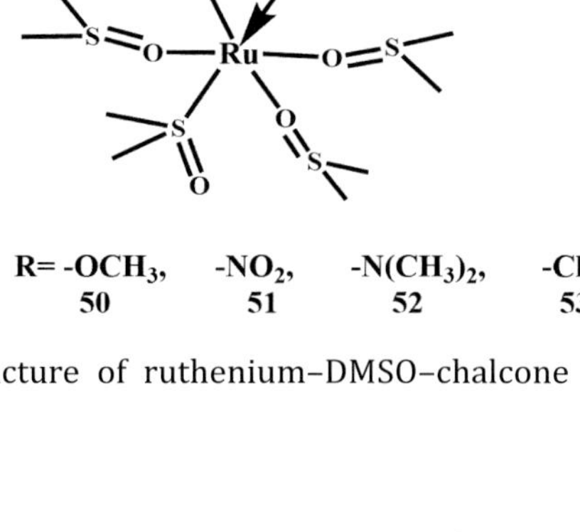

R= $-OCH_3$, $-NO_2$, $-N(CH_3)_2$, -Cl
50 51 52 53

Figure 10.13 Structure of ruthenium–DMSO–chalcone complexes (Singh, 2016).

10.4.3 Ruthenium–DMSO Complexes Containing N,O-Coordinated Ligands

In particular, *cis*-Ru(DMSO)$_4$Cl$_2$ had been used as a synthon for numerous compounds, i.e., *cis,cis,trans*-RuL$_2$(DMSO)$_2$Cl$_2$ (L = nitrofurylsemicarbazone ligands with increased conjugation in Fig. 10.14), although the conjugated products did not offer improved cytotoxicity (MCF-7, TK-10, HT-29) (Cabrera, 2004).

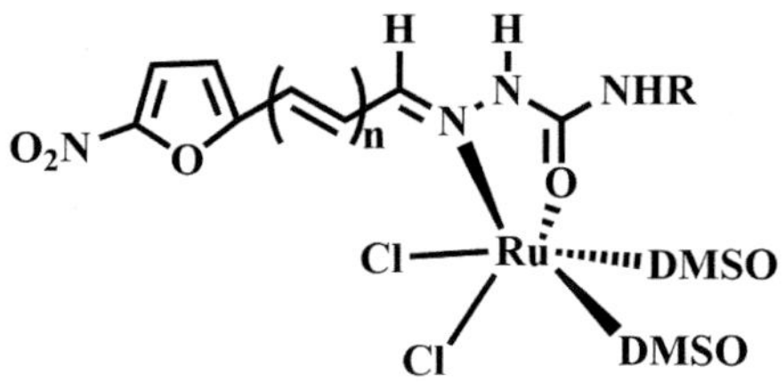

n= 0, R= H 54
n=1, R= H 55
n=0, R=butyl 56

54-56

Figure 10.14 Structure of ruthenium nitrofurylsemicarbazone complexes (Cabrera, 2004).

Chen et al. reported *cis*-[RuCl$_2$(liriodenine)(DMSO)$_2$]·1.5H$_2$O **57** where natural occurring liriodenine (isolated from *Zanthoxylum nitidum*), an active component of the anticancer traditional Chinese medicine (TCM) (Chen, 2009), shown in Fig. 10.15. The interactions of CT DNA and complex revealed that it adopts a classical interaction mode and binds covalently to DNA simultaneously because it easily hydrolyzes to give coordinate active sites.

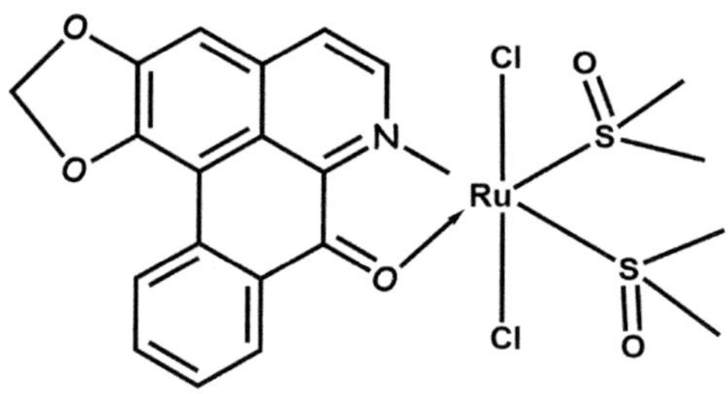

57

Figure 10.15 Structure of liriodenine-based *cis*-[RuCl$_2$(Liriodenine)(DMSO)$_2$]·1.5H$_2$O (Chen, 2009).

Alegesan et al. investigated the binding of biomolecules, i.e., calf thymus DNA (CT-DNA) and bovine serum albumin (BSA) protein with some bivalent ruthenium hydrazine DMSO complexes and revealed their intercalation mode of interactions (Fig. 10.16). The high cytotoxicity of these complexes toward cancer cells was attributed to the nature of the hydrazide moiety and impacts their biological activities (Alagesan, 2014).

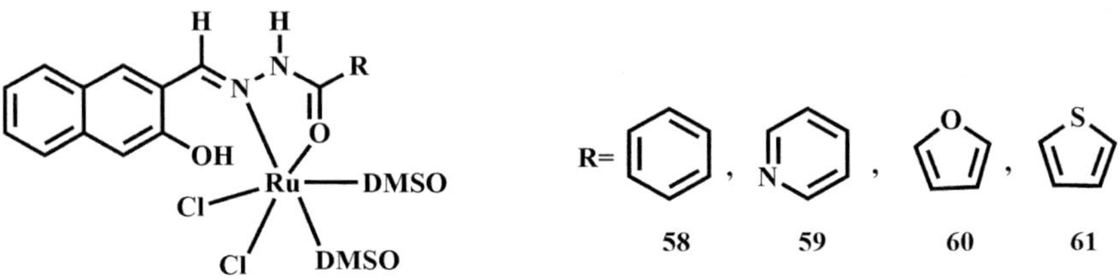

58-61

Figure 10.16 Molecular structure of ruthenium hydrazine DMSO complexes (Alagesan, 2014).

10.5 Mechanism of Action of Ru–DMSO Complexes

In view of the rapid coordination of Ru(II) to biomolecules, Clarks proposed "activation by reduction" hypothesis (Clarke, 1999). Since, tumor cells rapidly spread under hypoxic (poor in O_2) and acidic conditions than normal tissues, the low oxygen content together with lower extracellular pH and the presence of appreciable amounts of cellular reducing agents such as glutathione induce the reduction of Ru(III) to Ru(II) in vitro (Kelman, 1977). The reducing capability of ruthenium(III)-based drugs mostly depends on their ligand environment, and they serve as a prodrug. The prodrug is activated by this selective reduction and generates an active specie which rapidly reacts with target and ultimately leads to cell death (apoptosis). In accordance with this hypothesis, a large number of Ru(III) complexes were exploited for their anticancer activity. The first real breakthrough of this hypothesis was made by Alessio and Sava on the imidazole-dimethylsulfoxide ruthenium(III) complex named sodium *trans*-(dimethylsulfoxide) (imidazole) tetrachloro ruthenate-(III) Na[*trans*-$RuCl_4$(Im)

(DMSO-*S*)] (NAMI), which was specifically active against solid metastases tumors in mice. At doses active on lung metastases, NAMI represents the most extensively studied compound of a new class of ruthenium complexes endowed with an interesting antitumor activity (Alessio et al., 1991; Sava et al., 1992). Later on, Alessio and Sava also explored its analogue (ImH)[*trans*-$RuCl_4$(Im)(DMSO-*S*)] (NAMI-A) (Im = imidazole) which prevented tumor metastases via a mechanism not to involve Ru-DNA binding (Cocchietto, 2003). Hypoxia induced by drugs generally affects cancerous cells via iron-dependent processes. In particular, hypoxia-activated prodrugs were efficient to transferrin-mediated cellular uptake selectivity and induce elevated transferrin receptor expression in tumor cells (Tacchini, 1999). The compounds NAMI-A and KP1019 followed the anticancer mechanism by binding to its iron-binding pockets (Kratz et al., 1994; Sava et al., 1999; Bergamo et al., 2000).

NAMI-A is very well-studied ruthenium DMSO complex in clinical trials. It has also undergone phase I clinical trials and it was well tolerated in patients; reversible blister formation is the dose-limiting toxicity (Rademaker-Lakhai et al., 2004; Sava et al., 2002). NAMI-A coordinated with nucleic acids and directly affected the DNA of cancerous cells (Sava, 2004). It is documented that NAMI-A seems to be independent of direct cytotoxic effects but showed inhibitory effects in various types of primary tumors or metastases (Sava, 1999). Among Ru (III) complexes, NAMI-A is a leading compound that has the cooperative effects of anti-angiogenic and anti-invasive properties on tumor cells and blood vessels involving the inhibition of the processes of tumor invasiveness. The scavenging properties of NAMI-A influenced the NO produced by the endothelial cells probably control the angiogenesis in chick allantoic membrane (embryonic membrane) and the eye cornea model in rabbit (Vacca et al., 2002; Morbidelli et al., 2003). The molecular investigation showed that it blocks extracellular mitogen-activated protein kinase/mitogen and activates phosphokinase/extracellular signal-regulated kinase pathway in endothelial cells. This is probably due to the suppression of a membrane PKC (protein kinase C), leading to the arrest of c-myc transcription, capspase-3 activation, and apoptosis (Sanna, 2002). In view of anti-invasive effects, NAMI-A also reduces the spontaneous invasion of matrigel (gelatinous protein mixture) by tumor cells in a dose-dependent reduction of gelatinase release by the same cells and activation of adhesion molecules such as beta-integrins (Sava, 2004). According to further pharmacological studies, NAMI-A is also capable of changing cell shape or reduce metastases growth by combining to CD44 (cell membrane receptor modulator: cell surface glycoprotein involved in cell/cell and cell/matrix interactions) and tumor infiltrating lymphocyte recruitment (Pacor, 2004).

It is well established that albumin and transferrin are responsible for most of bindings in blood cells (Messori et al., 2000; Messori et al., 2000). Iron is essential for the fast growth of malignant cells and transferrin appears to induce entry into cells (Som, 1983) and transferrin is appropriate for the delivery of drugs into cancerous cells (Sun, 1999). The drugs, owing to strong binding with serum proteins, could be explored in selective delivery to cancer cells through receptor-mediated transferrin delivery (Bergamo, 2003). The iron-binding site of apo transferrin protein also binds with diaqua intermediates of Na[*trans*-$RuCl_4$(Im)(DMSO-*S*)] and its indazole analogue in 2:1 stoichiometry and Im-Ru-DMSO core can be reversibly detached with citrate (Messori et al., 1996; Kratz et al., 1994; Smith et al., 1996). NAMI-A mostly binds to surface accessible histidyl imidazole nitrogens on a number of proteins.

10.6 Conclusion

Finally, the evolution of a new therapeutic agent for the anticanacer activity is a multistep operation which involves design, synthesis, structural characterization, promotion to biological activity, mode of mechanism, and clinical testing. It is worth to emphasize that biologically active ligands together with ruthenium–DMSO complexes can improve the anticancer activity. The improvement of biological activity of ruthenium–DMSO complexes implicates primary molecular targets such as DNA and topoisomerase enzyme, activity against cultured cells (in vitro), or activity in animals (in vivo). Unanticipated side effects in clinical trials have generated a demanding situation for the emergence of new pharmaceuticals. Consequently, a large number of new chemical compounds are being produced. In this outlook, the challenge and prospects of ruthenium–DMSO-based therapeutics, activity against DNA binding, DNA cleavage, enzymes inhibition, low cytotoxicity, and apoptosis, can only be reliably identified in animal models.

In spite of extensive research work carried out in this area of research, it demands deeper investigation.

Abbreviations

BSA: Bovine Serum Albumin
DMSO: dimethylsulfoxide
CT DNA: Calf Thymus DNA
HSA: human serum albumin
MMPs: metalloproteinases
NAMI: Na[*trans*-$RuCl_4$(Im)(DMSO-*S*)]
NAMI-A: (ImH)[*trans*-$RuCl_4$(Im)(DMSO-*S*)]
PDT: photodynamic therapy
Ru: Ruthenium
Topo: Topoisomerase

Acknowledgements

The authors are grateful to the authorities of SERB grant no. SB/FT/CS-188/2013 (DST fast track young scientist fellowship to Ruchi Gaur) and UGC (BSR Faculty Fellowship to L.M.), New Delhi, India, for their financial support.

References

Alagesan M, Sathyadevi P, Krishnamoorthy P, Bhuvanesh NSP and Dharmaraj N (2014). DMSO containing ruthenium(II) hydrazone complexes: In vitro evaluation of biomolecular interaction and anticancer activity, *Dalton Trans*, 43: 15829–15840.

Alcindor T, Beauger N (2011). Oxaliplatin: A review in the era of molecularly targeted therapy, *Curr Oncol*, 18.1: 18–25.

Alderden RA, Hall MD, Hambley TW (2006). The discovery and development of cisplatin, *J Chem Educ*, 83: 728–734.

Allardyce CS, Dyson PJ (2016). Metal-based drugs that break the rules, *Dalton Trans*, 45: 3201–3209.

Alessio E, Balducci G, Calligaris M, Costa G, Attia WM, Mestroni G (1991). Synthesis, molecular structure, and chemical behavior of hydrogen trans-bis(dimethyl sulfoxide)tetrachlororuthenate(III) and mer-trichlorotris(dimethy1 sulfoxide)ruthenium(III): The first fully characterized chloride-dimethyl sulfoxide-ruthenium(III)complexes, *Inorg Chem*, 30: 609–618.

Alessio E, Iengo E, Zorzet S, Bergamo A, Coluccia M, Boccarelli A, Sava G (2000). Antimetastatic properties and DNA interactions of the novel class of dimeric Ru(III) compounds $Na_2[[trans\text{-}RuCl_4(Me_2SO)]_2$ (mu-L)] (L = ditopic, non-chelating aromatic N-ligand). A preliminary investigation, *J Inorg Biochem*, 79: 173–177.

Alessio E, Mestroni G, Lacidogna G, Nassi A, Giordano D, Coluccia M (1991). Interaction of $RuCl_2(DMSO)_4$ isomers with DNA, *Anticancer Res*, 11(4): 1549–1554.

Alessio E, Mestroni G, Nardin G, Attia WM, Calligaris M, Sava G, Zorzet S (1988). *cis*- and *trans*-Dihalotetrakis(dimethyl sulfoxide) ruthenium(II) Complexes ($RuX_2(DMSO)_4$; X = CI, Br): Synthesis, Structure, and Antitumor Activity, *Inorg. Chem*, 27: 4099–4106.

Alessio E, Milani B, Bolle M, Mestroni G, Faleschini P, Todone F, Geremia S, Calligaris M (1995). Carbonyl derivatives of chloride-dimethyl sulfoxide-ruthenium(II) complexes: Synthesis, structural characterization, and reactivity of $Ru(CO)_x(DMSO)_{4-x}Cl_2$Complexes (x = 1–3), *Inorg Chem*, 34: 4722–4734.

Alessio E, Xu Y, Cauci S, Mestroni G, Quadrifoglio F, Viglino P, Marzilli LG (1989). Novel diastereomers with opposite chirality at ruthenium formed by N^7, alpha-PO_4 chelation of 5′-dGMP to the antimetastatic agent *trans*-$RuCl_2(DMSO)_4$: NMR and CD evidence, *J Am Chem Soc*, 111: 7068–7071.

Allessio E (2004). Synthesis and Reactivity of Ru-, Os-, Rh-, and Ir-halide-sulfoxide Complexes, *Chem Rev*, 104: 4203–4242.

Anagnostopoulou A, Moldrheim E, Katsaros N, Sletten E (1999). Interaction of *cis*- and *trans*-$RuCl_2(DMSO)_4$ with the nucleotides GpA, d(GpA), ApG, d(ApG) and d(CCTGGTCC): High-field NMR characterization of the reaction products, *J Biol Inorg Chem*, 4: 199–208.

Ang WH, Dyson PJ (2006). Classical and non-classical ruthenium-based anticancer drugs: Towards targeted chemotherapy, *Eur J Inorg Chem*, 2006: 4003–4018.

Antonarakis ES, Emadi A (2010). Ruthenium-based chemotherapeutics: Are they ready for prime time?, *Cancer Chemother Pharmacol*, 66: 1–9.

Bailly C (2012). Contemporary challenges in the design of topoisomerase II inhibitors for cancer chemotherapy, *Chem Rev*, 112(7): 3611–3640.

Bailey VM, LaChance-Galang KJ, Doan PE, Clarke MJ (1997). ^{1}H and ^{31}P NMR and EPR of pentaammineruthenium(III) complexes of endocyclically coordinated nucleotides, nucleosides, and related heterocyclic bases. Autoxidation of $[(Guok^{N7}(NH_3)_5Ru^{III}](Guo=$ guanosine). Crystal structure of $[7MeGuak^{N9}(NH_3)_5Ru]Cl_3 \cdot 3H_2O$, *Inorg Chem*, 36: 1873–1883.

Barton JK, Lolis E (1985). Chiral Discrimination in the covalent binding of bis(phenanthroline)dichlororuthenium(II) to B-DNA, *J Am Chem Soc*, 107: 709–711.

Bergamo A, Messori L, Piccioli F, Cocchietto M, Sava G (2003). Biological role of adduct formation of the ruthenium(III) complex NAMI-A with serum albumin and serum transferrin, *Invest New Drugs*, 21: 401–411.

Bergamo A, Sava G (2007). Ruthenium complexes can target determinants of tumor malignancy, *Dalton Trans*, 10.1039/B617769G: 1267–1272.

Bergamo A, Stocco G, Casarsa C, Cocchietto M, Alessio E, Serli B, Zorzet S, Sava G (2004). Reduction of in vivo lung metastases by dinuclear ruthenium complexes is coupled to inhibition of in vitro tumor invasion, *Int J Oncol*, 24: 373–379.

Bergamo A, Stocco G, Gava B, Cocchietto M, Alessio E, Serli B, Iengo E, Sava G (2003). Distinct effects of dinuclear ruthenium(III) complexes on cell proliferation and on cell cycle regulation in human and murine tumor cell lines, *J Pharmacol Exp Ther*, 305: 725–732.

Bergamo A, Zorzet S, Gava B, Sorc A, Alessio E, Iengo E, Sava G (2000). Effects of NAMI-A and some related ruthenium complexes on cell viability after short exposure of tumor cells, *Anti Cancer Drugs*, 11: 665–672.

Bhargava A, Vaishampayan UN (2009). Satraplatin: Leading the new generation of oral platinum agents, *Expert Opin Invest Drugs*, 18: 1787–1797.

Bratsos I, Bergamo A, Sava G, Gianferrara T, Zangrando E, Alessio E (2008). Influence of the anionic ligands on the anticancer activity of Ru(II)-dmso complexes: Kinetics of aquation and in vitro cytotoxicity of new dicarboxylate compounds in comparison with their chloride precursors, *J Inorg Biochem*, 102: 606–617.

Bratsos I, Jedner S, Gianferrara T, Alessio E (2007). Ruthenium anticancer compounds: Challenges and expectations, *Chimia*, 61: 692–697.

Cabrera E, Cerecetto H, Gonzalez M, Gambino D, Noblia P, Otero L, Parajon-Costa B, Anzellotti A, Sanchez Delgado R, Azqueta A, de Cerain AL, Monge A (2004). Ruthenium (II) nitrofurylsemicarbazone complexes: New DNA binding agents, *Eur J Med Chem*, 39: 377–382.

Calligaris M, Carugo O (1996). Structure and bonding in metal sulfoxide complexes: An update, *Coord Chem Rev*, 153: 83–154.

Cao WD, Ferrance JP, Demas J, Landers, JP (2006). Quenching of the Electrochemiluminescence of *Tris*(2,2′-bipyridine)ruthenium(II) by Ferrocene and its potential application to quantitative DNA detection, *J Am Chem Soc*, 128: 7572–7578.

Carcell M, Mazzo P, Pelizzi C, Zani F (1995). Antimicrobial and genotoxic activity of 2,6-diacetylpyridine bis(acylhydrazones) and their complexes with some first transition series metal ions. X-ray crystal structure of a dinuclear copper(II) complex, *J Inorg Biochem*, 57: 43–62.

Chan PKI, James BR, Frost DC, Chan PKH, Hu H–L (1989). Effects of halide (X) and sulfoxide (R_2SO) replacement within the ruthenium(I1) nitroimidazole complexes, $RuX_2(R_2SO)_m$ (nitroimidazole)$_n$ m = 1–3, n = 1 or 2: Their characterization, solution chemistry, radiosensitizing activity, and related properties, *Can J Chem*, 67: 508–516.

Chen SH, Chan NL, Hsieh TS (2013). New mechanistic and functional insights into DNA topoisomerases, *Annu Rev Biochem*, 82: 139–70.

Chen ZF, Liu YC, Liu LM, Wang HS, Qin SH, Wang BL, Bian HD, Yang B, Fun HK, Liu HG, Liang H, Orvig C (2009). Potential new inorganic antitumor agents from combining the anticancer traditional Chinese medicine (TCM) liriodenine with metal ions, and DNA binding studies, *J Chem Soc Dalton Trans*, 0: 262–272.

Chikamori K, Grozav AG, Kozuki T, Grabowski D, Ganapathi R, Ganapathi MK (2010). DNA topoisomerase II enzymes as molecular targets for cancer chemotherapy, *Curr Cancer Drug Targets*, 10: 758–771.

Clarke MJ (1980). Oncological implications of the chemistry of ruthenium, *Met Ions Biol Syst*, 11: 231–283.

Clarke MJ, Zhu F, Frasca D (1999). Non-platinum chemotherapeutic metallopharmaceuticals, *Chem Rev*, 99: 2511–2533.

Cocchietto M, Zorzet S, Sorc A, Sava G (2003). Primary tumor, lung and kidney retention and antimetastasis effect of NAMI-A following different routes of administration, *Invest New Drugs*, 21: 55–62.

Coluccia M, Sava G, Loseto F, Nassi A, Boccarelli A, Giordano D, Alessio E, Mestroni G (1993). Antileukemic action of $RuCl_2(DMSO)_4$ isomers and prevention of brain involvement on P388 leukemia and on P388/DDP subline, *Eur J Cancer*, 29A: 1873–1879.

Davey JM, Moerman KL, Ralph SF, Kanitz R, Sheil MM (1998). Comparison of the reactivity of *cis*-$[RuCl_2(DMSO)_4]$ and *trans*-$[RuCl_2(DMSO)_4]$ towards nucleosides, *Inorg Chim Acta*, 281: 10–17.

Egger EA, Rappel C, Jakupec MA, Hartinger CG, Heffeter P, Keppler BK (2009). Development of an experimental protocol for uptake studies of metal compounds in adherent tumor cells, *J Anal At Spectrom*, 24: 51–61.

Erkkila KE, Odom DT, Barton JK (1999). Recognition and reaction of metallointercalators with DNA, *Chem Rev*, 99: 2777–2796.

Evans P, Spencer A, Wilkinson G (1973). Dichlorotetrakis(dimethyl sulphoxide)ruthenium(ii) and its use as a source material for some new ruthenium(ii) complexes, *J Chem Soc Dalton Trans*, 0: 204–209.

Gaur R, Khan RA, Tabassum S, Shah P, Siddiqi MI, Mishra L (2011). Interaction of a ruthenium(II)–chalcone complex with double stranded DNA: Spectroscopic, molecular docking and nuclease properties, *J Photochem Photobiol A*, 220: 145–152.

Gaur R, Mishra L (2012). Synthesis and characterization of Ru(II)–DMSO–Cl–chalcone complexes: DNA binding, nuclease, and topoisomerase II inhibitory activity, *Inorg Chem*, 51: 3059–3070.

Gaur R, Mishra L (2013). Bi-nuclear Ru(II) complexes of bis-chalcone and bisflavonol: Synthesis, characterization, photo cleavage of DNA and Topoisomerase I inhibition, *RSC Adv*, 3: 12210–12219.

Gopal YNV, Konuru N, Kondapi AK (2002). Topoisomerase II antagonism and anticancer activity of coordinated derivatives of $[RuCl_2(C_6H_6)(dmso)]$, *Arch Biochem Biophys*, 401: 53–62.

Gray HB, Winkler JR (1996). Electron transfer in proteins, *Ann Rev Biochem*, 65: 537–561.

Griffith D, Cecco S, Zangrando E, Bergamo A, Sava G, Marmion CJ (2008). Ruthenium(III) dimethyl sulfoxide pyridinehydroxamic acid complexes as potential antimetastatic agents: Synthesis, characterisation and in vitro pharmacological evaluation, *J Biol Inorg Chem*, 13: 511–520.

Haq I, Lincoln P, Suh D, Norden BZ, Chairs CB (1995). Interaction of DELTA- and LAMBDA-$[Ru(phen)_2DPPZ]^{2+}$ with DNA: A calorimetric and equilibrium binding study, *J Am Chem Soc*, 117: 4788–4796.

Heffeter P, Pongratz M, Steiner E, Chiba P, Jakupec MA, Elbling L, Marian B, Koerner W, Sevelda F, Micksche M, Keppler BK, Berger W (2005). Intrinsic and acquired forms of resistance against the anticancer ruthenium compound KP1019 [indazolium *trans*-[tetrachlorobis (1H-indazole)ruthenate (III)] (FFC14A), *J Pharmacol Exp Ther*, 312: 281–289.

Hernandez-Gil J, Llusar SF, Maldonado CR, Mareque-Rivas JC (2011). Synergy between quantum dots and 1,10-phenanthroline–copper(II) complex towards cleaving DNA, *Chem Commun*, 47: 2955–2957.

Hesek D, Inoue Y, Everitt SRL, Ishida H, Kunieda M, Drew MGB (1999). Conversion of a new chiral reagent Δ-[Ru(bpy)$_2$(dmso)Cl]PF$_6$ to Δ-[Ru(bpy)$_2$(dmbpy)]PF$_6$Cl with 96.8% retention of chirality (dmbpy = 4,4′-dimethyl-2,2′-bipyridine), *Chem Commun*, 0: 403–404.

Hu L, Bian Z, Li H, Han S, Yuan Y, Gao L, Xu G (2009). [Ru(bpy)$_2$dppz]$^{2+}$ electrochemiluminescence switch and its applications for DNA interaction study and label-free ATP aptasensor, *Anal Chem*, 81: 9807–9811.

Huxham LA, Cheu ELS, Patrick BO, James BR (2003). The synthesis, structural characterization, and in vitro anti-cancer activity of chloro(*p*-cymene) complexes of ruthenium(II) containing a disulfoxide ligand, *Inorg Chim Acta*, 352: 238–246.

Jakupec MA, Galanski M, Arion VB, Hartinger CG, Keppler BH (2008). Antitumor metal compounds: More than theme and variations, *Dalton Trans*, 0: 183–194.

Jakupec MA, Galanski M, Keppler BK (2003). Tumor-inhibiting platinum complexes–state of the art and future perspectives, *Rev Physiol Biochem Pharmacol*, 146: 1–54.

Jayaraju D, Gopal YNV, Kondapi AK (1999). Topoisomerase II is a cellular target for antiproliferative cobalt salicylaldoxime complex, *Arch Biochem Biophys*, 369: 68–77.

Ji L, Zheng W, Lin Y, Wang X, Lü S, Hao X, Luo Q, Li X, Yang L, Wang F (2014). Novel ruthenium complexes ligated with 4-anilinoquinazoline derivatives: Synthesis, characterisation and preliminary evaluation of biological activity, *Eur J Med Chem*, 77: 110–120.

Johnston DH, Glasgow KC, Thorp HH (1995). Electrochemical measurement of the solvent accessibility of nucleobases using electron transfer between DNA and metal complexes, *J Am Chem Soc*, 117: 8933–8938.

Jovanovic KK, Gligorijevic N, Gaur R, Mishra L, Radulovic S (2016). Anticancer activity of two ruthenium(II)-DMSO-chalcone complexes: Comparison of cytotoxic, pro-apoptotic and antimetastatic potential, *JBUON*, 21(2): 482–490.

Kelland, L (2007). The resurgence of platinum-based cancer chemotherapy, *Nat Rev Cancer*, 7: 573–584.

Kelman AD, Clarke MJ, Edmonds SD, Peresie HJ (1977). Biological activity of ruthenium purine complexes, *J Clin Hematol Oncol*, 7: 274–288.

Keppler BK, Rupp W, Juhl UM, Endres H, Niebl R, Balzer W (1987). Synthesis, molecular structure, and tumor-inhibiting properties of imidazolium *trans*-bis(imidazole)tetrachlororuthenate(III) and its methyl-substituted derivatives, *Inorg Chem*, 26: 4366–4370.

Kirin SI, Ott I, Gust R, Mier W, Weyhermüller T, Metzler- Nolte N (2008). Cellular uptake quantification of metalated peptide and peptide nucleic acid bioconjugates by atomic absorption spectroscopy, *Angew Chem Int Ed*, 47: 955–959.

Kratz F, Hartmann M, Keppler B, Messori L (1994). The binding properties of two antitumor ruthenium(III) complexes to apotransferrin, *J Biol Chem*, 269: 2581–2588.

Kratz F, Keppler BK, Messori L, Smith C, Baker EN (1994). Protein-binding properties of two antitumor Ru(III) complexes to human apotransferrin and apolactoferrin, *Metal Based Drugs*, 1: 169–173.

Lang DR, Davis JA, Lopes LGF, Ferro AA, Vasconcellos LCG, Franco DW, Tfouni E, Wieraszko A, Clarke MJ (2000). A controlled NO-releasing compound: Synthesis, molecular structure, spectroscopy, electrochemistry, and chemical reactivity of (*R,R,S,S*)*trans*-$[RuCl(NO)(cyclam)]^{2+}$ (1,4,8,11-tetraazacyclotetradecane), *Inorg Chem*, 39: 2294–2300.

Larsen AK, Escargueil AE, Skladanowski A (2003). From DNA damage to G2 arrest: The many roles of topoisomerase II, *Prog Cell Cycle Res*, 5: 295–300.

Larsen AK, Escargueil AE, Skladanowski A (2003). Catalytic topoisomerase II inhibitors in cancer therapy, *Pharmacol Ther*, 99 (2): 167–181.

Levina A, Mitra A, Lay PA (2009). Recent developments in ruthenium anticancer drugs, *Metallomics*, 1: 458–470.

Li TK, Liu LF (2001). Tumor cell death induced by topoisomerase-targeting drugs, *Annu Rev Pharmacol Toxicol*, 41: 53–77.

Lippert B (2000). Multiplicity of metal ion binding patterns to nucleobases, *Coord Chem Rev*, 200: 487–516.

Liu CL, Zhou JY, Xu HB (1998). Interaction of the copper(II) macrocyclic complexes with DNA studied by fluorescence quenching of ethidium, *J Inorg Biochem*, 71: 1–6.

Mahadevan S, Palaniandavar M (1997). Spectroscopic and voltammetric studies of copper (II) complexes of bis (pyrid-2-yl)-di/trithia ligands bound to calf thymus DNA, *Inorg Chim Acta*, 254: 291–302.

Malina J, Novakova O, Keppler BK, Alessio E, Brabec V (2001). Biophysical analysis of natural, double-helical DNA modified by anticancer heterocyclic complexes of ruthenium(Ill) in cell-free media, *J Biol Inorg Chem*, 6: 435–445.

Mandal PC (2004). Reactions of the nitro radical anion of metronidazole in aqueous and mixed solvent: A cyclic voltammetric study, *J Electroanal Chem*, 570: 55–61.

Messori L, Kratz F, Alessio E (1996). The interaction of the antitumor complexes Na[*trans*-$RuCl_4$(DMSO)(Im)] and Na[*trans*-$RuCl_4$(DMSO)(Ind)] with apotransferrin: A spectroscopic study, *Met Based Drugs*, 3: 1–9.

Messori L, Orioli P, Vullo D, Alessio E, Iengo E (2000). A spectroscopic study of the reaction of NAMI, a novel ruthenium(III)anti-neoplastic complex, with bovine serum albumin, *Eur J Biochem*, 267: 1206–1213.

Messori L, Vilchez, FG, Vilaplana R, Piccioli F, Alessio E, Keppler B (2000). Binding of antitumor ruthenium(III) complexes to plasma proteins, *Met Based Drugs*, 7: 335–342.

Mestroni G, Alessio E, Calligaris M, Attia WM, Quadrifoglio F, Cauci S, Sava G, Zorzet S, Pacor S, Monti-Bragadin C, Tamaro M, Dolzani L (1989). Chemical, biological and antitumor properties of ruthenium (II) complexes with dimethylsulfoxide, in *Progress in Clinical Biochemistry and Medicine*, Clarke MJ, ed, Springer Verlag, Heidelberg, Germany, pp. 71–87.

Metcalfe C, Thomas JA (2003). Kinetically inert transition metal complexes that reversibly bind to DNA, *Chem Soc Rev*, 32: 215–224.

Mitsopoulou CA, Dagas CE, Makedonas C (2008). Characterization and DNA interaction of the Pt(II) (pq)(bdt) complex. A theoretical and experimental research, *Inorg Chim Acta*, 361: 1973–1982.

Morbidelli L, Donnini S, Filippi S, Messori L, Piccioli F, Orioli P, Sava G, Ziche M (2003). Antiangiogenic properties of selected ruthenium(III) complexes that are nitric oxide scavengers, *Br J Cancer*, 88: 1484–1491.

Montgomery P, Murray BS, New EJ, Pal R, Parker D (2009). Cell-penetrating metal complex optical probes: Targeted and responsive systems based on lanthanide luminescence, *Acc Chem Res*, 42: 925–937.

Mura P, Camalli M, Messori L, Piccioli F, Zanello P, Corsini M (2004). Synthesis, structural characterization, solution chemistry, and preliminary biological studies of the ruthenium(III) complexes TzH *trans*-$RUCl_4(Tz)_2$ and TzH *trans*-$RuCl_4(DMSO)(Tz)\cdot(DMSO)$, the thiazole analogues of antitumor ICR and NAMI-A, *Inorg Chem*, 43: 3863–3870.

Novakova O, Hofr C, Brabec V (2000). Modification of natural, double-helical DNA by antitumor *cis*- and *trans*-$[Cl_2(Me_2SO_4)_4Ru]$ in cell-free media, *Biochem Pharmacol*, 60: 1761–1771.

Nitiss JL (2009). Targeting DNA topoisomerase II in cancer chemotherapy, *Nat Rev Cancer*, 9: 338–350.

Pacor S, Zorzet S, Cocchietto M, Bacac M, Vadori M, Turrin C, Gava B, Castellarin A, Sava G (2004). Intratumoral NAMI-A treatment triggers metastasis reduction, which correlates to CD44 regulation and tumor infiltrating lymphocyte recruitment, *J Pharmacol Exp Ther*, 310: 737–744.

Pluim D, van Waardenburg RC, Beijnen JH, Schellens JH (2004). Cytotoxicity of the organic ruthenium anticancer drug NAMI-A is correlated with DNA binding in four diVerent human tumor cell lines, *Cancer Chemother Pharmacol*, 54: 71–78.

Prajapati R, Dubey SK, Gaur R, Koiri RK, Maurya BK, Trigun SK, Mishra L (2010). Structural characterization and cytotoxicity studies of ruthenium(II)–dmso–chloro complexes of chalcone and flavone derivatives, *Polyhedron*, 29: 1055–1061.

Puckett CA, Barton JK (2007). Methods to explore cellular uptake of ruthenium complexes, *J Am Chem Soc*, 129: 46–47.

Qu Y, Farrell N (1995). Synthesis and chemical properties of a hetero-dinuclear (Pt,Ru) DNA-DNA and DNA-protein cross-linking agent, *Inorg Chem*, 34: 3573–3576.

Rademaker-Lakhai JM, Van Den Bongard D, Pluim D, Beijnen JH, Schellens JHM (2004). A Phase I and pharmacological study with imidazolium-*trans*-DMSO-imidazole-tetrachlororuthenate, a novel ruthenium anticancer agent, *Clin Cancer Res*, 10: 3717–3727.

Reedijk J (2009). Platinum anticancer coordination compounds: Study of DNA binding inspires new drug design, *Eur J Inorg Chem*, 0: 1303–1312.

Roecker L, Dobson JC, Vining WJ, Meyer TJ (1987). Oxygen atom transfer in the oxidations of dimethyl sulfide and dimethyl sulfoxide by $[(bpy)_2(py)Ru(O)]^{2+}$, *Inorg Chem*, 26: 779–781.

Rosenberg B, Van Camp L, Trosko JE, Mansour VH (1969). Platinum compounds: A new class of potent antitumor agents, *Nature*, 222: 385–386.

Sanna B, Debidda M, Pintus G, Tadolini B, Posadino AM, Bennardini F, Sava G, Ventura C (2002). The antimetastatic agent imidazolium *trans*-imidazole tetrachlororuthenate induces endothelial cell apoptosis by inhibiting the mitogen-activated protein kinase/extracellular signal-regulated kinase signaling pathway, *Arch Biochem Biophys*, 403: 209–218.

Sava G, Bergamo A, Zorzet S, Gava B, Casarsa C, Cocchietto M, Furlani A, Scarcia V, Serli B, Iengo E, Alessio E, Mestroni G (2002). Influence of chemical stability on the activity of the antimetastasis ruthenium compound NAMI-A, *Eur J Cancer*, 38: 427–435.

Sava G, Capozzi I, Bergamo A, Gagliardi R, Cocchietto M, Masiero L, Onisto M, Alessio E, Mestroni G, Garbisa S (1996). Down-regulation of tumor gelatinase/inhibitor balance and preservation of tumor endothelium by an anti-metastatic ruthenium complex, *Int J Cancer*, 68: 60–66.

Sava G, Frausin F, Cocchietto M, Vita F, Podda E, Spessotto P, Furlani A, Scarcia V, Zabucchi G (2004). Actin-dependent tumor cell adhesion after short-term exposure to the antimetastasis ruthenium complex NAMI-A, *Eur J Cancer*, 40: 1383–1396.

Sava G, Gagliardi R, Bergamo A, Alessio E, Mestroni G (1999). Treatment of metastases of solid mouse tumors by NAMI-A: Comparison with cisplatin, cyclophosphamide and dacarbazine, *Anticancer Res*, 19: 969–972.

Sava G, Pacor S, Coluccia M, Mariggio M, Cocchietto M, Alessio E, Mestroni G (1994). Response of MCa mammary carcinoma to cisplatin and to Na[*trans*-$RuCl_4$(DMSO)(Im)], *Drug Invest*, 8: 150–161.

Sava G, Pacor S, Mestroni G, Alessio E (1992). Effects of the Ru(III) complexes [mer-$RuCl_3(DMSO)_2$Im] degrees and Na[trans-$RuCl_4$(DMSO)Im] on solid mouse tumors, *Anti Cancer Drugs*, 3: 25–31.

Sava G, Pacor S, Zorzet S, Alessio E, Mestroni G (1989). Antitumor properties of ruthenium(II)-dimethylsulfoxide complexes in the Lewis lung carcinoma system, *Pharmacol Res*, 21: 617–628.

Schmid WF, Zorbas-Seifried S, John RO, Arion VB, Jakupec MA, Roller A, Galanski M, Chiorescu I, Zorbas H, Keppler BK (2007). The first ruthenium-based paullones: Syntheses, X-ray diffraction structures, and spectroscopic and antiproliferative properties in vitro, *Inorg Chem*, 46: 3645–3656.

Serli B, Zangrando E, Gianferrara T, Yellowlees L, Alessio E (2003). Coordination and release of NO by ruthenium-dimethylsulfoxide complexes: Implications for anti-metastases activity, *Coord Chem Rev*, 245: 73–83.

Serli B, Zangrando E, Iengo E, Mestroni G, Yellowlees L, Alessio E (2002). Synthesis and structural, spectroscopic, and electrochemical characterization of new ruthenium dimethyl sulfoxide nitrosyls, *Inorg Chem*, 41: 4033–4043.

Singh AK, Saxena G, Dixit S, Hamidullah, Singh SK, Singh SK, Arshad M, Konwar R (2016). Synthesis, characterization and biological activities of some Ru(II) complexes with substituted chalcones and their applications as chemotherapeutics against breast cancer, *J Mol Struct*, 1111: 90–99.

Slocik JM, Ward MS, Somayajula KV, Shepherd RE (2001). Coordination of $RuCl_3(NO)(H_2O)_2$ by imidazole, histidine and iminodiacetate ligands: A study of complexation of 'Caged NO' by simple bio-cellular donors, *Trans Met Chem*, 26: 351–364.

Smith CA, Sutherland-Smith AJ, Keppler BK, Kratz F, Baker EN (1996). Binding of ruthenium(III) anti-tumor drugs to human lactoferrin probed by high resolution X-ray crystallographic structure analyses, *J Biol Inorg Chem*, 1: 424–431.

Som P, Oster ZH, Matsui K, Gugliemi G, Persson B, Pellettieri ML, Srivastava SC, Richards P, Atkins HL, Brill AB (1983). ^{97}Ru-transferrin uptake in tumor and abscess, *Eur J Nucl Med*, 8: 491–494.

Spence JM, Fournier REK, Oshimura M, Regnier V, Farr CJ (2005). Topoisomerase II cleavage activity within the human D11Z1 and DXZ1 alpha-satellite arrays, *Chromosome Res*, 13: 637–648.

Sun H, Li H, Sadler PJ (1999). Transferrin as a metal ion mediator, *Chem Rev*, 99: 2817–2842.

Tacchini L, Bianchi L, Bernelli-Zazzera A, Cairo G (1999). Transferrin receptor induction by hypoxia, *J Biol Chem*, 274: 24142–24146.

Tan C, Liu J, Chen L, Shi S, Ji L (2008). Synthesis, structural characteristics, DNA binding properties and cytotoxicity studies of a series of Ru(III) complexes, *J Inorg Biochem*, 102: 1644–1653.

Tsao YP, Russo A, Nyamuswa G, Silber R, Liu LF (1993). Interaction between replication forks and topoisomerase I-DNA cleavable complexes: Studies in a cell-free SV40 DNA replication system, *Cancer Res*, 53: 5908–5914.

Tomita A, Sano M (1994). Linkage isomerizations of (sulfoxide)ammine ruthenium complexes induced by electrochemical processes, *Inorg Chem*, 33: 5825–5830.

Turel I, Pecanac M, Golobic A, Alessio E, Serli B, Bergamo A, Sava G (2004). Solution, solid state and biological characterization of ruthenium(III)-DMSO complexes with purine base derivatives, *J Inorg Biochem*, 98: 393–401.

Vacca A, Bruno M, Boccarelli A, Coluccia M, Ribatti D, Bergamo A, Garbisa S, Sartor L, Sava LG (2002). Inhibition of endothelial cell functions and of angiogenesis by the metastasis inhibitor NAMI-A, *Br J Cancer*, 86: 993–998.

Van Houten B, Illenye S, Qu Y, Farrell N (1993). Homodinuclear (Pt,Pt) and heterodinuclear (Ru,Pt) metal compounds as DNA-protein cross-linking agents: Potential suicide DNA lesions, *Biochemistry*, 32: 11794–11801.

Velders AH, Bergamo A, Alessio E, Zangrando E, Haasnoot JG, Casarsa C, Cocchietto M, Zorzet S, Sava G (2004). Synthesis and chemical-pharmacological characterization of the antimetastatic NAMI-A-TypeRu(III)complexes(Hdmtp)[*trans*-$RuCl_4$(dmso-S)(dmtp)], (Na)[*trans*-$RuCl_4$(dmso-*S*)(dmtp)], and [mer-$RuCl_3$(H_2O)(dmso-*S*)(dmtp)] (dmtp) 5,7-Dimethyl[1,2,4]triazolo[1,5-a]pyrimidine), *J Med Chem*, 47: 110–1121.

Wang JC (2009). A journey in the world of DNA rings and beyond, *Annu Rev Biochem*, 78: 31–54.

Wang F, Bella J, Parkinson JA, Sadler PJ (2005). Competitive reactions of a ruthenium arene anticancer complex with histidine, cytochrome c and an oligonucleotide, *J Biol Inorg Chem*, 10: 147–155.

Wescott BL, Enemark JH (1999). In: Solomon EI, Lever ABP, (eds.). Transition metal nitrosyls, *In Inorganic Electronic Structure and Spectroscopy, Volume II: Applications and Case Studies*, vol. II, Wiley, New York, pp. 403–450.

Wu J, Liu LF (1997). Processing of topoisomerase I cleavable complexes into DNA damage by transcription, *Nucleic Acid Res*, 25: 4181–4186.

Yapp DTT, Rettig SJ, James BR, Skov KA (1997). Bidentate sulfoxide complexes of ruthenium(II) and their preliminary biological assessment in vitro, *Inorg Chem*, 36: 5635–5641.

Zeglis BM, Divilov V, Lewis JS (2011). Role of metalation in the topoisomerase IIα inhibition and antiproliferation activity of a series of α-heterocyclic-N^4-substituted thiosemicarbazones and their Cu(II) complexes, *J Med Chem*, 54(7): 2391–2398.

Index

#0120 - 060918 - C18 - 229/152/18 [20] - CB - 9789814774390